2/27/89

Real Variables

Alberto Torchinsky

University of Indiana, Bloomington

Addison-Wesley Publishing Company, Inc.

The Advanced Book Program

Redwood City, California • Menlo Park, California • Reading, Massachusetts
New York • Amsterdam • Don Mills, Ontario • Sydney
Bonn • Madrid • Singapore • Tokyo • Bogotá • Santiago
San Juan • Wokingham, United Kingdom

Publisher: *Allan M. Wylde*
Production Administrator: *Karen L. Garrison*
Editorial Coordinator: *Pearline Randall*
Electronic Production Consultant: *Mona Zeftel*
Promotions Manager: *Celina Gonzales*

Library of Congress Cataloging-in-Publication Data

Torchinsky, Alberto.
Real variables / Alberto Torchinsky.
p. cm.
Includes index.
ISBN 0-201-15675-x : $39.95
1. Functions of real variables. I. Title
QA331.5.T588 1987 87-18629
515.8–dc19 CIP

This book was typeset in MicroTEX using a Leading Edge Model D computer. Camera-ready output from an Apple Laserwriter Plus Printer.

ABCDEFGHIJ-AL-8987
0-201-15675-x

To Massi, Kurosh, and Darius

Author's Foreword

During the academic year 1985–1986 I gave a course on Real Variables at Indiana University. The main source of reference for the course was a set of class notes prepared by the students as we went along; this book is based on those notes. One of the purposes in those lectures was to present to students who are beginning a deeper study of the fairly esoteric subject of Real Variables an overview of how the familiar results covered in Advanced Calculus develop into a rich theory. Motivation is an essential ingredient in this endeavour, as are convincing examples and interesting applications.

Now, teaching a course at this level two facts become quickly apparent, to wit: (i) The background of the students is quite varied, as first year graduate and upper division undergraduate Math students, as well as various science and economics majors, enroll in it, and, (ii) Even those students with a strong background are not entirely at ease with proofs involving either an abstract new concept or an ε-δ argument. My idea of a course at this level is one that presents to the students a modern introduction to the theory of real variables without subjecting them to undue stress.

Although the material is not presented here in a radically different way than in other textbooks, this book offers a conceptually different approach. First, it takes into account, both in placement and content of the topics discussed, the uneven nature of the background of the students. Second, an attempt has been made to motivate the material discussed, and always the most "natural" rather than the most elegant proof of a result is given. We also stress the unity of the subject matter rather than individual results. Third, we go from the particular to the general, discussing each definition and result rather carefully, closer to the way a mathematician first thinks about a new concept. Finally, students are not "talked down," but rather feel that the issues at hand are addressed in a forthright manner and in a direct language, one they can understand. It is important that readers have no difficulty in following the actual arguments presented and spend their time instead in considering questions such as: What is the role or roles of a given result? What is it good for? What are the important ideas, and which are the secondary ones? What are the basic problems in this area and how are they approached and solved?

In fact, we expect the serious students at this level to learn to ask these questions and this text will serve as a guide to ask them at the appropriate time.

How does the text present the material? An important consideration is that the students see the "big picture" rather than isolated theorems, and basic ideas rather than generality are stressed. Each chapter starts with a short reader's guide stating the goals of the chapter. Specific examples are discussed, and general concepts are developed through particular cases. There are 599 problems and questions that are used to motivate the material as well as to round out the development of the subject matter. The reader will be pleasantly surprised to find out that problems are in fact problems, and not further theorems to be proved. Problems are thought-provoking, and there is a mixture of routine to difficult, and concrete to theoretical.

Because I wanted this book to be essentially self-contained for those students with a good Advanced Calculus background as well as an elementary knowledge of the theory of metric spaces, the point of departure is an informal discussion of the theory of sets and cardinal numbers in Chapter I, and ordinal numbers and Zorn's Lemma in Chapter II. These topics give the student the opportunity to work with abstract, possible new, concepts. Chapter III introduces the Riemann-Stieltjes integral and the limitations of the Riemann integral become quickly apparent; ε–δ proofs are discussed here. At the completion of these chapters the background of the students has been essentially equalized. Chapter IV is the exception that proves the rule. It develops the abstract concept of measure, a particular case of which, the Lebesgue measure on R^n, is discussed in Chapter V. Anyone objecting to this treatment can plainly, and almost painlessly, read these chapters in the opposite order. The construction of the Lebesgue measure is a favorite among the students, as it allows them to discover where measures come from and how they are constructed.

In Chapter VI we return to a somewhat abstract setting, although for reasons of simplicity Lusin's theorem is presented in the line where all the difficulties are already apparent. An important feature of this chapter is working with "good" and "bad" sets; this is an indispensable tool in other areas, including the Calderón-Zygmund decomposition of integrable functions discussed in Chapter VIII. The proof of Egorov's theorem illustrates our point of view: It is longer than the usual proof, but it is clear and understandable. In Chapter VII we introduce the notion of the integral and the role of almost everywhere convergence. I am confident that the path that leads to the various convergence theorems is direct and motivational. The material described thus far constitutes a solid first semester of a yearlong course.

Chapter VIII presents new properties of integrable functions, including the Lebesgue Differentiation Theorem. The proof given here makes use of the Hardy-Littlewood maximal function, and is one that most experts agree should have worked its way into the standard treatment of this topic by now. Chapter IX constructs important new examples of measures on the line, the Borel measures. The correspondence between these measures and their distribution functions, a subject that lies at the heart of the theory of Probability, is established in an elementary and computational manner. Chapter X discusses properties of absolutely continuous functions, including the Lebesgue decomposition of functions of bounded variation and the characterization of those functions on the line that may be recovered by integrating their derivatives. The abstract setting of these results is presented in detail in Chapter XI, where the Radon-Nikodým theorem is discussed. The basic theory of the Lebesgue L^p spaces, including duality and the notion of weak convergence is covered in Chapter XII. Chapter XIII deals with product measures and Fubini's theorem in the following manner: In the first section we discuss the version dealing with Lebesgue integrals in Euclidean space; the second section discusses some important applications, including convolutions and approximate identities; and, finally, the third section presents Fubini's theorem in an abstract setting. This is a concrete example on how to proceed from the particular to the general. However, if preferred, the third and second sections can be covered, and the first section assigned for reading.

Chapter XIV deals with normed linear spaces, an abstraction of the notion of the L^p spaces, and the Hahn-Banach theorems. Students are happy to see both the geometric and analytic forms of this result and their applications. Chapter XV covers the basic principles of Functional Analysis, to wit, the Uniform Boundedness Principle, the Closed Graph Theorem, and the Open Mapping Theorem; each principle is given individual attention. In Chapter XVI we consider those Banach spaces whose norm comes from an inner product, or Hilbert spaces. The discussion of the geometry of Hilbert spaces and the spectral decomposition of compact self-adjoint operators are some of the features of this chapter.

Brief historical references concerning the origin of some of the concepts introduced in the text have been made throughout the text, and Chapter XVII presents these remarks in their natural setting, namely, the theory of Fourier series. Finally, Chapter XVIII contains suggestions and comments to some of the problems and questions posed in the book; they are not meant, however, to make the learning of the material effortless.

The notations used throughout the book are either standard or else they are explained as they are introduced. "Theorem 3.2" means that the result alluded to appears as the second item in Section 3 of the present chapter, and "Theorem 3.2 in Chapter X" means that it appears as the second item of the third section in Chapter X. The same convention is used for formulas and problems.

A word about where the text fits into the existing literature. It is more advanced than Rudin's book *Principles of Mathematical Analysis*, a good references for the material on Advanced Calculus and metric spaces. It is also more abstract than the treatise *Measure and Integral* by Wheeden and Zygmund. I learned much of the material on integration from Antoni Zygmund, and some of the topics discussed, including the construction of the Lebesgue measure and the outlook on the Euclidean version of Fubini's theorem, have his imprint. Then, there are the classics. They include Natanson's *Theory of Functions of a Real Variable*, Saks' *Theory of the Integral*, F.Riesz and Sz.-Nagy's *Leçons d'Analyse Fonctionnelle*, Halmos' *Measure Theory*, Hewitt and Stromberg's *Real and Abstract Analysis*, and Dunford and Schwartz's *Linear Operators.* Anyone consulting these books will gain the perspective of the masters.

Where do we go from here? I am confident that the reading of this book will adequately prepare the student to venture into diverse fields of Mathematics. Specifically, books such as Billingsley's *Probability and Measure*, Conway's *A Course in Functional Analysis*, Stein's *Singular Integrals and Differentiabilty Properties of Functions* and Zygmund's *Trigonometric Series* are now within reach.

Acknowledgments

It is always a pleasure to acknowledge the contribution of those who make a project of this nature possible. My friends and colleagues Hari Bercovici and Ron Kerman read the complete manuscript and made valuable suggestions and comments. The opportunity to create this manuscript with the MicroTEX version of TEX was an unexpected pleasure and challenge. Elena Fraboschi and George Springer were my mentors in this endeavour, and I owe them much. Pam Cunningham Pierce contributed with the illustrations. My largest debt, though, is to the students who attended the course and kept a keen interest in learning throughout the ordeal. Many examples and solutions to the problems are due to them, particularly to Steve Rowe. Steve Blakeman, Nick Kernene, and Shilin Wang were also very helpful. The manuscript was cheerfully typed by Storme Day. The staff at Addison-Wesley handled all my questions efficiently. Mona Zeftel provided the much needed technical assistance, and Allan Wylde was the best publisher this ambitious project could have had.

Contents

CHAPTER I

Cardinal Numbers

We open our discussion by introducing, in a naive fashion, the notion of set. We are particularly interested in operating with sets and in the concept of "number of elements" in a set, or cardinal number. We consider various cases of infinite cardinals and do some cardinal arithmetic.

1. SETS

What is a set? According to G. Cantor (1845–1918), who initiated the theory of sets in the last part of the nineteenth century: "A set is a collection into a whole of definite, distinct objects of our intuition or our thought. The objects are called the elements (members) of the set." The origin of the theory of sets, like that of many of the basic notions and results that are covered in this book, can be traced back to the theory of trigonometric and Fourier series. The theory of sets was created by Cantor to address the problem of uniqueness for trigonometric series.

We refer to the "whole of distinct objects" in Cantor's definition as the universal set. We denote sets by capital letters $A, \ldots$ and elements by small letters $a, \ldots$, say. The notation $a \in A$, which reads a belongs to A, indicates the fact that a is a member of A. Most of the sets we consider are of the following form: If X is the universal set, then A is the set of those x in X for which the property $P(x)$ is true. The convenient, and descriptive, notation we adopt in this instance is $A = \{x \in X : P(x)\}$, or plainly $A = \{x : P(x)\}$ or even $A = \{P(x)\}$.

N or Z_+ is the set of natural numbers $\{1, 2, \ldots\}$, $Z = \{\ldots, -1, 0, 1, \ldots\}$ is the set of integers, $Q = \{r : r = m/n, m, n \in Z, n \neq 0\}$ is the set of rational numbers, I is the set of irrational numbers and R is the (universal) set of real numbers. $Q_+ = \{r \in Q : r \geq 0\}$ and Q_- denote the sets of nonnegative and negative rational numbers respectively; similarly for I_+, I_-, R_+ and R_-.

If a is not a member of A we write $a \notin A$, which reads a does not belong to A. The complement $B \setminus A$ of a set A relative to a set B is defined as

$$B \setminus A = \{b \in B : b \notin A\}.$$

We call $X \setminus A$ the complement of A. For instance, in the universal set R, the complement of Q is I and that of I is Q. It is not clear at this point what the complement of the universal set X should be. For this, and other important reasons, we postulate the existence of a particular set. We say that $\emptyset$ is the empty set if $x \in \emptyset$ holds for no element x. For instance, for every set A, $A \setminus A = \emptyset$.

If every element of a set A also belongs to a set B we say that A is a subset of B and we write $A \subseteq B$ or $B \supseteq A$; these expressions read A is contained in or equal to B and B contains or is equal to A, respectively. For instance, $Z \subseteq Q \subseteq R$ and $\emptyset \subseteq A$ for any A. We say that sets A and B are equal, and we write $A = B$, if $A \subseteq B$ and $B \subseteq A$. Although this definition seems a bit cumbersome, it often represents the only practical way we have to determine whether two sets are equal. To emphasize that A is a proper subset of B, i.e., $A \subseteq B$ and $A \neq B$, we write $A \subset B$ or $B \supset A$.

Given a set A, we let $\mathcal{P}(A)$, or parts of A, be the set consisting of all the subsets of A, i.e., $\mathcal{P}(A) = \{B : B \subseteq A\}$. For instance, if $A = \{a, b\}$, then $\mathcal{P}(A) = \{\emptyset, \{a\}, \{b\}, \{a, b\}\}$.

What operations can we perform with sets, and what new sets are generated? We begin by introducing the union and intersection. Let A and B be any two sets. By the union $A \cup B$ of A and B we mean the set consisting of those elements which belong to either A or B. Thus $A \cup B = \{x : x \in A \text{ or } x \in B\}$. By the intersection $A \cap B$ of A and B we mean the set consisting of all elements which belong to both A and B, i.e., $A \cap B = \{x : x \in A \text{ and } x \in B\}$. In case $A \cap B = \emptyset$ we say that the sets A and B are disjoint. For instance, $Q \cup I = R$ and $Q \cap I = \emptyset$.

How do we operate with more than two sets? A set whose elements are sets is referred to as a collection, a class or a family. Families are denoted by script letters $\mathcal{A}, \ldots$ For the question we posed it often suffices to consider a family $\mathcal{A}$ of indexed sets. More precisely, if I is a nonempty set and $\mathcal{A} = \{A_i : i \in I\}$, then we put

$$\bigcup_{i \in I} A_i = \{x : x \in A_i \text{ for some } i \text{ in } I\}$$

and

$$\bigcap_{i \in I} A_i = \{x : x \in A_i \text{ for all } i \text{ in } I\}\,.$$

It is quite straightforward to operate with these concepts, cf. 5.1 below.

If $\mathcal{A} = \{A_i : 1 \leq i \leq n\}$ is a family of n sets, we define the Cartesian product $\prod_{i=1}^{n} A_i$, or product, of the A_i's as the set of ordered n-tuples

$$\prod_{i=1}^{n} A_i = \{(a_1, \ldots, a_n) : a_i \in A_i, 1 \leq i \leq n\}\,.$$

This set is named after Descartes (1596–1650), who introduced the rectangular coordinates for the plane; the analogy of the concepts is clear. A familiar product is $R^n = \{(x_1, \ldots, x_n) : x_i \in R, 1 \leq i \leq n\}$. A product of two sets A and B, say, is denoted by $A \times B$.

A useful application of the notion of product is the following: If $x \notin A \cup B$, then the sets $A \times \{x\}$ and $\{x\} \times B$ look essentially like A and B, and yet are disjoint.

2. FUNCTIONS AND RELATIONS

Various fields of human endeavour have to do with relationships that exist between sets of objects. Graphs and formulas, for instance, are devices for describing special relations in a quantitative way. We start by defining a particular kind of relation, namely, a function. The terminology goes back to Leibniz (1646–1716) who used the term primarily to refer to certain kinds of mathematical formulas. The notion of function generally accepted today was first formulated in 1837 by Dirichlet (1805–1859) in a memoir dealing with the convergence of Fourier series.

Given two sets A and B, say, a function f from A into B is a correspondence which associates with each element a of A, in some manner, an element, and only one, b in B, which we denote by $f(a)$. We refer to f as a function (or map, mapping, correspondence or transformation) of A into B. A is called the domain of f and those elements of B of the form $f(a)$ form a subset of B, denoted by $f(A)$, called the range of f. Any letter in the English or Greek alphabets, capital or small, may be used to denote a function.

The symbol $f : A \to B$ means that f is a function with domain A and range contained in B. If $f : A \to B$ and $g : B \to C$, then the mapping $g \circ f : A \to C$ is defined by $g \circ f(a) = g(f(a))$ for a in A. The function $g \circ f$ is called the composition of f and g. A function F is said to be an extension of a function f, and f a restriction of the function F, if the domain of F contains that of f and $F(a) = f(a)$ for every a in the domain of f. The restriction of F to a subset A of its domain is denoted by $F|A$.

The function f is said to map A onto B if $f(A) = B$; we also say that f is surjective. The function f is said to be a one-to-one mapping of A into B, or plainly one-to-one or injective, if $f(a_1) \neq f(a_2)$ whenever $a_1 \neq a_2$ for all a_1, a_2 in A.

Suppose $f: A \to B$ is one-to-one and onto. Then we can define the mapping $g: B \to A$ by means of $g(b) = a$ whenever $f(a) = b$. The function g is called the inverse of f and is denoted by f^{-1}. For example, the function $f: (-1,1) \to R$ given by $f(x) = \tan(\pi x/2)$ is one-to-one and onto, and its inverse $f^{-1}: R \to (-1,1)$ is $f^{-1}(x) = 2\arctan(x)/\pi$.

Although somewhat inconsistent, we conform to tradition and adopt the following notation: If $f: A \to B$ and $C \subseteq B$, the set $\{a \in A : f(a) \in C\}$ is called the inverse image of C by f and is denoted by $f^{-1}(C)$. This set should not be confused with $(f^{-1})(C) = \{a : a = f^{-1}(b), b \in C\}$ which is only defined when f^{-1} exists.

Two particular functions have a specific name. They are the identity function $1: A \to A$, $1(a) = a$ for all a in A, and the characteristic function χ_E of a set E, i.e., the function defined by the equation $\chi_E(x) = 1$ if $x \in E$ and 0 otherwise.

We often work with families of functions. The collection of all the functions $f: A \to B$ from a set A into a set B is denoted by B^A. For example, R^N denotes the family of all real sequences $\{r_1, r_2, \ldots\}$.

We visualize a function f from A into B as a particular subset of $A \times B$. Indeed, we think of f as the subset of $A \times B$ consisting of the ordered pairs $(a, f(a))$; in other words there is a natural identification between f and its graph. This notion can be extended considerably. An arbitrary subset R of $A \times B$ is called a relation. To emphasize this correspondence we often write aRb to indicate that $(a,b) \in R$. In addition to functions, an important instance of relations are the so-called equivalence relations. In this particular case we have $A = B$ and the equivalence relation R satisfies the following three properties:

R(reflexivity) aRa, all a in A,
S(symmetry) aRb iff(if and only if) bRa,
T(transitivity) If aRb and bRc, then aRc.

The equivalence class $\mathcal{R}(a)$ of an element $a \in A$ is the set $\mathcal{R}(a) = \{b \in A : aRb\}$; A is then the disjoint union of these equivalence classes.

For instance, let A be the collection of all the straight lines L in R^2. Then the relation $L_1 R L_2$ iff L_1 and L_2 are parallel is an equivalence relation, and the equivalence class $\mathcal{R}(L)$ of any line L consists precisely of all the lines parallel to it.

3. EQUIVALENT SETS

Suppose A and B are two sets for which there is a function $f: A \to B$ which is one-to-one and onto. Intuitively, the sets A and B are interchangeable provided we are interested in some property that does not concern the specific nature of their elements. Therefore, in this case we say that A and B are equivalent, with equivalence function f, and we write $A \sim B$. It is readily seen that $\sim$ is an equivalence relation among sets. Indeed, $\sim$ verifies the following three properties:

R. $A \sim A$,

S. If $A \sim B$, then $B \sim A$,

T. If $A \sim B$ and $B \sim C$, then $A \sim C$.

First, in the case of R, the identity equivalence function will do. As for S, if $f: A \to B$ is an equivalence function, then $f^{-1}: B \to A$ establishes an equivalence between B and A. Finally, if $f: A \to B$ and $g: B \to C$ are equivalence functions, so is $g \circ f: A \to C$, cf. 5.10 below.

By means of this equivalence relation we are able to sort sets as follows: A finite set is any set that is either empty or equivalent to $\{1, \ldots, n\}$ for some $n \in N$. Any set that is not finite is called infinite. For instance N, and any set equivalent to N, is infinite. Sets equivalent to N are called countable; it is easy to see why. If A is a countable set and $f: N \to A$ is an equivalence function, then each element $a \in A$ is of the form $a = f(n)$, $n \in N$, and can be identified with n. Thus A can be explicitly written as the sequence $(a_1, a_2, \ldots)$, where $a_n = f(n), n \in N$. A set which is either finite or countable is said to be at most countable. An uncountable set is one which is not at most countable.

It is not hard to see that there are uncountable sets. Indeed, let $I_0 = [0,1]$ be the unit interval of the real line; we claim that I_0 is uncountable. Suppose not, then I_0 can be expressed as $r_1, r_2, \ldots$, say. Dividing I_0 into three closed intervals, each of length $1/3$ (they may have common endpoints), it is clear that one of the intervals, I_1 say, does not contain r_1; if there is more than one interval just choose any. Next we divide I_1 into three closed intervals of equal length and choose a second subinterval, I_2 say, which does not contain r_2. Proceeding in this fashion we construct a nested sequence I_n of closed intervals, each of which is one-third of the preceding in length and such that $r_n \notin I_n$, all n in N. By the well-known nested interval principle, the intersection $\bigcap_{n \in N} I_n$ is not empty and consists of a single real number r, say. Clearly $0 \leq r \leq 1$. Since our assumption is that all the real numbers in I_0 are listed in the sequence $r_1, r_2, \ldots,$ r must be one of the r_n's. But since by construction $r_n \notin I_n$, then $r \neq r_n$ for all n, and we have reached a contradiction. This result is so interesting that it deserves another proof, cf. 5.15 below.

It is often helpful to "picture" a proof and once this is achieved to translate this proof into one that can be written out. For instance, let us show that any two closed, bounded intervals are equivalent. If the intervals are $[a,b]$ and $[c,d]$, say, and if $(b-a) < (d-c)$, then an equivalence function can be readily obtained as indicated in Figure 1. In fact, the picture hints that an explicit expression for f might be $f(x) = ((d-c)/(b-a))(x-a)+c$.

A similar picture can be used to establish that any two bounded, open intervals in the line are equivalent. Combining this result with the above observation that $(-1,1) \sim R$, it readily follows that any two open intervals, bounded or not, are equivalent.

It is natural to consider whether $[0,1]$ and $[0,1)$ are equivalent. This is a slightly more complicated question since a proof by pictures is not easy to come by. Let $A = \{r_1, r_2, \ldots\}$ consist of a decreasing sequence of distinct points in $[0,1]$ such that $r_1 = 1$ and $\lim_{n\to\infty} r_n = 1/2$. Then the function $f\colon [0,1] \to [0,1)$ given by $f(r) = r$ if $r \notin A$ and $f(r_n) = r_{n+1}, r_n \in A$, establishes the desired equivalence.

Now, that $[0,1)$ and $[0,1]$ are equivalent is not so surprising since $[0,1/2] \subseteq [0,1) \subseteq [0,1]$ and $[0,1/2] \sim [0,1]$. The remarkable fact is that a similar result, conjectured by Cantor, is true for arbitrary sets.

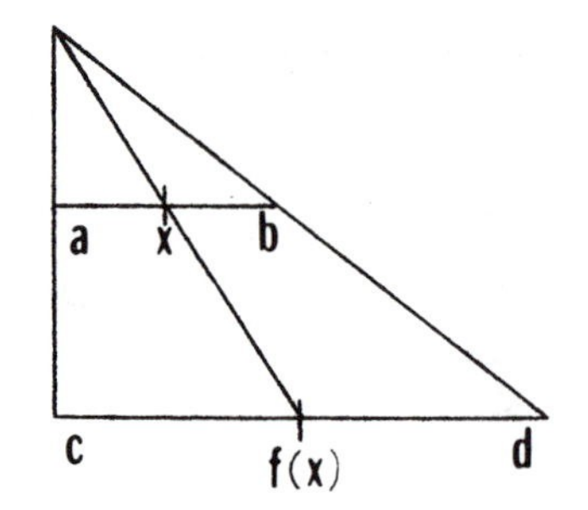

Figure 1

Theorem 3.1 (Schröder-Bernstein). Let A_0, A_1, A_2 be distinct sets such that $A_2 \subset A_1 \subset A_0$ and suppose that $A_0 \sim A_2$. Then also $A_0 \sim A_1$.

Proof. Let $f: A_0 \to A_2$ be an equivalence function and consider $f|A_1$, the restriction of f to A_1. If we let $A_3 = f(A_1)$, then clearly $(f|A_1): A_1 \to A_3$ is one-to-one and onto and it establishes an equivalence between A_1 and A_3. By a proof by pictures it also follows that $A_0 \setminus A_1 \sim A_2 \setminus A_3$ and that $f|(A_0 \setminus A_1)$ is an equivalence function in this case.

Next we put $g = f|A_1$ and we repeat the above argument with A_2 and g in place of A_1 and f. That is, let $A_4 = g(A_2)$ and note that $g|A_2: A_2 \to A_4$ is an equivalence function. It is also readily seen that $g|(A_1 \setminus A_2)$ establishes the equivalence of $A_1 \setminus A_2$ and $A_3 \setminus A_4$.

We repeat this procedure and thus obtain a decreasing sequence $\{A_n\}$ of subsets of A_0 which satisfies the following properties:

(i) $A_1 \sim A_3 \sim A_5 \sim \dots$

(ii) $A_2 \sim A_4 \sim A_6 \sim \dots$

(iii) $A_n \setminus A_{n+1} \sim A_{n+2} \setminus A_{n+3}$, all $n \geq 0$.

Furthermore, note that

$$A_0 = (A_0 \setminus A_1) \cup (A_1 \setminus A_2) \cup (A_2 \setminus A_3) \cup \dots \cup (\textstyle\bigcap_{n=0}^{\infty} A_n). \qquad (3.1)$$

and

$$A_1 = (A_1 \setminus A_2) \cup (A_2 \setminus A_3) \cup (A_3 \setminus A_4) \cup \dots \cup (\textstyle\bigcap_{n=0}^{\infty} A_n). \qquad (3.2)$$

The equivalence of A_0 and A_1 follows now readily since the sets on the right-hand side of (3.1) and of (3.2) are pairwise disjoint, the sets located at the odd spots in (3.1) are equivalent to the sets located at the even spots in (3.2) and the remaining sets are the same. ∎

This result has many important consequences, and we mention some.

Corollary 3.2. Let A, B be arbitrary sets, and let $A_1 \subseteq A$ and $B_1 \subseteq B$ be such that $A_1 \sim B$ and $B_1 \sim A$. Then $A \sim B$.

Proof. Simply observe that by assumption $B_1 \subseteq B \sim A_1 \subseteq A$, and $B_1 \sim A$. Then (a simple variant of) Theorem 3.1 applies with $A_2 \sim B_1$ and $A_0 = A$. ∎

An interesting application of Theorem 3.1 is to show that Q_+ is countable. Since $N \subseteq Q_+$ it is enough to show that Q_+ is equivalent to a subset of N. But this is not hard: If $r = m/n, m, n \in N$, is the relatively prime expression of $r \in Q_+$, then put $f(r) = f(m/n) = 2^m 3^n$. It is clear that f is one-to-one, and consequently, it is an equivalence between Q_+ and a subset of N, as we wanted to show. There are at least two other ways to verify that Q_+ is countable. For instance, we may exhibit Q_+, including repetitions, as the sequence $1/1, 2/1, 1/2, 1/3, 2/2, 3/1, 4/1, 3/2, 2/3, 1/4, \ldots$, ordered by the increasing magnitude of the sum of the numerator and denominator of each rational number. A proof by pictures leading to the above sequence is also available; we leave it to the reader to set it up.

These observations may be cast in a more general setting.

Proposition 3.3. Let $\mathcal{A} = \{A_n\}$ be a family of countable sets. Then $A = \bigcup_{n \in N} A_n$ is also countable.

Proof. List the elements of each $A_n = \{a_{n,1}, a_{n,2}, \ldots\}$ and introduce the mapping $f: A \to N$ given by $f(a_{n,m}) = 2^n 3^m$. Since f is one-to-one, A is at most countable. Also, since $A \supseteq A_1$, say, A is actually countable.

The argument can be readily modified to show that a finite, or countable, union of at most countable sets is again at most countable. ∎

4. CARDINALS

As pointed out above, sets which are equivalent cannot be told apart by purely set-theoretic properties. This observation leads to the following definition. Given a set A, we associate with it its cardinal number, with the property that any two sets A and B have the same cardinal number, or cardinality, provided that they are equivalent. We denote the cardinal number of A by $\operatorname{card} A$ and it is clear that $\operatorname{card} A = \operatorname{card} B$ whenever $A \sim B$. This definition is somewhat imprecise, but it will do for the applications we have in mind.

The cardinal number of the class of sets equivalent to $\emptyset$ is denoted by 0, that of $\{1, \ldots, n\}$ by n, and that of N by $\aleph_0$. Thus $\aleph_0$ is the first infinite cardinal. The cardinal number of the uncountable set $[0,1]$, or that of R for that matter, is denoted by c (for continuum). Small letters often are used to denote cardinal numbers.

The inclusion relation between sets translates into a comparison relation for cardinal number. More precisely, given cardinal numbers, or plainly cardinals, a and b, we say that a precedes b, or that a is less than or equal to b, and we write $a \leq b$, if there are sets A and B and a function $f: A \to B$ such that $\operatorname{card} A = a$, $\operatorname{card} B = b$ and f is one-to-one. In other words, and with the above notation, $a \leq b$ if and only if $A \sim B_1$, where $B_1 \subseteq B$. It is clear that $\aleph_0 \leq c$, and that $n \leq m$ (in the cardinal sense) iff $n \leq m$ (in the usual sense).

Inspired by the concept of equivalent sets we say that the cardinals a and b are equal, and we write $a = b$, if $a \leq b$ and $b \leq a$. We say that $a < b$ if $a \leq b$ and $a \neq b$. For instance, $\aleph_0 < c$.

Next we develop the arithmetic of cardinal numbers, including the operations of addition, multiplication and exponentiation. We do addition first. Given cardinals a,b, we define the sum $a + b$ of a and b as the cardinal number obtained as follows: Let A,B be disjoint sets such that $\text{card}\, A = a$ and $\text{card}\, B = b$. Then put $a + b = \text{card}\,(A \cup B)$. It is not hard to see that addition is commutative (since $A \cup B = B \cup A$) and associative (since $A \cup (B \cup C) = (A \cup B) \cup C$). For example, if n, m are finite cardinals then $n + m$ is, as it should be, $(n + m)$ (let $A = \{1, \ldots, n\}, B = \{n+1, \ldots, n+m\}$). On the other hand, $n + \aleph_0 = \aleph_0$ (choose $A = \{1, \ldots, n\}, B = \{n+1, \ldots\}$ and note that $A \cap B = \emptyset$ and $A \cup B = N$) and $\aleph_0 + \aleph_0 = \aleph_0$ (A = even natural numbers, B = odd natural numbers). Also $\aleph_0 + c = c$, cf. 5.18 below, and $c + c = c\,(A = [0,1/2), B = [1/2,1])$.

As for the multiplication of cardinal numbers, given cardinals a and b, we define the product ab of a and b as the cardinal obtained as follows: Let A, B be sets such that $\text{card}\, A = a$ and $\text{card}\, B = b$. Then put $ab = \text{card}\,(A \times B)$. Multiplication of cardinal numbers is commutative and associative, and distributive with respect to addition, cf. 5.3 below. For example, in the case of finite cardinals n and m, the product nm is, as it should be, (nm), i.e., the cardinal of $\{1, \ldots, n, \ldots, n2, \ldots, nm\}$, and that of $\aleph_0 \aleph_0$ is $\aleph_0$ (Put $A = N, B = \{1/n : n \in N\}$).

Finally we consider exponentiation. Given cardinals a and b, we define the cardinal b^a as follows: Let A,B be sets with $\text{card}\, A = a$ and $\text{card}\, B = b$. Then we set $b^a = \text{card}\, B^A$. The usual properties of exponentiation are not hard to check, cf. 5.21, 5.22 below.

There is at least one exponential that is readily computed, and it corresponds to the case $b = 2$, since it is not hard to identify 2^A. More precisely, we have

Proposition 4.1. Given any set A, $2^A \sim \mathcal{P}(A)$.

Proof. Let $\psi: 2^A \to \mathcal{P}(A)$ be defined as follows: If $f: A \to \{0, 1\}$, then let $\psi(f)$ be the subset of A corresponding to $f^{-1}(\{1\})$, i.e., put $\psi(f) = f^{-1}(\{1\})$. We claim that ψ is an equivalence function. First note that if $\psi(f) = \psi(g)$, then $f^{-1}(\{1\}) = g^{-1}(\{1\})$, and consequently also $f^{-1}(\{0\}) = g^{-1}(\{0\})$ and $f = g$; thus ψ is one-to-one. Next suppose that

$B \in \mathcal{P}(A)$ and let $f = \chi_B$. Then $\psi(f) = f^{-1}(\{1\}) = B$ and ψ is also onto. Thus ψ is an equivalence function . ■

This result explains why $\mathcal{P}(A)$ is also referred to as the power set of A, and it can be used to show that there is no largest cardinal number.

Proposition 4.2. For any set A, $\operatorname{card} A < \operatorname{card} 2^A$.

Proof. Since all the singletons of A belong to $\mathcal{P}(A)$ it is clear that $\operatorname{card} A \leq \operatorname{card} \mathcal{P}(A)$. Let ψ be a (one-to-one if you wish) map from A into $\mathcal{P}(A)$, we show that ψ cannot be onto. This is not hard; suppose that ψ is onto. Now, for each $x \in A$, $\psi(x)$ is a subset of A and consequently the set $B = \{x \in A : x \notin \psi(x)\}$ is well defined. Since by assumption ψ is onto, there exists $a \in A$ such that $\psi(a) = B$. Now, if $a \in B$, then by the definition of B, $a \notin \psi(a) = B$, and this cannot happen. If, on the other hand, $a \notin B$, then also $a \notin \psi(a) = B$ and consequently, by the definition of B, $a \in B$, which is also a contradiction. In other words, ψ cannot be onto. ■

Proposition 4.1 in particular implies that for finite cardinals n, 2^n is as expected. How about $2^{\aleph_0}$? This requires a new idea. Each real number r in [0,1] can be expressed as

$$r = \sum_{n=0}^{\infty} a_n 2^{-n} = .a_1 a_2 \ldots , \quad a_n = 0, 1, \quad \text{all } n .$$

This is the so-called dyadic expansion of r. A minor inconvenience arises since expansions are not necessarily unique. For instance $1/2 = .011\ldots = .100\ldots$, one dyadic expansion terminating in 0's and one in 1's. But the set of such r's is countable, cf. 5.14 below. In other words, if we consider all dyadic expansions, there are, counting repetitions, $c + \aleph_0 = c$ of them. Furthermore, the set of dyadic expansions is clearly equivalent to the set A of all sequences which assume the values 0 and 1, and this set in turn is equivalent to 2^N, $2 = \{0,1\}$. Now, by definition, $\operatorname{card} A = \operatorname{card} 2^N = 2^{\aleph_0}$, and by the above remarks $\operatorname{card} A = c$. Thus $2^{\aleph_0} = c$. A similar argument allows us to compute cc. On the other hand, to compute this product it suffices to note that $cc = 2^{\aleph_0} 2^{\aleph_0} = 2^{2\aleph_0} = 2^{\aleph_0} = c$.

One point remains open. Given two cardinals a and b, we cannot be sure that they are comparable. In order to answer this question we need a new concept, namely that of an ordered set, which we discuss in the next chapter.

5. PROBLEMS AND QUESTIONS

5.1 Show that union and intersection are distributive with respect to intersection and union respectively. In other words, show that

$$B \cap \left(\bigcup_{i \in I} A_i\right) = \bigcup_{i \in I}(B \cap A_i),$$

and

$$C \cup \left(\bigcap_{i \in I} A_i\right) = \bigcap_{i \in I}(C \cup A_i).$$

In addition the de Morgan's laws also hold, to wit,

$$B \setminus \left(\bigcup_{i \in I} A_i\right) = \bigcap_{i \in I}(B \setminus A_i) \text{ and } C \setminus \left(\bigcap_{i \in I} A_i\right) = \bigcup_{i \in I}(C \setminus A_i).$$

5.2 Let $\mathcal{A} = \{A_n : n \in N\}$ be a family of sets and let $A = \bigcup_{n \in N} A_n$. Show that there is a family $\mathcal{B}$ consisting of pairwise disjoint sets, $\mathcal{B} = \{B_n : n \in N\}$, such that $B_n \subseteq A_n$ and $A = \bigcup_{n \in N} B_n$.

5.3 Show that $A \times (B \cup C) = (A \times B) \cup (A \times C)$ and that, in general, $A \cup (B \times C) \neq (A \cup B) \times (A \cup C)$.

5.4 Show that if $A \neq \emptyset$ and $A \times B = A \times C$, then $B = C$.

5.5 Suppose that $B \cap C = \emptyset$ and show that $A^{B \cup C} \sim A^B \times A^C$.

5.6 Suppose that $A^B = B^A$ and show that $A = B$.

5.7 Suppose that A is a set of n elements, how many relations are there in $A \times A$?

5.8 Let $f: A \to B$ and suppose that for all $i \in I, A_i \subseteq A$ and $B_i \subseteq B$. Discuss the (inclusion) relations between

$$f\left(\bigcup_{i \in I} A_i\right) \quad \text{and} \quad \bigcup_{i \in I} f(A_i),$$

and do the same for

$$f\left(\bigcap_{i \in I} A_i\right) \quad \text{and} \quad \bigcap_{i \in I} f(A_i).$$

Also, what are the inclusion relations between

$$f^{-1}\left(\bigcup_{i \in I} B_i\right) \quad \text{and} \quad \bigcup_{i \in I} f^{-1}(B_i),$$

and between

$$f^{-1}\left(\bigcap_{i \in I} B_i\right) \quad \text{and} \quad \bigcap_{i \in I} f^{-1}(B_i)?$$

5.9 Let $f: A \to B$. Show that

$$f(f^{-1}(B)) \subseteq B \quad \text{and} \quad f^{-1}(f(A)) \supseteq A.$$

By means of examples show that the inclusions may indeed be proper.

5.10 Show that the composition of one-to-one functions is one-to-one, and that of onto functions is onto.

5.11 Let $f: A \to B$ and $g: B \to A$, and suppose that $g \circ f = 1$ (identity in A) and $f \circ g = 1$ (identity in B). Show that f and g are one-to-one and onto and that $g = f^{-1}$.

5.12 Suppose that $A \sim B$ and show that for any set C, $A^C \sim B^C$.

5.13 Show that if $f: [0,1] \times [0,1] \to [0,1]$ is one-to-one and onto, then it cannot be continuous.

5.14 Show that the set of real numbers in $[0,1]$ which have two decimal expansions (one terminating in 9's and one in 0's) is countable.

5.15 This is a sketch of a proof that $[0,1]$ is uncountable: Any countable listing $.a_{11}a_{12}a_{13}\ldots, .a_{21}a_{22}a_{23}\ldots, .a_{31}a_{32}a_{33}\ldots, \ldots$ of the real numbers in $[0,1]$ can not be complete. Indeed, put

$$r = .b_1 b_2 b_3 \ldots, \quad b_n \neq a_{nn}, \quad \text{all } n,$$

and note that the real number r cannot be in the above listing. The actual proof requires some care; 5.14 is relevant here. The method of proof uses a "Cantor diagonal selection process."

5.16 Prove the following restatement of the corollary to the Schröder-Bernstein theorem: If a and b are cardinal numbers such that $a \leq b$ and $b \leq a$, then $a = b$. This result does not require the Axiom of Choice as do the results of Chapter II, but it merely asserts that both $a < b$ and $b < a$ cannot occur.

5.17 Show that if A is an infinite set, then A contains a countable subset.

5.18 Prove these two corollaries to 5.17: (i) A set is infinite iff it is equivalent to a proper subset, and, (ii) If a is an infinite cardinal, then $a + \aleph_0 = a$.

5.19 Suppose a_1, a_2 are infinite cardinals, and that $a_1 < a_2$. Does it follow that for every cardinal b, $a_1 + b < a_2 + b$?

5.20 If a and b are cardinal numbers so that $a < c$ and $a + b = c$, does it follow that $b = c$?

5.21 Given cardinals a,b and d, show that $(a^b)^d = a^{bd}$.

5.22 Show that for arbitrary cardinals $a, b, 2^{a+b} = 2^a 2^b$.

5.23 Compute $\aleph_0 \aleph_0, c\aleph_0, cc, c^c$ and $\aleph_0^{\aleph_0}$.

5.24 Show that the set of all polynomials with rational coefficients is countable. Also the set of roots of such polynomials is countable; these are known as algebraic numbers.
A real number is said to be transcendental if it is not algebraic. Show that the set of transcendental numbers is uncountable.

5.25 What is the cardinal number of the set of all the real-valued functions defined on [0,1]? What is the cardinality of $C([0,1])$, the set of all the continuous real-valued functions defined on [0,1]?

5.26 Let S be the set of those sequences of natural numbers which are eventually 0; show that S is countable.

5.27 Let $\mathcal{A}$ be the family of all the finite subsets of R and let $\mathcal{B}$ be the family of all the at most countable subsets of R. Compute card $\mathcal{A}$ and card $\mathcal{B}$.

5.28 What is the cardinality of the family of all the open subsets of R? of R^2?

5.29 Let $\mathcal{A}$ be the family of all the convex subsets of R^2, and let $\mathcal{B}$ be the family of all the connected subsets of R^2. Evaluate card $\mathcal{A}$ and card $\mathcal{B}$.

5.30 Given a family of sets $\mathcal{A} = \{A\}$ such that card $\mathcal{A} \leq c$, let $\mathcal{B} = \{B : B = \bigcup_{k=1}^{\infty}(A_k \setminus A'_k), A_k, A'_k \in \mathcal{A}\}$. Prove that card $\mathcal{B} \leq c$.

5.31 If $\{A_i\}_{i \in I}$ is an indexed family of sets with card $A_i \leq c$ for all $i \in I$, and card $I \leq c$, show that card $\left(\bigcup_{i \in I} A_i\right) \leq c$.

5.32 Let A be a subset of the plane with the property that the (usual Euclidean) distance $d(x,y)$ is a rational number for any pair of elements $x, y \in A$. Show that A is at most countable.
Is the result valid for subsets of R^n with the same property?

5.33 Suppose $A_1 \subseteq A_2 \subseteq \ldots, A \subseteq \bigcup_{n=1}^{\infty} A_n$, and for each infinite subset B of A there exists A_N such that $B \cap A_N$ is an infinite set. Is it true that for some integer n_0, $B \subseteq A_{n_0}$?

5.34 Decide whether the following statements are true: (a) If $A \subseteq B$, then card $B =$ card $(B \setminus A)+$card A, and, (b) card $(A \cap B) =$ card A iff $A \subseteq B$.

5.35 Let a be an infinite cardinal such that $a = 2a$. Show that if $a < b$, then $a + b = b$.

5.36 Let a, b be cardinal numbers, $b \geq 1$. Show that $a + b \leq ab$.

5.37 (Russell's Paradox). A set is either a member of itself, or it is not. Let R denote the set of all sets which are not members of themselves. Then if $R \in R$ it follows that $R \notin R$. If $R \notin R$, it follows that $R \in R$. Hence it cannot be that $R \in R$ or that $R \notin R$.
It is clear from this and other paradoxes, that there is a need for the axiomatization of intuitive set theory. These paradoxes are avoided in axiomatic set theory by the elimination of "sets" that are "too large." To develop the theory of sets from the axiomatic point of view is a long and difficult process, far removed from Real Analysis, which is the main subject of our text. For this reason we have made no effort to be rigorous in dealing with sets, but have rather appealed to intuition.

CHAPTER II

Ordinal Numbers

In Chapter I we identified sets according to their cardinal number; an attribute we want to consider next is that of order. For instance, we want to distinguish between the sets $A = \{1, 2, \ldots\}$ ordered by the usual "$\leq$" relation and $B = \{\ldots, 2, 1\}$ ordered by "$\geq$." Although these sets are equivalent in the sense of cardinality, the same is not true if we take the order into account: A has a first, but not a last, element, and B has a last, but not a first, element.

1. ORDERED SETS

A partial order on a set M is a special type of relation. More precisely, we say that the relation $\prec$ on $M \times M$ is a partial ordering on M if it satisfies the following three properties:

R. $m \prec m$ for every $m \in M$.

AS. (antisymmetry) If $m_1 \prec m_2$ and $m_2 \prec m_1$, then $m_1 = m_2$.

T. If $m_1 \prec m_2$ and $m_2 \prec m_3$, then $m_1 \prec m_3$.

To stress that M is partially ordered by $\prec$ we denote a partially ordered set by $(M, \prec)$. Also $(\{a_1, a_2, \ldots\}, \prec)$ denotes that the a_n's are ordered by the relation $\prec$ in the order they are listed.

Although the symbol $\prec$ reads "precedes", this terminology by no means assigns to $\prec$ an intuitive meaning. For instance, Z ordered by $0, 1, -1, 2, -2, \ldots$, i.e., integers ordered by increasing magnitude with the positive integers coming first, satisfies the three order properties listed above. Another important example of a partial order is $(\mathcal{P}(A), \subseteq)$, namely, the family of all the subsets of a fixed set A partially ordered by inclusion; in this case not all the elements of $\mathcal{P}(A)$ are comparable.

If $M_1 \subseteq M$ and $(M, \prec)$ is a partially ordered set, then we may consider M_1 as a partially ordered set by simply restricting $\prec$ to M_1, i.e., by ordering elements in M_1 as they were ordered in M. This partial ordering is denoted by $(M_1, \prec |M_1)$.

As we saw above, a partial order does not insure that all the elements of a set are comparable. A partially ordered set $(M, \prec)$ is said to be totally ordered (also called linearly ordered or ordered), provided that for every $m \neq m_1$ in M, either $m \prec m_1$ or $m_1 \prec m$. $(Z, \leq)$ is totally ordered.

In identifying sets according to their order properties, when do we say that $(M, \prec)$ and $(M^*, \prec^*)$ are order equivalent? First M and M^* must be equivalent, and in addition we require that there exists an equivalence function $f: M \to M^*$ which is order preserving, i.e., for all m_1, m_2 in M,

$$m_1 \prec m_2 \quad \text{iff} \quad f(m_1) \prec^* f(m_2).$$

For example, $(N, \leq)$ and $(\{1,3,\ldots,2,4,\ldots\}, \prec)$ are not order equivalent, whereas $(N, \leq)$ and $(\{a_1, a_2, \ldots\}, \prec)$ are.

Given the ordered sets $(M, \prec)$ and $(M^*, \prec^*)$, we say that they have the same order type provided they are order equivalent. The order type of $(\{1,2,\ldots,n\}, \leq)$ is denoted by n, that of $(\{1,2.\ldots\}, \leq)$ by ω, that of $(\{\ldots,2,1\}, \geq)$ by ω^*, that of $([0,1], \leq)$ by λ and that of $([0,1] \cap Q, \leq |Q)$ by η. Although it is apparent that not all order types are comparable, we can not compare ω and ω^*, it still is possible to do some arithmetic with order types. We do addition and multiplication as examples.

Given order types μ_1 and μ_2, we define the sum $\mu_1 + \mu_2$ as the order type obtained as follows: Let $(M_1, \prec_1)$ and $(M_2, \prec_2)$, $M_1 \cap M_2 = \emptyset$, be ordered sets with order types μ_1 and μ_2 respectively. On $M = M_1 \cup M_2$ introduce the order $\prec$ by means of the following relations: If $m, m' \in M_1$, then $m \prec m'$ iff $m \prec_1 m'$, if $m, m' \in M_2$, then $m \prec m'$ iff $m \prec_2 m'$, and finally, if $m \in M_1$ and $m' \in M_2$, then let $m \prec m'$. It is readily seen that $(M, \prec)$ is an ordered set and $\mu_1 + \mu_2$ is defined to be the order type μ corresponding to $(M, \prec)$.

It is clear that $\mu_1 + \mu_2$ is well defined, i.e., it is independent of the sets $(M_1, \prec_1)$ and $(M_2, \prec_2)$ chosen above, and that, in general, $\mu_1 + \mu_2 \neq \mu_2 + \mu_1$. For instance, we have $n + \omega = \omega$ for all n, and $1 + \omega \neq \omega + 1$. Also, addition of order types is associative.

As for the multiplication of order types, given order types μ_1 and μ_2, we define the product $\mu_1\mu_2$ as the order type obtained as follows: Let $(M_1, \prec_1)$ and $(M_2, \prec_2)$ be ordered sets with order type μ_1 and μ_2 respectively. On $M_1 \times M_2$ consider the relation $\prec$ given by $(m_1, m_2) \prec (m_1', m_2')$ iff either $m_1 \prec_1 m_1'$, or $m_1 = m_1'$ and $m_2 \prec_2 m_2'$. It is not hard to see that $(M_1 \times M_2, \prec)$ is an ordered set, and $\mu_1\mu_2$ is defined as the order type μ corresponding to its order equivalence class. $\prec$ is called the lexicographic order on $M_1 \times M_2$, and the reason for this is apparent. Note that in general $\mu_1\mu_2 \neq \mu_2\mu_1$. However, as the reader can verify, addition is distributive with respect to multiplication, i.e., $\mu(\mu_1 + \mu_2) = \mu\mu_1 + \mu\mu_2$.

2. WELL-ORDERED SETS AND ORDINALS

Given a countable set A it is possible to assign to it various nonequivalent order types. For instance, there are countable sets with order type $\omega, \omega+\omega, \omega^*, \omega+\omega^*, \ldots$ In fact, there are $2^{\aleph_0}$ nonequivalent order types corresponding to sets of cardinal $\aleph_0$. Inspired by the basic model $(N, \leq)$ we will focus on the theory of well-ordered sets. We need some definitions first.

Given an ordered set $(M, \prec)$, we say that $m \in M$ is the first element of M if m precedes any other element of M, i.e., $m \prec m'$ for every m' in M. By AS the first element is unique.

Similarly, we say that the ordered set $(M, \prec)$ has a last element if there is $m \in M$ which is preceded by any other element of M, i.e., $m' \prec m$ for every m' in M. For instance, an ordered set with order type ω has a first, but no last element, and one with the order type $\omega + 1$ has both a first and a last element.

Closely related to these notions are the concepts of lower and upper bound, and minimal and maximal element. Given an ordered set $(M, \prec)$ and a subset M' of M, we say that $m \in M$ is a lower bound for M' if $m \prec m'$ for every $m' \in M'$; we say that m_1 is an upper bound for M' if $m' \prec m_1$ for every m' in M'. For instance, in $(\mathcal{P}(A), \subseteq)$, $\emptyset$ is a lower bound and A is an upper bound for any family of $\mathcal{P}(A)$.

An element m is said to be a minimal element of M iff there is no $m \neq m' \in M$ so that $m' \prec m$. Similarly, an element $m \in M$ is said to be maximal iff there is no $m' \in M$ so that $m \prec m'$. For instance, if $M = \{a, b, c\}$ is partially ordered by $a \prec b$ and $a \prec c$, then a is the minimal element and b and c are maximal elements.

Finally, we say that an ordered set $(M, \prec)$ is well-ordered if M and any of its nonempty subsets, ordered by the restriction order, has a first element. The order type of a well-ordered set is called an ordinal number, or plainly an ordinal. For instance, $(N, \leq)$ and $(N+N, \leq)$ are well-ordered sets with ordinal ω and $\omega + \omega$, respectively. On the other hand, the order types λ and η are not ordinals.

It is also clear that if $M_1 \subseteq M$ and $(M, \prec)$ is well-ordered, then so is $(M_1, \prec |M_1)$. More generally, if $(M, \prec)$ is well-ordered and if A is equivalent to a subset M_1 of M, then A can also be well-ordered. To see this let $f: M_1 \to A$ be an equivalence function and rewrite $A = \{a_m\}$ where $a_m = f(m)$, $m \in M_1$. It is then readily seen that the relation $\prec^*$ on $A \times A$ given by $a_m \prec^* a_{m'}$ iff $m \prec m'$ is an order relation on A, and that $(A, \prec^*)$ is well-ordered. Although η is not an ordinal, by the preceeding discussion $Q \cap (0,1)$, or any other countable set for that matter, can be well-ordered by means of an order induced by $(N, \leq)$.

The next step is, then, to consider whether arbitrary sets may be well-ordered. In the early 1900's Zermelo showed that this was the case provided we assume the validity of the so-called Axiom of Choice. Again, as in the case of the rationals, if the set under consideration is already ordered, the well-ordering that Zermelo's theorem induces does not, in general, coincide with the existing order. We discuss this in more detail later, but first we consider the Axiom of Choice. It was introduced by Zermelo in 1904 and, in one of its many equivalent formulations, it states:

Axiom of Choice. Given an arbitrary family $\mathcal{A} = \{A_i : i \in I\}$ of nonempty sets indexed by (the nonempty) set I, there exists a function $f: I \to \bigcup_{i \in I} A_i$, called a choice or selection function, such that for each $i \in I$, $f(i)$ is an element of A_i.

Where does the Axiom of Choice stand in comparison to the more familiar (Zermelo-Fraenkel) axioms of set theory? Gödel (1906-1978) established the fact that the Axiom of Choice and the generalized continuum hypothesis, which we do not discuss here, are consistent with the remaining axioms of set theory, provided that these axioms are consistent themselves. On the other hand, some 60 years after its introduction, P.J. Cohen showed that the Axiom of Choice can neither be proved nor refuted from the usual axioms of set theory. The Axiom of Choice is thus a new principle of set formation.

Zermelo's theorem is actually equivalent to the Axiom of Choice. We leave to the reader to prove that the Axiom of Choice implies Zermelo's theorem, cf. 5.14 below. Conversely, suppose that any set can be well-ordered and let $\mathcal{A} = \{A_i : i \in I\}$ be an arbitrary family of nonempty sets indexed by I. Well-order each A_i and observe that by choosing the first element a_i, say, in each nonempty subset A_i of $\bigcup_{i \in I} A_i$ we get the required selection function f by simply putting $f(i) = a_i \in A_i$.

The Axiom of Choice plays an important role in Analysis. As we shall have the opportunity to discover as we go along, in addition to the results discussed in the next section, the existence of a non Lebesgue-measurable set, the existence of a maximal orthonormal set in a Hilbert space and the proof of the Hahn-Banach Theorem, all rely on the Axiom of Choice. In applications we often prefer to work with an equivalent principle, namely Zorn's Lemma, because it only requires that we deal with partially ordered sets. To state Zorn's Lemma we need a definition.

We say that a subset A of a partially ordered set $(M, \prec)$ is a chain in M if $(A, \prec |A)$ is totally ordered. For instance, if $M = \{a, b\}$, then $A = \{\emptyset, \{a\}\}$ and $B = \{\emptyset, \{a\}, M\}$ are chains in $(\mathcal{P}(A), \subseteq)$.

The stage is now set for

Zorn's Lemma. A partially ordered set $(M, \prec)$ has a maximal element if every chain A in M has an upper bound.

As we stated above, Zorn's Lemma is equivalent to the Axiom of Choice. To illustrate a typical use of Zorn's Lemma we show that it actually implies the Axiom of Choice. Suppose, then, that $\mathcal{A} = \{A_i : i \in I\}$ is a family of nonempty sets indexed by a nonempty set I; we must find a selection function. Let $\mathcal{F} = \{f\}$ be the collection of those functions f which satisfy the following three properties:

1. Domain of $f \subseteq I$.
2. Range of $f \subseteq \bigcup_{i \in I} A_i$.
3. $f(i)$ is an element of A_i for each i in the domain of f.

It is readily seen that $\mathcal{F} \neq \emptyset$. Indeed, choose i_0 in the nonempty set I, and since A_{i_0} is nonempty, pick an element $a_{i_0} \in A_{i_0}$. Then, the function $f: \{i_0\} \to A_{i_0}$ given by $f(i_0) = a_{i_0}$ is in $\mathcal{F}$. $\mathcal{F}$ may be partially ordered as follows: Given f, g in $\mathcal{F}$, we say that $f \prec g$ iff J = domain of $f \subseteq$ domain of g and $g|J = f$.

Next we check that any chain $\mathcal{C} = \{g\}$ in $\mathcal{F}$ has an upper bound. Indeed, let $J = \bigcup_{g \in \mathcal{C}}(\text{domain of } g)$ and $A = \bigcup_{g \in \mathcal{C}}(\text{range of } g)$; by assumption $J \subseteq I$ and $A \subseteq \bigcup_{i \in I} A_i$. A candidate for an upper bound is the function $f: J \to A$ defined as follows: If $i \in I$, i is in the domain of some g in $\mathcal{C}$; then put $f(i) = g(i) \in A_i$. There may be more than one g in $\mathcal{C}$ so that $i \in$ (domain of g), but since $\mathcal{C}$ is a chain it readily follows that all functions g in $\mathcal{C}$ for which $g(i)$ is defined assume a common value at i, and consequently, f is well-defined. It is now clear that $f \in \mathcal{F}$ and that $g \prec f$ for every g in $\mathcal{C}$. In other words, f is an upper bound for $\mathcal{C}$, and every chain in $\mathcal{F}$ has an upper bound. By Zorn's Lemma $\mathcal{F}$ has a maximal element f, say. We claim that domain of $f = I$, and if this is the case then clearly f is a choice function. Suppose that the domain of f is strictly contained in I and let $i \in I \setminus (\text{domain of } f)$. Then since

$a_i \in A_i \neq \emptyset$, the function g defined on (domain of f) $\cup \{i\}$ by $g|$(domain of f) $= f$ and $g(i) = a_i$, satisfies properties 1, 2, and 3 listed above and $f \prec g$, thus contradicting the maximality of f in $\mathcal{F}$. Whence the domain of f is I, f is a selection function and the Axiom of Choice follows.

3. APPLICATIONS OF ZORN'S LEMMA

It is readily seen that any continuous function $f\colon R \to R$ which satisfies $f(x+y) = f(x) + f(y)$ is of the form $f(x) = cx$, for some real number $c(= f(1))$. Are there, necessarily discontinuous, functions $f(x) \neq cx$ which verify the same functional relation? Yes, there are, and to construct such functions we need some definitions.

We say that a nonempty subset M of R is l.i. (linearly independent) over Q if every finite subset $\{x_1, \ldots, x_n\}$ of M is l.i. over Q, i.e., there is no nontrivial linear combination $r_1x_1 + \cdots + r_nx_n = 0$, $r_i \in Q$. A Hamel basis $H = \{x_i\}$ for R is a maximal l.i. set. More precisely, H is a set of nonzero real numbers which satisfies the following three properties:

1. H is l.i. over Q.
2. Every x in R can be written as a finite linear combination, with rational coefficients, of the x_i's.
3. If $H \subset H'$, then H' is not l.i. over Q.

Suppose we have a Hamel basis H for R and observe that for each x in R, the decomposition given in 2 above is unique. We may then construct a function f as follows: If $x = r_1x_1 + \cdots + x_nr_n$, $x_1, \ldots, x_n$ in H, put

$$f(x) = r_1 f(x_1) + \cdots + r_n f(x_n)\,.$$

Since the decomposition for each x is unique, f is well-defined and it satisfies the relation $f(x+y) = f(x) + f(y)$ for all x, y in R. Since H has at least two elements x_1 and x_2, say, we may set $f(x_1) = 1$, $f(x_2) = 0$ and take arbitrary values for the other x_i's in H. Furthermore, since $x_2 \neq 0$, f cannot be of the form $f(x) = cx$ unless $c = 0$. But in this case we have $1 = f(x_1) = cx_1 = 0$, which is a contradiction. Thus f is not linear and our construction will be accomplished once we exhibit a Hamel basis for R. It is at this juncture that we invoke Zorn's Lemma.

Proposition 3.1. There is a Hamel basis for R.

Proof. The class $\mathcal{M}$ of those subsets of R which are l.i. over Q is nonempty since it contains any nonzero real number r. Let $\mathcal{C}$ be a chain in the partially ordered set $(\mathcal{M}, \subseteq)$, we check that $\mathcal{C}$ has an upper bound.

This is not hard to do since the union of all the real numbers in $\mathcal{C}$ is an element of $\mathcal{M}$ which is an upper bound. By Zorn's Lemma, $\mathcal{M}$ has a maximal element M, say. It is readily seen that M is a Hamel basis for R. Indeed, let Y denote the span of M in R. If $Y = R$ we are done, otherwise let $r \in R \setminus Y$ and observe that the set $M \cup \{r\}$ is l.i. over Q and it properly contains M, contrary to the latter's maximality. Therefore, H is a Hamel basis for R. ■

We close this section with two interesting results that also follow from Zorn's Lemma, namely that cardinals are comparable, and that ordinals are comparable, by the relation less than or equal to.

Theorem 3.2. Any two cardinal numbers are comparable.

Proof. Let a, b be nonzero cardinals and let A and B be (nonempty) sets so that card $A = a$ and card $B = b$. Furthermore let $\mathcal{F}$ be the family of functions f defined from a subset A_0 of A, into a subset B_0 of B, which are one-to-one and onto. Given $x \in A$ and $y \in B$, $f(x) = y$ establishes an equivalence between $A_0 = \{x\}$ and $B_0 = \{y\}$, and $\mathcal{F} \neq \emptyset$. Also $\mathcal{F}$ may be partially ordered by the usual extension relation between functions. Next we check that any chain $\mathcal{C}$ in $\mathcal{F}$ has an upper bound. Let $X = \bigcup(\text{domain of } f) \subseteq A, Y = \bigcup(\text{range of } f) \subseteq B$, where the unions are taken over those f's in $\mathcal{C}$. Since it is readily seen that the function $g: X \to Y$ given by $g|(\text{domain of } f) = f$ is an upper bound for $\mathcal{C}$ (here we use the fact that $\mathcal{C}$ is a chain to insure that g is well-defined), by Zorn's Lemma $\mathcal{F}$ has a maximal element $f: X \to Y$, say. There are four possibilities, to wit: (i) $X = A$ and $Y = B$, in this case $a = b$, (ii) $X = A$ and $Y \subset B$, then $a \leq b$, (iii) $X \subset A$ and $Y = B$, then $b \leq a$, and finally (iv) $X \subset A$ and $Y \subset B$, which we show cannot occur. For, if case (iv) holds, then there are elements $x \in A \setminus X$ and $y \in B \setminus Y$, and the function $g: X \cup \{x\} \to Y \cup \{y\}$ given by $g|X = f$ and $g(x) = y$ is in $\mathcal{F}$, and it properly extends f, contradicting its maximality. ■

In order to prove our next result we need a concept that will enable us to work with the first, or initial, elements of a well-ordered set. Given a well-ordered set $(M, \prec)$ and m in M, let

$$I_m = \{m' \in M : m' \prec m, m' \neq m\};$$

I_m is called the initial segment of M corresponding to m.

An important property that initial segments satisfy is

Proposition 3.3. No nonempty well-ordered set is order equivalent to any of its initial segments.

Proof. Let $(M, \prec)$ be a well-ordered set, suppose there is an element m in M so that M is order equivalent to I_m, and let $f: M \to I_m$ be a function which establishes the order equivalence. Then $f(m) \in I_m$ and by the definition of I_m, $m_1 = f(m) \prec m = m_0$. We iterate this step and by setting $m_k = f(m_{k-1})$, $k = 1, 2, \ldots$, we obtain a sequence of elements of M such that for all k, $m_k \prec m_{k-1} \prec \ldots \prec m_0$. Thus the subset of M consisting precisely of the m_k's has no first element, contradicting the fact that M is well-ordered. Whence M is not order equivalent to any I_m, as we wanted to show. ■

We are now ready to compare ordinal numbers. Given ordinal numbers a and b, we say that a precedes b, and we write $a \le b$, provided the following holds: If A, B are sets with ordinal number a and b respectively, then A is order equivalent to B or an initial segment of B. If A is order equivalent to an initial segment of B we write $a < b$. Also if $a \le b$ and $b \le a$, then $a = b$.

The stage is now set for

Theorem 3.4. Any two ordinal numbers are comparable.

Proof. Let a, b be nonzero ordinal numbers and let $(A, \prec)$ and $(B, \prec^*)$ be well-ordered sets with ordinal a and b respectively. Inspired by Theorem 3.2, let $\mathcal{F}$ denote the family of all order preserving mappings which establish an order equivalence between an initial segment of A, or A itself, onto an initial segment of B or B. It is clear that $\mathcal{F} \ne \emptyset$. Indeed, if x is the first element of A and y is the first element of B, then the function $f: \{x\} \to \{y\}$ defined by $f(x) = y$ is in $\mathcal{F}$. As in the proof of Theorem 3.2 it is readily seen that when $\mathcal{F}$ is partially ordered by extension, any chain C in $\mathcal{F}$ has an upper bound and consequently, $\mathcal{F}$ has a maximal element f, say. If $f: A_1 \to B_1$ is an order equivalence and A_1, B_1 are initial segments of A and B, and if $A_1 \subset A$ and $B_1 \subset B$, then let x, y be the first elements of $A \setminus A_1$ and $B \setminus B_1$ respectively. Then the mapping $g: A_1 \cup \{x\} \to B_1 \cup \{y\}$ defined by $g|A_1 = f$ and $g(x) = y$ is an order equivalence of an initial segment of A, or A, onto an initial segment of B, or B, which contradicts the maximality of f. Thus, either $A_1 = A$ or $B_1 = B$, or both. In the first case $a \le b$ and actually by Proposition 3.3 $a < b$; in the second case $b < a$ and in the last case $a \le b$ and $b \le a$, i.e., $a = b$. ■

4. THE CONTINUUM HYPOTHESIS

We close this chapter with the discussion of a particular set A, namely one that is well-ordered, uncountable, and such that all its initial segments are at most countable. That uncountable well-ordered sets exist follows from Zermelo's theorem, so let $(X, \prec)$ be one such set. If X has all the desired properties we are finished, otherwise the set $X_0 = \{x \in X : I_x \text{ is uncountable }\} \neq \emptyset$; let x_0 be the first element of X_0. Then I_{x_0} is well-ordered, uncountable, and for every $x \prec x_0$, I_x is at most countable. This is precisely the set A we wanted to construct. Also observe that if B is another set with similar properties, then clearly B is not order equivalent to an initial segment of A (since they are all at most countable), nor for the same reason is A order equivalent to any initial segment of B. Thus by Theorem 3.4 the ordinal of A is equal to that of B, and A is unique.

The ordinal of A is special, it is denoted by Ω and it is referred to as the first uncountable ordinal. The cardinal of A is denoted by $\aleph_1$, and since R is uncountable it is clear that $\aleph_1 \leq c$. The question is whether $\aleph_1 = c$ or not.

The Continuum Hypothesis was formulated by Cantor and it asserts that $\aleph_1 = c$. In 1900, Hilbert (1862–1943) included this hypothesis as the first item in a list of open problems that he presented to the Second International Congress of Mathematicians in Paris. In 1939 Gödel showed that the Continuum Hypothesis is consistent with the axioms of set theory, that is, using the usual axioms of set theory, including the Axiom of Choice, the Continuum Hypothesis cannot be proved false. On the other hand, in 1963 P.J. Cohen showed that the Continuum Hypothesis is independent of these axioms; this means that from these axioms one cannot prove the Continuum Hypothesis true.

5. PROBLEMS AND QUESTIONS

5.1 By means of an example show that not every AS relation may be extended to an order.

5.2 Let $\mathcal{P} = \{p\}$ be the family of all real polynomials and let R be the following relation on $\mathcal{P} \times \mathcal{P}$: $p_1 R p_2$ iff there is a real number x_0 such that $p_1(x) \leq p_2(x)$ for any $x \geq x_0$.
Is R a partial order on $\mathcal{P}$? Is it a total order? Is it a well-ordering?

5.3 Let $I = [0,1]$, $C(I) = \{f: I \to R: f \text{ is continuous}\}$, and let R be the following relation on $C(I) \times C(I)$: $f_1 R f_2$ iff $f_1(x) \leq f_2(x)$ for every x in I. Is R a partial order on $C(I)$? Is it a total order? Is it a well-ordering?

5.4 Let $\mathcal{A} = \{\prec\}$ be the collection of all partial orders on a given set M. $\mathcal{A}$ can be partially ordered by "extension" of orders; let $(\mathcal{A}, \sqsubseteq)$ denote this partially ordered set. Show that $\prec^*$ is a maximal element of $(\mathcal{A}, \sqsubseteq)$ iff $(M, \prec^*)$ is totally ordered.

5.5 Let $(M, \prec)$ be totally ordered. Show that $(M, \prec)$ is well-ordered iff there exists no infinite strictly decreasing sequence in M.

5.6 Let $(M, \prec)$ be totally ordered. Show that if every initial segment I_m is well-ordered, then M itself is well-ordered.

5.7 Suppose every countable subset of a totally ordered set $(M, \prec)$ is well-ordered, does it follow that $(M, \prec)$ is well-ordered?

5.8 Is the lexicographic order on $N \times N$ a well-ordering?
How about the lexicographic order on $N \times R$?

5.9 Suppose $(M, \prec)$ is totally ordered and introduce a relation R on $(M \times M) \times (M \times M)$ as follows: $(m, m_1) R (m', m_1')$ iff (i) $(m, m_1) \prec \max(m', m_1')$, or (ii) $\max\,(m, m_1) = \max\,(m', m_1')$ and $m \prec m'$, or (iii) $\max\,(m, m_1) = \max\,(m', m_1')$, $m = m'$, and $m_1 \prec m_1'$. Is R a total order? If $(M, \prec)$ is well-ordered, is $(M \times M, R)$ well-ordered?

5.10 (Principle of Transfinite Induction) Given a nonempty well-ordered set $(M, \prec)$, let $P(m)$ be a statement which is formulated for each m in M. Further suppose that (i) $P(m)$ is true for the first element of M, (ii) If $P(m)$ is true for each $m \prec m^*$, then $P(m^*)$ is also true. Show that $P(m)$ is true for every $m \in M$.

5.11 Suppose M_1 is a subset of a well-ordered set M and that the following holds: If $m \in M$ and the initial segment $I_m \subseteq M_1$, then $m \in M_1$. Show that $M_1 = M$.

5.12 An interesting application of Transfinite induction is to show the various properties concerning operations with order types. For instance, prove that the addition is associative and left distributive with respect to multiplication.

5.13 Does the conclusion of the Principle of transfinite induction hold if, instead of (ii) there, we assume: (ii') If $P(m)$ is true, then $P(m+1)$ is also true?

5.14 Prove that the Axiom of Choice implies Zermelo's theorem.

5.15 Let $\mathcal{A}$ be a family of nonempty sets of nonnegative integers. Can you construct, without invoking the Axiom a Choice, a selection function on $\mathcal{A}$?

5.16 Use the Axiom of Choice to prove the following result: A mapping $f: A \to B$ is onto iff there is a mapping $g: B \to A$ such that $f \circ g = 1$(identity on B). Moreover, any such function g is one-to-one.

5.17 Use Zorn's Lemma to show that for every infinite cardinal a,

$$a + a = aa = a\,.$$

5.18 Show that any two Hamel bases for R have the same cardinality.

5.19 Prove that Zermelo's theorem implies Zorn's Lemma.

5.20 Prove that every set of cardinals is well-ordered.

5.21 Given an ordinal a, let $W_a = \{\text{ordinals } b : b < a\}$. Show that W_a is well-ordered and find its ordinal.

5.22 Is $\omega + \omega = \omega 2$? Is $\omega + \omega = 2\omega$?

5.23 Discuss which of the following statements is true: Given ordinals a, b, d, (i) If $a + b = a + d$, then $b = d$, (ii) If $b + a = d + a$, then $b = d$, (iii) If $b \neq 0$, then $a < a + b$, and, (iv) If $b \neq 0$, then $b < a + b$.

5.24 Show that given an ordinal a, there exists no ordinal b such that $a < b < a + 1$.

5.25 If a, b are ordinals and $a < b$, show that there exists a unique ordinal d such that $a + d = b$.

5.26 Show that every set of ordinals is well-ordered.

5.27 Prove that the Axiom of Choice is equivalent to the following principle: Let $\mathcal{A}$ be a family of pairwise disjoint nonempty sets. Then there exists a set C such that $A \cap C$ contains exactly one element, for each $A \in \mathcal{A}$.

5.28 Suppose a discontinuous function $f: R \to R$ satisfies $f(x + y) = f(x) + f(y)$ for all $x, y \in R$. Show that $f^{-1}(\{0\})$ is dense in R.

5.29 Show that there exists no infinite cardinal a such that $\aleph_0 < a < \aleph_1$.

5.30 Prove that the set A constructed in Section 4 has the following three properties: (i) A has no largest element, (ii) For every ordinal $x_0 < \Omega$, the set $\{x : x_0 < x < \Omega\}$ is uncountable, and, (iii) Let A_0 denote the subset of A consisting of all elements x such that x has no immediate predecessor, i.e., for no $y \in A$ we have $x = y + 1$. Then, A_0 is uncountable.

5.31 Can a subset S of real numbers ordered by the order induced by $(R, \leq)$ be (order) equivalent to the set of all countable ordinals?

5.32 Let Ω be the first uncountable ordinal and suppose that f is a real-valued monotone increasing function defined for all ordinals $a < \Omega$. Show that f is bounded, and that in fact it is eventually constant.

5.33 Let A be a set with ordinal Ω. Show that every countable subset of A has an upper bound.

5.34 (The Burali-Forti Paradox) For every infinite set A of ordinals there is an ordinal number that exceeds every ordinal in A. Can we then consider the set of all ordinals?

CHAPTER **III**

The Riemann-Stieltjes Integral

In this chapter we introduce the Riemann-Stieltjes integral and study some of its basic properties. We also describe a natural class of functions related to this notion of integral, namely the class BV of functions of bounded variation.

1. FUNCTIONS OF BOUNDED VARIATION

What do questions in Physics which involve mass distributions that are partly discrete and partly continuous, problems in Probability that consider continuous and discrete random variables simultaneously, and the determination of the arc length of a curve have in common? They all involve the Riemann-Stieltjes integral. It was to solve the "moment problem", that is, to find a distribution of mass whose moments of all orders are known, that led Stieltjes (1856–1894) to introduce this integral, and in the process to give respectability to discontinuous functions.

We begin by discussing the question of finding the arc length of a plane curve $\Gamma: I = [a,b] \to R^2$ with graph $\{(x,y): x = \phi(t), y = \psi(t), t \in I\}$. Note that the graph may have self-intersections and is not necessarily continuous or bounded. Consider a partition $\mathcal{P} = \{a = t_0 < t_1 < \ldots < t_n = b\}$ of I and the corresponding points $P_i = (\phi(t_i), \psi(t_i))$, $1 \le i \le n$, on the graph of Γ. If we draw the polygonal line segments connecting P_0 to P_1, etc., we can obtain an approximation to the length of the graph of the curve Γ by means of the expression

$$L(\Gamma, \mathcal{P}) = \left(\sum_{j=1}^{n} (\phi(t_j) - \phi(t_{j-1}))^2 + (\psi(t_j) - \psi(t_{j-1}))^2\right)^{1/2}, \quad (1.1)$$

which corresponds to the length of the polygonal line segment. The next step is to take finer partitions of I and to define the length $L(\Gamma)$ of Γ as

$L(\Gamma) = \lim L(\Gamma, \mathcal{P})$, where the limit is taken as the norm of the partition, i.e., $\sup(t_j - t_{j-1})$, $1 \le j \le n$, goes to 0. Rectifiable curves are those Γ's for which $L(\Gamma) < \infty$.

There are many physical quantities whose calculation takes a similar form. For instance, to compute the mass of a slab of unit thickness we must consider expressions of the form $\sum_j \varrho(x_j)(A(x_j) - A(x_{j-1}))$, where ϱ denotes density and A denotes area, or $\sum_j \varrho(x_j)(V(x_j) - V(x_{j-1}))$, where V denotes volume, and their limits as the norm of the partition goes to 0.

So, in general, we are led to consider, for arbitrary functions f, g defined on I, expressions of the form

$$\sum_{j=1}^{n} g(\eta_j)\,(f(t_j) - f(t_{j-1}))\,, \quad \eta_j \in [t_{j-1}, t_j]\,, \tag{1.2}$$

and their limits as the norm of the partition goes to 0. If this limit is to exist we expect that it will satisfy properties similar to those of the Riemann integral, which corresponds to the choice $f(t) = t$ in (1.2).

It then becomes quickly apparent that step functions, i.e., those g's of the form $\sum_{j=1}^{n} c_j \chi_j$, where the c_j's are arbitrary real constants and the χ_j's are the characteristic functions of a partition $\{I_j\}$ of I consisting of nonoverlapping closed intervals (their interiors are disjoint), should be incorporated into the theory. In particular, if the c_j's are ± 1, we note that not only is each step function Riemann integrable, but also that

$$\left| \int_a^b g(t)\, dt \right| = \left| \sum\nolimits_{j=1}^{n} c_j (t_j - t_{j-1}) \right| \le b - a\,. \tag{1.3}$$

A moment's thought will convince the reader that an appropriate general version of (1.3), now with a constant that also depends on f on the right-hand side there, should hold as well. Thus, with the familiar notation

$$\operatorname{sgn} x = \begin{cases} 1 & \text{if } x > 0 \\ 0 & \text{if } x = 0 \\ -1 & \text{if } x < 0, \end{cases}$$

if we put $g = \sum_j c_j \chi_j$, where $c_j = \operatorname{sgn}(f(t_j) - f(t_{j-1})) = \pm 1$ (or 0, but then the term is unimportant) in (1.2) above, and if $c = c_{I,f}$ denotes the uniform bound in the estimate analogous to (1.3), it follows that

$$\sum_{j=1}^{n} |f(t_j) - f(t_{j-1})| \le c\,, \quad \text{all partitions } \mathcal{P}\,. \tag{1.4}$$

Functions that satisfy (1.4) are called of bounded variation and the class of such functions is denoted by BV. These functions were discovered by Jordan (1838–1921) in 1881 while working out the proof of Dirichlet concerning the convergence of Fourier series.

Before we continue with the discussion of the integral we investigate what BV functions are like. A useful notation in what follows is this: For a BV function f defined on $I = [a,b]$ and a subinterval $J = [c,d]$ of I, put $V(f;c,d) =$ sup of the left-hand side of (1.4), where the sup is taken over all finite partitions $\mathcal{P}$ of J. So, BV functions f on I are precisely those with $V(f;a,b) < \infty$.

What are some BV functions? Monotone functions f are BV and it is readily seen that

$$V(f;a,b) \le |f(b) - f(a)| .$$

On the other hand, there are continuous functions which are not BV. Rather than looking for a function in our repertoire, cf. 4.12 below, we construct one. Let $\{a_k\}, \{d_k\}$ be two sequences in (0,1], $a_1 = 1$, which decrease to 0, and suppose that $\sum d_k$ diverges. Define f as follows: On each subinterval $[a_{k+1}, a_k]$ the graph of f is the isosceles triangle with base $[a_{k+1}, a_k]$ and height d_k. Thus $f(a_k) = 0$ and $f((a_k + a_{k+1})/2) = d_k$. If we are careful to set $f(0) = 0$, f is a continuous function and

$$V(f;0,1) \ge \sum_{k=1}^{n} d_k \to \infty , \quad \text{as } n \to \infty .$$

Nevertheless, as we shall see below, the continuity of a BV function is reflected by that of its "variation function" $V(x) = V(f;a,x), a \le x \le b$.

Some of the properties of V are readily obtained. For instance, $V(0) = 0$, V is nondecreasing on I,

$$|f(x+h) - f(x)| \le V(f;x,x+h) , \quad a \le x < x+h \le b \tag{1.5}$$

and

$$V(x) + V(f;x,x+h) = V(x+h)., \quad a \le x < x+h \le b . \tag{1.6}$$

We sketch a proof of the identity (1.6). Given $\varepsilon > 0$, let $\mathcal{P}$ be a partition of $[a,x]$ such that

$$s = \sum_{\text{over } \mathcal{P}} |f(x_j) - f(x_{j-1})| > V(f;a,x) - \varepsilon/2 ,$$

and let $\mathcal{P}_1$ be a partition of $[x,x+h]$ such that

$$s_1 = \sum_{\text{over } \mathcal{P}_1} |f(x_j) - f(x_{j-1})| > V(f;x,x+h) - \varepsilon/2 .$$

Since $\mathcal{P}_2 = \mathcal{P} \cup \mathcal{P}_1$, i.e., the partition consisting of the points in $\mathcal{P}$ or $\mathcal{P}_1$, is a partition of $[a,x+h]$ it follows that

$$\begin{aligned} V(f;a,x) + V(f;x,x+h) &\le s + \varepsilon/2 + s_1 + \varepsilon/2 \\ &\le \sum_{\text{over } \mathcal{P}_2} |f(x_j) - f(x_{j-1})| + \varepsilon \\ &\le V(f;a,x+h) + \varepsilon\,. \end{aligned} \tag{1.7}$$

Since $\varepsilon > 0$ is arbitrary, (1.7) gives one inequality ($\le$) in (1.6).

In order to show the opposite inequality, given $\varepsilon > 0$, let $\mathcal{P}$ be a partition of $[a,x+h]$ with the property that $V(f;a,x+h) \le s = \sum_{\text{over } \mathcal{P}} |f(x_j) - f(x_{j-1})| + \varepsilon$. Clearly we may assume that x is a point in $\mathcal{P}$ since s only increases when we add points to it. But then, breaking $\mathcal{P}$ up into the partitions $\mathcal{P}_1 = \{a = x_0 < \ldots < x_n = x\}$ and $\mathcal{P}_2 = \{x = x_n < \ldots < x_m = x+h\}$, it readily follows that

$$V(f;a,x+h) \le s + \varepsilon \le V(f;a,x) + V(f;x,x+h) + \varepsilon. \tag{1.8}$$

Again, since $\varepsilon > 0$ is arbitrary, (1.8) gives the other inequality ($\ge$) in (1.6) and equality holds there, as asserted.

We are now ready to prove

Proposition 1.1. Suppose f is BV on $I = [a,b]$ and let $x \in [a,b)$. Then f is right-continuous at x iff V is right-continuous at x.

Proof. From the inequality (1.5) and the relation (1.6) it follows that if V is right-continuous at x, so is f. Conversely, suppose that f is right-continuous at x and observe that since V is nondecreasing, $\lim_{h\to 0^+} V(x+h) = L$ exists, cf. 4.3 below. Suppose that

$$L - V(x) = \eta > 0\,, \tag{1.9}$$

and choose $\delta > 0$ so that $x + \delta < b$ and

$$|f(y) - f(x)| < \eta/2\,, \quad 0 < y - x < \delta\,. \tag{1.10}$$

From (1.9), (1.6) and the monotonicity of V it readily follows that $V(x+\delta) - V(x) = V(f;x,x+\delta) > \eta$, and consequently, there is a partition $\mathcal{P}_0 = \{x < x_1 = x + \delta_1 < x_2 < \ldots < x_n = x + \delta\}$ of $[x, x+\delta]$ for which also $\sum_{j=1}^{n} |f(x_{j+1}) - f(x_j)| > \eta$. Now, as before, by (1.10), there exists a partition $\mathcal{P}_1$ of $[x, x+\delta_1]$ consisting of more than one intermediate point such that

$$\sum_{\text{over } \mathcal{P}_1} |f(x_{j+1}) - f(x_j)| > \eta\,.$$

Therefore combining $\mathcal{P}_0$ and $\mathcal{P}_1$ we get a partition $\mathcal{P}$ of $[x, x+\delta]$ so that $\sum_{\text{over } \mathcal{P}} |f(x_{j+1}) - f(x_j)| > 2\eta$. We may now repeat this procedure, i.e., subdivide the first interval in each partition and thus obtain, for any given k, a partition $\mathcal{P}$ of $[x, x+h]$ so that

$$\sum_{\text{over } \mathcal{P}} |f(x_{j+1}) - f(x_j)| > k\eta .$$

But this implies that f is not of bounded variation over I, which is a contradiction. In other words, $\eta = 0$, V is right-continuous at x, and we have finished. ■

A similar, yet simpler, argument shows that for x in $(a,b]$, f is left-continuous at x iff V is left-continuous at x. Thus combining these results we get that for $x \in (a,b)$, f is continuous at x iff V is continuous at x.

To complete the description of BV functions we note the following properties of V.

Proposition 1.2. Let f be BV on I. Then $V \pm f$ are nondecreasing on I.

Proof. Observe that for $a \le x < y < b$ we have

$$(V(y) - f(y)) - (V(x) - f(x)) = V(f; x, y) - (f(y) - f(x)) \ge 0 ,$$

and $V - f$ is nondecreasing. A similar argument applies to $V + f$. ■

The difference, or any linear combination for that matter, of nondecreasing functions is BV. The interesting fact is that the converse is also true.

Theorem 1.3 (Jordan). Suppose f is BV on I. Then f can be written as a difference of two nondecreasing functions.

Proof. A decomposition that works is $f = V - (V - f)$. ■

The decomposition in Jordan's theorem is not unique; in fact if $f = f_1 - f_2$ and if g is any increasing function, then we also have $f = (f_1 + g) - (f_2 + g)$. In particular, if we need to, we may assume that the nondecreasing functions in Jordan's theorem are actually strictly increasing. Also, if f is right-continuous on $[a,b)$, the increasing functions in the Jordan decomposition may be assumed to be right-continuous there.

2. EXISTENCE OF THE RIEMANN-STIELTJES INTEGRAL

We return to the task at hand, namely that of defining the Riemann-Stieltjes integral. Since BV functions are differences of nondecreasing functions we do first the theory for nondecreasing functions, and then discuss the general case.

Given a bounded function $g: I \to R$ and a partition $\mathcal{P}$ of $I = [a,b]$ consisting of the points $a = x_0 < \ldots < x_n = b$, put $I_k = [x_k, x_{k+1}]$ and set $m_k = \inf_{I_k} g$ and $M_k = \sup_{I_k} g$. Further, if f is a nondecreasing function defined on I, let $\Delta_k f = f(x_{k+1}) - f(x_k)$, $0 \le k \le n-1$.

The lower and upper sums of g corresponding to $\mathcal{P}$ with respect to f are then defined by the expressions

$$s(g,f,\mathcal{P}) = \sum_{k=1}^{n} m_k \Delta_k f\,, \quad \text{and} \quad S(g,f,\mathcal{P}) = \sum_{k=1}^{n} M_k \Delta_k f\,,$$

respectively.

It is readily seen that as the partitions get finer, the lower sums increase and the upper sums decrease. Indeed, suppose that $\mathcal{P}$ is a partition of I, and that we have refined it by adding a point t, say, between x_k and x_{k+1}. Then

$$\inf_{[x_k,t]} g\,,\ \inf_{[t,x_{k+1}]} g \ge m_k\,, \quad \text{and} \quad \sup_{[x_k,t]} g\,,\ \sup_{[t,x_{k+1}]} g \le M_k\,,$$

and our assertion follows at once from this.

An argument using similar ideas, more precisely, working with a common refinement, shows that for any partitions $\mathcal{P}$ and $\mathcal{P}_1$ of I, we have

$$s(g,f,\mathcal{P}) \le S(g,f,\mathcal{P}_1)\,. \tag{2.1}$$

It is therefore natural to define the lower and upper Riemann-Stieltjes integrals of g with respect to f by

$$L(g,f,I) = \sup_{\mathcal{P}} s(g,f,\mathcal{P})\,, \quad \text{and} \quad U(g,f,I) = \inf_{\mathcal{P}} S(g,f,I)\,, \tag{2.2}$$

where the sup and inf are taken over all finite partitions $\mathcal{P}$ of I. By (2.1) it follows that $L(g,f,I) \le U(g,f,I)$.

If $L(g,f,I) = U(g,f,I)$ we say that g is Riemann-Stieltjes integrable with respect to f over I, and we write $g \in \mathcal{R}(f,I)$; this common value is denoted by $\int_a^b g(x)\,df(x)$, or plainly by $\int_a^b g\,df$.

When $f(x) = x$ this definition coincides, as it should, with that of $\int_a^b g(x)\,dx$, the Riemann integral of g over I; the class of Riemann integrable functions is denoted by $\mathcal{R}(I)$.

Whereas we are interested in finding sufficient conditions for g to be integrable, it is easier to find functions which are not integrable. For instance, on $I = [0,1]$, if g is the Dirichlet function, i.e., if g takes the value 1 on the irrationals and 0 on the rationals there, we have $s(g,f,\mathcal{P}) = 0$ and $S(g,f,\mathcal{P}) = f(b) - f(a)$ for any partition $\mathcal{P}$, and, unless f is constant, $g \notin \mathcal{R}(f,I)$.

As for the integrability of g, a useful criterion is the following result which deals with a single partition at a time.

Proposition 2.1. Suppose g is a bounded real-valued function on I, and let f be nondecreasing there. Then $g \in \mathcal{R}(f,I)$ iff given $\varepsilon > 0$, there is a partition $\mathcal{P}$ of I such that

$$S(g,f,\mathcal{P}) - s(g,f,\mathcal{P}) \le \varepsilon\,. \tag{2.3}$$

Note that if (2.3) holds, then it also holds for every partition finer than $\mathcal{P}$.

Proof. To show that the condition is necessary, given $\varepsilon > 0$, let $\mathcal{P}_1$ and $\mathcal{P}_2$ be partitions of I such that $S(g,f,\mathcal{P}_1) \le U(g,f,I) + \varepsilon/2 = \int_a^b g\,df + \varepsilon/2$, and $s(g,f,\mathcal{P}_2) \ge L(g,f,I) - \varepsilon/2 = \int_a^b g\,df - \varepsilon/2$, and let $\mathcal{P} = \mathcal{P}_1 \cup \mathcal{P}_2$. Since the upper sums are nonincreasing and the lower sums are nondecreasing, we have

$$\begin{aligned} S(g,f,\mathcal{P}) - s(g,f,\mathcal{P}) &\le S(g,f,\mathcal{P}_1) - s(g,f,\mathcal{P}_2) \\ &\le \left(\int_a^b g\,df + \varepsilon/2\right) - \left(\int_a^b g\,df - \varepsilon/2\right) \le \varepsilon\,, \end{aligned}$$

and (2.3) holds for $\mathcal{P}$.

Conversely, if (2.3) holds, by the definition of $L(g,f,I)$ and $U(g,f,I)$, it readily follows that $0 \le U(g,f,I) - L(g,f,I) \le \varepsilon$, and, since ε is arbitrary, we actually have $U(g,f,I) = L(g,f,I)$. ∎

Corollary 2.2. Suppose $g \in \mathcal{R}(f,I)$ and let $J \subseteq I$. Then $g \in \mathcal{R}(f,J)$. Moreover, if $I = [a,b]$, and if $a < c < b$, then

$$\int_a^b g\,df = \int_a^c g\,df + \int_c^b g\,df\,.$$

The conclusion of Corollary 2.2 should not be confused with its converse, which is false. More precisely, if $g \in \mathcal{R}(f,[0,1])$ and $g \in \mathcal{R}(f,[1,2])$, then it is not necessarily true that $g \in \mathcal{R}(f,[0,2])$. To see this just take $g(x) = 0$ if $0 \leq x \leq 1$ and $g(x) = 1$ if $1 < x \leq 2$, and $f(x) = 0$ if $0 \leq x < 1$ and $f(x) = 1$ if $1 \leq x \leq 2$. Then $\int_0^1 g\,df = 0$ (since g vanishes on [0,1]) and $\int_1^2 g\,df = 0$ (since f is constant on [1,2]). However, $g \notin \mathcal{R}(f,[0,2])$, since for any partition $\mathcal{P}$ which does not contain the point 1 we have $s(g,f,\mathcal{P}) = 0$ and $S(g,f,\mathcal{P}) = 1$, and consequently, $U(g,f,I) = 1 > L(g,f,I) = 0$. The reader may have noticed that f and g have a common point of discontinuity, and it is a general fact that under these circumstances g cannot be integrable with respect to f, cf. 4.17 below.

Corollary 2.3. Suppose that $g \in \mathcal{R}(f_1,I) \cap \mathcal{R}(f_2,I)$ and that λ is a nonnegative real number. Then $g \in \mathcal{R}(f_1 + \lambda f_2, I)$, and

$$\int_a^b g\,d(f_1 + \lambda f_2) = \int_a^b g\,df_1 + \lambda \int_a^b g\,df_2 \,. \tag{2.4}$$

Also, if $g_1, g_2 \in \mathcal{R}(f,I)$, then $g_1 + \lambda g_2 \in \mathcal{R}(f,I)$, and

$$\int_a^b (g_1 + \lambda g_2)\,df = \int_a^b g_1\,df + \lambda \int_a^b g_2\,df \,.$$

The converse to both the statements in Corollary 2.3 is false.

Proposition 2.1 can be used to show that important classes of functions are Riemann-Stieltjes integrable, and Corollary 2.3 to show that the integral with respect to a BV function is well-defined.

Proposition 2.4. Suppose that g is continuous on I, and that f is nondecreasing there. Then $g \in \mathcal{R}(f,I)$.

Proof. If f is constant, all upper and lower sums of g are 0 and $\int_a^b g\,df = 0$. Otherwise, let $\varepsilon > 0$ be given, and put $\eta = \varepsilon/(f(b) - f(a))$. Since I is closed and bounded and g is continuous on I, g is uniformly continuous there and there exists $\delta > 0$ so that

$$|g(x) - g(y)| < \eta \quad \text{whenever} \quad |x - y| < \delta\,, \quad \text{all} x, y \in I\,.$$

Put then $x_0 = a, x_1 = a + \delta, \ldots, x_{n-1} = a + (n-1)\delta$, where n is chosen so that $a + n\delta \geq b$, and, if necessary, complete a partition $\mathcal{P}$ of I by setting

$x_n = b$. It is clear that

$$\begin{aligned} S(g,f,\mathcal{P}) - s(g,f,\mathcal{P}) &= \sum_{k=1}^{n}(M_k - m_k)\Delta_k f \\ &\le \eta \sum_{k=1}^{n} \Delta_k f = \eta(f(b) - f(a)) = \varepsilon\,. \end{aligned}$$

That is, (2.3) holds for this partition $\mathcal{P}$, and Proposition 2.1 gives that $g \in \mathcal{R}(f,I)$. ■

If f is BV on an interval $I = [a,b]$, then by Jordan's theorem we have $f = f_1 - f_2$, where f_1 and f_2 are nondecreasing functions defined on I. Thus, if the integral of g with respect to both f_1 and f_2 over I is defined, we put

$$\int_a^b g\,df = \int_a^b g\,df_1 - \int_a^b g\,df_2\,. \tag{2.5}$$

It is not hard to check that the left-hand side of (2.5) above is a well-defined quantity. Specifically, if $f = f_3 - f_4$ is another decomposition of f as a difference of nonincreasing functions, and if $g \in \mathcal{R}(f_3, I) \cap \mathcal{R}(f_4, I)$, then the left-hand side of (2.5) also equals $\int_a^b g\,df_3 - \int_a^b g\,df_4$. Indeed, since it readily follows that $f_1 + f_4 = f_2 + f_3$, by Corollary 2.3 we get that $\int_a^b g\,df_1 + \int_a^b g\,df_4 = \int_a^b g\,df_2 + \int_a^b g\,df_4$, and our observation follows by subtraction.

Corollary 2.5. Suppose that g is continuous on I and that f is a BV function there. Then $g \in \mathcal{R}(f, I)$.

Proof. Since f can be written as the difference of nondecreasing functions, the assertion follows readily from Proposition 2.4 and Corollary 2.3. ■

An interesting point to consider is whether in Corollary 2.5 the roles of g and f can be interchanged. Of course this question is loosely posed since the integral of a BV function with respect to a continuous function has not yet been defined. We proceed as follows: First we introduce a notion of a Riemann-Stieltjes integral that makes sense for arbitrary functions f, g, then show that when restricted to our old setting, i.e., $f \in$ BV and g continuous, both definitions coincide, and finally we prove an "integration by parts" formula that will enable us to answer the question we posed.

Given bounded functions f, g defined on I, we say that they satisfy the property $(\mathcal{RS})$ if there is a number L so that for any $\varepsilon > 0$ there exists a partition of $\mathcal{P}_\varepsilon$ of I with the property that for every partition $\mathcal{P}$ of I finer than $\mathcal{P}_\varepsilon$ and every choice of points t_k in $[x_k, x_{k+1}]$,

$$\left|\sum_{k=1}^{n} g(t_k)\Delta_k f - L\right| < \varepsilon . \tag{2.6}$$

Note that the number L, when it exists, is uniquely determined.

Theorem 2.6. Suppose g is a bounded function defined on $I = [a,b]$, and f is nondecreasing there. Then the following are equivalent:
(i) $g \in \mathcal{R}(f,I)$
(ii) f, g satisfy property $(\mathcal{RS})$ on I.

Proof. (i) implies (ii). Since $g \in \mathcal{R}(f,I)$, given $\varepsilon > 0$, there are partitions $\mathcal{P}'_\varepsilon$ and $\mathcal{P}''_\varepsilon$ of I such that $S(g,f,\mathcal{P}'_\varepsilon) \le \int_a^b g\,df + \varepsilon$, and $s(g,f,\mathcal{P}''_\varepsilon) \ge \int_a^b g\,df - \varepsilon$. Put $\mathcal{P}_\varepsilon = \mathcal{P}'_\varepsilon \cup \mathcal{P}''_\varepsilon$ and note that if $\mathcal{P} = \{a = x_0 < \ldots < x_n = b\}$ is finer than $\mathcal{P}_\varepsilon$, and if t_k is any choice of points in $[x_k, x_{k+1}]$, then

$$\begin{aligned}-\varepsilon + \int_a^b g\,df \le s(g,f,\mathcal{P}''_\varepsilon) \le s(g,f,\mathcal{P}) \le \sum g(t_k)\Delta_k f \\ \le S(g,f,\mathcal{P}) \le S(g,f,\mathcal{P}'_\varepsilon) \le \int_a^b g\,df + \varepsilon .\end{aligned}$$

This chain of inequalities gives that $\left|\sum g_k \Delta_k f - \int_a^b g\,df\right| < \varepsilon$, i.e., (ii) holds with $L = \int_a^b g\,df$.
(ii) implies (i). Given $\varepsilon > 0$, let $\mathcal{P}_\varepsilon = \{a = x_0 < \ldots < x_n = b\}$ be the partition corresponding to the choice $\varepsilon/4$ in the definition of $(\mathcal{RS})$. Now, since

$$M_k - m_k = \sup\{g(x) - g(x') : x, x' \in I_k\}$$

it follows that given $\eta > 0$, there is a choice of points t_k and t'_k in I_k so that $M_k - m_k \le g(t_k) - g(t'_k) + \eta$. Whence

$$\begin{aligned}S(g,f,\mathcal{P}_\varepsilon) - s(g,f,\mathcal{P}_\varepsilon) &= \sum (M_k - m_k)\Delta_k f \\ &\le \left(\sum g(t_k)\Delta_k f - L\right) + \left(L - \sum g(t'_k)\Delta_k f\right) + \eta(f(b) - f(a)),\end{aligned}$$

and consequently, by the choice of $\mathcal{P}_\varepsilon$,

$$S(g,f,\mathcal{P}_\varepsilon) - s(g,f,\mathcal{P}_\varepsilon) \le \varepsilon/2 + \eta(f(b) - f(a)) .$$

If $f(b) = f(a)$, then f is constant and there is not much to do. Otherwise let $\eta = \varepsilon/2(f(b) - f(a))$ and observe that we have

$$S(g,f,\mathcal{P}_\varepsilon) - s(g,f,\mathcal{P}_\varepsilon) \le \varepsilon ,$$

and by Proposition 2.1, $g \in \mathcal{R}(f,I)$, as we wanted to show. In fact, now that the integrability of g has been established, we may refer to the proof of the first implication and note that also $L = \int_a^b g\,df$. ■

Because of Theorem 2.6, if f and g satisfy property $(\mathcal{RS})$ we also write $g \in \mathcal{R}(f,I)$ and denote L by $\int_a^b g\,df$.

With this notation we have

Theorem 2.7. Let f, g be real-valued bounded functions defined on I. If $g \in \mathcal{R}(f,I)$, then $f \in \mathcal{R}(g,I)$ and

$$\int_a^b g\,df + \int_a^b f\,dg = f(b)g(b) - f(a)g(a)\,. \tag{2.7}$$

Proof. Let $\varepsilon > 0$ be given, and let $\mathcal{P}_\varepsilon$ be the partition of I corresponding to ε in the definition of the integrability of g. Let $\mathcal{P} = \{a = x_0 < \ldots < x_n = b\}$ be a partition of I finer than $\mathcal{P}_\varepsilon$ and finally let t_k be any choice of points in $[x_k, x_{k+1}]$. Since $A = f(b)g(b) - f(a)g(a)$ can also be written as

$$A = \sum_{k=1}^{n} f(x_k)g(x_k) - \sum_{k=1}^{n} f(x_{k-1})g(x_{k-1})\,,$$

it readily follows that $A - \sum f(t_k)\Delta_k g$ equals

$$\sum_{k=1}^{n} g(x_k)(f(x_k) - f(t_k)) + \sum_{k=1}^{n} g(x_{k-1})(f(t_k) - f(x_{k-1}))\,. \tag{2.8}$$

The sum in (2.8) corresponds to the partition of I obtained by combining the x_k's and the t_k's, and, since this partition is finer than $\mathcal{P}_\varepsilon$, we have that

$$\left| A - \int_a^b g\,df - \sum f(t_k)\Delta_k g \right| < \varepsilon\,.$$

Since $\mathcal{P}$ and the t_k's are arbitrary it follows that $f \in \mathcal{R}(g,I)$, and that (2.7) holds. ■

Theorem 2.7 and the results we have covered thus far provide examples of integrable functions; our aim, however, is to characterize these functions. For instance, when $f(x) = x$ experience indicates that it is reasonable to expect that the set of discontinuities of a Riemann integrable function can be covered by intervals whose total length is small. To make this precise Hankel (1839–1873) introduced the notion of "content" of a set (contained in R). Shortly after that Cantor and Stoltz extended the results to subsets of R^n, although there were some problems when the sets in question were not closed. To deal with this inconvenience, a few years later, Peanno (1858–1932) and Jordan introduced the concepts of inner and outer content. The notion of content accepted today was introduced by Borel (1871–1956) in 1898 and Lebesgue (1875–1941) in 1902. It is in terms of these concepts, namely the Lebesgue and Borel measures, that we characterize the Riemann integrable functions in Chapter VII, and the Riemann-Stieltjes integrable functions in Chapter IX.

3. THE RIEMANN-STIELTJES INTEGRAL AND LIMITS

In this section we explore how successful we are in operating with the Riemann-Stieltjes integral, in other words what we can, and cannot, do with it. First note that if g is a bounded function defined on I and if f is nondecreasing there, then $g \in \mathcal{R}(f,I)$ implies $|g| \in \mathcal{R}(f,I)$. This fact is a simple consequence of Proposition 2.1. Indeed, given $\varepsilon > 0$, by Proposition 2.1 there is a partition $\mathcal{P}$ of I so that $S(g,f,I) - s(g,f,I) < \varepsilon/2$. Also, for any $\eta > 0$, there exist points x'_k, x''_k with the property that

$$\sup_{I_k} |g| - \inf_{I_k} |g| \leq |g(x'_k)| - |g(x''_k)| + \eta\,, \quad \text{all } k\,.$$

But since $|g(x'_k)| - |g(x''_k)| \leq |g(x'_k) - g(x''_k)|$, rearranging if necessary x'_k and x''_k to remove the absolute value, it follows that

$$\sup_{I_k} |g| - \inf_{I_k} |g| \leq g(x'_k) - g(x''_k) + \eta \leq M_k - m_k + \eta\,, \quad \text{all } k\,.$$

Multiplying these inequalities through by $\Delta_k f$ and adding them up we get

$$\begin{aligned} S(|g|,f,I) - s(|g|,f,I) &\leq S(g,f,I) - s(g,f,I) + \eta(f(b) - f(a)) \\ &\leq \varepsilon/2 + \eta(f(b) - f(a))\,, \end{aligned}$$

and this quantity can be made less than or equal to ε by choosing η small enough. Whence by Proposition 2.1, $|g| \in \mathcal{R}(f,I)$ as we wanted to show.

The converse to this fact is not true: There is a bounded function g so that $|g| \in \mathcal{R}(I)$ and yet $g \notin \mathcal{R}(I)$; on [0,1] the function $g(x) = 1$ if x is irrational and $g(x) = -1$ if x is rational will do.

If g and f are as above, then $|g| \pm g$ are nonnegative Riemann-Stieltjes integrable functions, and consequently, $\int_a^b (|g| \pm g)\, df \ge 0$. By the linearity of the integral we get that $\int_a^b g\, df \le \mp \int_a^b |g|\, df$, or

$$\left| \int_a^b g\, df \right| \le \int_a^b |g|\, df\,. \tag{3.1}$$

A moment's thought will convince the reader that it is possible to extend (3.1) to include the BV functions f as well. The estimate in this case proceeds as follows: Since for arbitrary partitions of I we have

$$|s(g, f, I)| \le \sup_I |g| \sum |\Delta_k f| \le \sup_I |g|\, V(f; a, b)\,,$$

it is not difficult to see that

$$\left| \int_a^b g\, df \right| \le \sup_I |g|\, V(f; a, b)\,. \tag{3.2}$$

Inequality (3.2) allows us to address the following question: Suppose $g_n \to g$, and both the g_n's and g are integrable, is it then true that $\int_a^b g_n\, df \to \int_a^b g\, df$? A simple example shows that this is not always the case: If $I = [0,1]$ and $g_n(x) = n\chi_{(0,1/n)}(x)$, then $g_n(x) \to g(x) = 0$ for x in I, but $\int_0^1 g_n(x)\, dx = 1 \not\to \int_0^1 g(x)\, dx = 0$. Nevertheless, this situation can be remedied.

Proposition 3.1. Suppose that the sequence of bounded functions $\{g_n\}$ converges uniformly to g on I, that f is a BV function there, and that $g_n, g \in \mathcal{R}(f, I)$. Then $\lim_{n\to\infty} \int_a^b g_n\, df = \int_a^b g\, df$.

Proof. By the linearity of the integral and (3.2) it follows that

$$\left| \int_a^b g_n\, df - \int_a^b g\, df \right| = \left| \int_a^b (g_n - g)\, df \right| \le \sup_I |g_n - g|\, V(f; a, b)\,.$$

Now, since the g_n's converge uniformly to g on I, $\sup_I |g_n - g| \to 0$ as $n \to \infty$, the right-hand side of the above inequality goes to 0 as n tends to ∞, and so does the left-hand side. ■

Some remarks concerning this result are in order. First, we assumed that $g \in \mathcal{R}(f, I)$. It would be more interesting if we could derive the integrability of g from that of the g_n's. Next, we assumed that the g_n's converge uniformly to g, and this is quite restrictive. However, these conditions cannot essentially be relaxed. Consider, for instance, the following situation: Enumerate the rationals in $I = [0,1]$, $r_1, r_2, \ldots$, say, and let $g_n(x) = 1$ if $x = r_1, \ldots, r_n$ and $g_n(x) = 0$ otherwise. Then $g_n \in \mathcal{R}(I)$, $\int_0^1 g_n(x)\,dx = 0$ for all n, $\lim_{n\to\infty} g_n(x) = g(x)$ exists everywhere in I, but, as we have seen above, g is not Riemann integrable. Although the g_n's do not converge uniformly to g, they do converge boundedly, and the example points out an unhappy state of affairs that will be corrected, in Chapter VII, by the Lebesgue integral.

There is yet another way to interpret this last observation. Suppose the distance between $f, g \in \mathcal{R}(I)$ is measured by the quantity

$$d(f,g) = \int_0^1 |f(x) - g(x)|\,dx\,.$$

Now, endowed with this distance $\mathcal{R}(I)$ becomes a metric space. In this metric, the sequence $\{g_n\}$ introduced above is Cauchy, but it does not converge. Thus, this metric space is not complete. This shortcoming will also be corrected by the Lebesgue integral.

4. PROBLEMS AND QUESTIONS

4.1 Assume $\{f_n\}$ is a sequence of real-valued nondecreasing functions defined on $I = [a,b]$, and suppose $f(x) = \lim_{n\to\infty} f_n(x)$ exists for $x \in I$. Is f necessarily nondecreasing?

4.2 (F. Riesz's Rising Sun Lemma) Assume f is a bounded real-valued function defined on $I = [a,b]$, and let $\mathcal{F} = \{g : g$ is defined on I, g is nonincreasing and $g(x) \ge f(x)$ for x in $I\}$. Show that

$$f^*(x) = \sup\{f(y) : x \le y \le 1\}, \quad x \in I\,,$$

belongs to $\mathcal{F}$, and in fact it is the smallest element there. Moreover, if f is continuous at x, so is f^*.

4.3 Let f be a real-valued nondecreasing function defined on $I = [a,b]$. Show that for $x \in [0,1)$,

$$f(x^+) = \lim_{h\to 0} f(x+h)\,, \quad h > 0$$

exists. Also, for $x \in (0,1]$,

$$f(x^-) = \lim_{h\to 0} f(x+h)\,, \quad h < 0$$

exists. Show that the functions

$$f_l(x) = f(x^-), \quad 0 < x \le 1, \quad f_l(0) = f(0)$$

and

$$f_r(x) = f(x^+), \quad 0 \le x < 1, \quad f_r(1) = f(1)$$

are nondecreasing on I. Furthermore, f_l is left-continuous, and f_r is right-continuous.

4.4 Show that a monotone function $f\colon [a,b] \to R$ has, at most, countably many discontinuities, and that all are of the first kind. Conversely, if D is an at most countable subset of $[a,b]$, there is a monotone function $f\colon [a,b] \to R$ such that $D = \{x \in [a,b] : f \text{ is discontinuous at } x\}$.

4.5 Let A be a nonempty subset of R, and let f be a bounded nondecreasing real-valued function defined on A. Show that f can be extended to R as a nondecreasing function.

4.6 Let $f\colon [a,b] \to R$ be nondecreasing, and for x in $[a,b]$ put

$$s(x) = \sum_{a \le y < x} \left(f(y^+) - f(y^-)\right) + f(x) - f(x^-).$$

Show that $s(x)$ is a well-defined function, called the "saltus" function of f. Show that s and $f - s$ are nondecreasing and that $f - s$ is continuous.

4.7 A real-valued function f defined on $I = [a,b]$ is said to be Lipschitz there if there is a constant c so that $|f(x) - f(x')| \le c\,|x - x'|$ for all x, x' in I. Show that if f is Lipschitz on I it is BV there.

4.8 Let f, g be BV on $I = [a,b]$. Show that f, g are bounded on I, and that for any real number η, $f + \eta g$ is BV on I and

$$V(f + \eta g; a, b) \le V(f; a, b) + |\eta| V(g; a, b).$$

4.9 Let f, g be BV on $I = [a,b]$. Show that fg is BV on I, and that if $|g(x)| \ge \varepsilon > 0$ for $x \in I$, then also f/g is BV on I. Estimate $V(fg; a, b)$ and $V(f/g; a, b)$ in terms of $V(f; a, b), V(g; a, b)$ and ε.

4.10 Let f, g be BV on $I = [a,b]$. Show that $(f \vee g)(x) = \max(f(x), g(x))$ and $(f \wedge g)(x) = \min(f(x), g(x))$ are BV on I.

4.11 Let f, g be real-valued functions defined on $I = [a,b]$, and suppose that f and g differ at finitely many values. Show that f is BV on I iff g is BV on I, and that $V(f; a, b) = V(g; a, b)$.

4.12 Characterize those real numbers η, ε for which $f(x) = x^\eta \sin^\varepsilon(1/x)$, $x \neq 0$, $f(0) = 0$, is BV on [0,1]. Verify that the choice $\eta = 2, \varepsilon = 3/2$ gives an example of a function f which is BV on I, differentiable there, and yet f' is unbounded.

4.13 Show that the plane curve $\Gamma: I \to R^2$ with graph $\{(\phi(t), \psi(t)): t \in I\}$ is rectifiable iff ϕ and ψ are BV on I. What is $L(\Gamma)$ in this case?

4.14 Let f be BV and continuous on $I = [a,b]$. Show that for $a \leq x \leq b$ we have

$$V(x) = \lim_{\text{norm}(\mathcal{P}) \to 0} \sum_{\text{over } \mathcal{P}} |f(x_j) - f(x_{j-1})| \,.$$

The above statement is understood as follows: Given $\varepsilon > 0$, there exists $\eta > 0$ such that for any partition $\mathcal{P}$ of $[a,x]$ with norm less than or equal to η we have $\sum_{\text{over } \mathcal{P}} |f(x_j) - f(x_{j-1})| > V(x) - \varepsilon$.

4.15 Assume that f is BV on $I = [a,b]$ and for $a \leq x \leq b$ let $\mathcal{P} = \{a = x_0 < \ldots < x_n = x\}$ be a partition of $[a,x]$. Let

$$P(\mathcal{P}) = \{k : \Delta_k f > 0, 0 \leq k \leq n-1\}\,,$$

and

$$N(\mathcal{P}) = \{k : \Delta_k f < 0, 0 \leq k \leq n-1\}\,,$$

and put

$$P(x) = \sup\left\{\sum\nolimits_{k \in P(\mathcal{P})} \Delta_k f\right\}, \; N(x) = \sup\left\{\sum\nolimits_{k \in N(\mathcal{P})} \Delta_k f\right\},$$

where the sup in each expression above is taken over all finite partitions $\mathcal{P}$ of $[a,x]$; P and N are called the positive and the negative variations of f on I. Prove they satisfy the following properties: (i) P, N are nonnegative and nondecreasing, (ii) $P(x)+N(x) = V(x)$ and $P(x) - N(x) = f(x) - f(a)$, and, (iii) Every point of continuity of f is also a point of continuity of P and N.

4.16 Recall that for any real number r we have

$$r^+ = \begin{cases} r & \text{if } r > 0 \\ 0 & \text{if } r \leq 0\,, \end{cases} \qquad r^- = \begin{cases} 0 & \text{if } r > 0 \\ r & \text{if } r \leq 0\,. \end{cases}$$

These are called the positive and negative parts of r and satisfy the relations $r^+, r^- \geq 0$, $r = r^+ - r^-$, and $|r| = r^+ + r^-$.

Show that if f is BV on $I = [a,b]$, and if $f \in \mathcal{R}(I)$ and $f(x) = \int_a^x f'(t)\,dt$ for $x \in I$ (this condition is not redundant), then for $a \le x \le b$, we have

$$P(x) = \int_a^x (f'(t))^+\,dt\,, \quad N(x) = \int_a^x (f'(t))^-\,dt\,,$$

and

$$V(x) = \int_a^x |f'(t)|dt\,.$$

4.17 Assume that g, f are bounded functions defined on $I = [a,b]$ which are discontinuous from the right at $x \in (a,b)$. Show that $g \notin \mathcal{R}(f,I)$.

4.18 Assume that f is a nondecreasing real-valued function defined on an interval I, and that $g \in \mathcal{R}(f,I)$. Show that $g^2 \in \mathcal{R}(f,I)$.
Show that the converse is not true, i.e., there are functions f, g such that $g^2 \in \mathcal{R}(f,I)$ but $g \notin \mathcal{R}(f,I)$.

4.19 Assume that f is a nondecreasing real-valued function defined on an interval I and that $g, h \in \mathcal{R}(f,I)$. Show that $gh \in \mathcal{R}(f,I)$.

4.20 Suppose that $g \in \mathcal{R}(f,I)$, and that f has a bounded derivative f' on $I = [a,b]$. Show that $gf' \in \mathcal{R}(I)$ and

$$\int_a^b g\,df = \int_a^b g(x)f'(x)\,dx\,.$$

4.21 Let f be BV on $I = [a,b]$, and let, as usual, V denote its variation on I, $V(a) = 0$. Prove that if g is bounded on I and $g \in \mathcal{R}(f,I)$, then $g \in \mathcal{R}(V,I)$.

4.22 Suppose f is BV on $I = [a,b]$ and the bounded function $g \in \mathcal{R}(f,I)$. For $x \in I$ put $G(x) = \int_a^x g\,df$, and show that G is BV on I and continuous at those points of I where f is continuous.

4.23 Let f be a nondecreasing bounded function on $I = [a,b]$ and let $g \in \mathcal{R}(f,I)$, $m \le g(x) \le M$ for all $x \in I$. Show that there is a real number c, $m \le c \le M$, so that

$$\int_a^b g\,df = c(f(b) - f(a))\,.$$

4.24 Let f, f_1 be nondecreasing functions defined on an interval $I = [a,b]$ of the line with the property that $f(a) = f_1(a)$ and

$$\int_a^b g\,df = \int_a^b g\,df_1\,, \quad \text{all } g \text{ continuous on } I\,.$$

Prove that if $x \in I$ is a point of continuity of both f and f_1, then $f(x) = f_1(x)$.

4.25 Let f_1, f_2 be two real-valued nondecreasing functions defined on $I = [a,b]$ and suppose there is a value $c \in R$ such that the set $D = \{x \in I : f_1(x) = f_2(x) + c\}$ is dense in I. Show that

$$\int_a^b g\,df_1 = \int_a^b g\,df_2\,, \quad \text{all } g \text{ continuous on } I\,.$$

4.26 Let $I = [0,1]$ and suppose f is a BV function defined on I. Let h be the function defined on I as follows: $h(0) = 0$, $h(x) = f(x+0) - f(0)$ if $0 < x < 1$, and $h(1) = f(1) - f(0)$. Show that h is BV on I, and that for each continuous function g we have $\int_0^1 g\,df = \int_0^1 g\,dh$.

4.27 Let f be a continuous function defined on $I = [a,b]$ and suppose that g is nondecreasing there. Show that there is a point $x_0 \in I$ such that

$$\int_a^b g\,df = g(a)\int_a^{x_0} df + g(b)\int_{x_0}^b df\,.$$

4.28 (Change of variable) Let f, g be bounded on $I = [a,b]$ and suppose that $g \in \mathcal{R}(f,I)$. Furthermore, assume there are an interval $J = [c,d]$ and a continuous, strictly monotone function ϕ such that $I = \phi(J)$, $\phi(c) = a$, $\phi(d) = b$. Show that the functions $F(x) = f(\phi(x))$ and $G(x) = g(\phi(x))$ are well-defined on J, $G \in \mathcal{R}(F,J)$, and that

$$\int_a^b g\,df = \int_c^d G\,dF\,.$$

4.29 Let $\{f_n\}$ be a sequence of BV functions on $I = [a,b]$ and suppose there exists a BV function f defined on I such that the variation $V(f - f_n; a, b)$ tends to 0 as $n \to \infty$. Assume also that $f_n(a) = f(a) = 0$ for each $n = 1, 2, \ldots$ If g is continuous on I, prove that $g \in \mathcal{R}(f,I)$ and

$$\lim_{n\to\infty} \int_a^b g\,df_n = \int_a^b g\,df\,.$$

CHAPTER **IV**

Abstract Measures

In this chapter we study the notions of measure and of sets of "content" zero. These concepts are essential to measure the level sets of the new class of functions to be integrated and in the characterization of Riemann-Stieltjes integrable functions. A successful approach to these problems requires that we operate freely with sets, including taking limits. This we achieve with the introduction of algebras and σ-algebras of sets.

1. ALGEBRAS AND σ-ALGEBRAS OF SETS

A class $\mathcal{A}$ of subsets of a (universal) set X is called an algebra of sets, or plainly an algebra, provided the following three properties hold:

(i) $\mathcal{A}$ is nonempty.

(ii) If $E \in \mathcal{A}$, then $X \setminus E \in \mathcal{A}$.

(iii) If $\{E_k\}_{k=1}^{n} \subseteq \mathcal{A}$, then $\bigcup_{k=1}^{n} E_k \in \mathcal{A}$.

Some sets of an algebra $\mathcal{A}$ are easily identified, namely $\emptyset$ and X. In fact, $\mathcal{A} = \{\emptyset, X\}$ is the most economical algebra. On the other hand, $\mathcal{A} = \mathcal{P}(X)$ is also an algebra. Another interesting example is $\mathcal{E} = \{E \subseteq R :$ E can be written as a finite pairwise disjoint union of half-open intervals $(a,b]$, with a,b in $R\}$. Also, it is not hard to check that if $\{\mathcal{A}_i\}_{i\in I}$ is a collection of algebras, then $\mathcal{A} = \bigcap_{i\in I} \mathcal{A}_i$ is an algebra.

If $\mathcal{A}$ is an algebra of subsets of X and $E \subseteq X$, then the family $\mathcal{A}_E = \{E \cap A : A \in \mathcal{A}\}$ is an algebra of subsets of E.

What operations can we perform with the sets of an algebra $\mathcal{A}$ and still remain in $\mathcal{A}$?

Proposition 1.1. Suppose $\mathcal{A}$ is an algebra of sets, and $E_1, E_2 \in \mathcal{A}$. Then $E_1 \cap E_2$ and $E_1 \setminus E_2$ belong to $\mathcal{A}$.

Proof. Since by 5.1 in Chapter I

$$X \setminus (E_1 \cap E_2) = (X \setminus E_1) \cup (X \setminus E_2), \tag{1.1}$$

by (ii) and (iii) the set on the right-hand side of (1.1) is in $\mathcal{A}$ and consequently, the complement of $E_1 \cap E_2$ belongs to $\mathcal{A}$. By (ii) again, $E_1 \cap E_2 \in \mathcal{A}$.

Moreover, since $E_1 \setminus E_2 = E_1 \cap (X \setminus E_2)$, by (ii) and the first part of the proof, we have $E_1 \setminus E_2 \in \mathcal{A}$. ■

In applications it is often convenient to replace (iii) by the seemingly weaker condition that $\mathcal{A}$ be closed under the union of pairwise disjoint sets, namely:

(iii') If $\{E_k\}_{k=1}^n$ is a collection of pairwise disjoint subsets of $\mathcal{A}$, then $\bigcup_{k=1}^n E_k \in \mathcal{A}$.

However, as an argument using 5.2 in Chapter I and Proposition 1.1 readily shows, (iii) and (iii') are actually equivalent.

We consider the taking of limits next. Given a sequence $\{A_n\}$, we define the sets $\limsup A_n = \{x : x \text{ belongs to infinitely many } A_n\text{'s}\}$ and $\liminf A_n = \{x : x \text{ belongs to all but finitely many } A_n\text{'s}\}$. It is not hard to see that

$$\limsup A_n = \bigcap_{m=1}^{\infty} \left(\bigcup_{n=m}^{\infty} A_n \right), \tag{1.2}$$

and

$$\liminf A_n = \bigcup_{m=1}^{\infty} \left(\bigcap_{n=m}^{\infty} A_n \right). \tag{1.3}$$

For instance, if $A_n = [0,1]$, n odd, and $A_n = [1,2]$, n even, then $\liminf A_n = \{1\}$, and $\limsup A_n = [0,2]$.

When the limits are equal we say that the sequence $\{A_n\}$ converges and the common value is denoted by $\lim A_n$.

From the expressions for the limits it is apparent that limiting operations are not necessarily closed in an algebra of sets; we are thus led to the concept of σ-algebra.

We say that an algebra $\mathcal{A}$ of subsets of X is a σ-algebra of sets, or plainly a σ-algebra, if it satisfies the additional property

(iv) If $\{E_k\}_{k=1}^{\infty} \subseteq \mathcal{A}$, then $\bigcup_{k=1}^{\infty} E_k \in \mathcal{A}$.

As before, (iv) is equivalent to the condition obtained by requiring that the E_k's be pairwise disjoint.

$\mathcal{P}(X)$ is a σ-algebra, and the algebra $\mathcal{E}$ introduced above is not. Also, if $\{\mathcal{A}_i\}_{i\in I}$ is a family of σ-algebras, then $\mathcal{A} = \bigcap_{i\in I} \mathcal{A}_i$ is a σ-algebra as well.

If $\mathcal{A}$ is a σ-algebra of subsets of X and $E \subseteq X$, the collection $\mathcal{A}_E = \{A \cap E : A \in \mathcal{A}\}$ is a σ-algebra of subsets of E.

As for the consideration of the limits we have

Proposition 1.2. Suppose $\{A_n\}$ is a sequence of sets of a σ-algebra $\mathcal{A}$. Then $\liminf A_n$ and $\limsup A_n$ belong to $\mathcal{A}$.

Proof. Since $A_n = X \setminus (X \setminus A_n)$, by the relations 5.1 in Chapter I we have that $\limsup A_n$ equals

$$\begin{aligned}\bigcap_{m=1}^{\infty} \left(\bigcup_{n=m}^{\infty} (X \setminus (X \setminus A_n))\right) &= \bigcap_{m=1}^{\infty} \left(X \setminus \left(\bigcap_{n=m}^{\infty} (X \setminus A_n)\right)\right) \\ &= X \setminus \bigcup_{m=1}^{\infty} \left(\bigcap_{n=m}^{\infty} (X \setminus A_n)\right) ,\end{aligned}$$

i.e., $\limsup A_n = X \setminus \liminf (X \setminus A_n)$.

Thus, if we prove the conclusion for the $\limsup A_n$, it will follow for the $\liminf A_n$, and vice versa. Now, since $\mathcal{A}$ is a σ-algebra, $B_m = \bigcup_{n=m}^{\infty} A_n \in \mathcal{A}$ for all m, and by the countable version of Proposition 1.1 (which holds for σ-algebras and which is proved in a similar fashion), $\lim A_n = \bigcap_{m=1}^{\infty} B_m \in \mathcal{A}$, and we are done. ■

That the notion of σ-algebra is the natural one to deal with limits is also expressed by

Proposition 1.3. Suppose $\mathcal{A}$ is an algebra. Then $\mathcal{A}$ is a σ-algebra iff for every sequence $\{A_k\} \subseteq \mathcal{A}$, $\limsup A_k \in \mathcal{A}$.

Proof. Since Proposition 1.2 gives the necessity of the condition we only do the sufficiency. We must only show that property (iv) holds. Let $\{A_k\} \subseteq \mathcal{A}$, and set $B_n = \bigcup_{k=1}^{n} A_k$. Since $\mathcal{A}$ is an algebra, $B_n \in \mathcal{A}$ for all n, and consequently, by assumption, $\limsup B_n \in \mathcal{A}$. But since every x in $\bigcup_{k=1}^{\infty} A_k$ belongs to infinitely many B_n's, in fact, if $x \in A_k$, then $x \in B_n$ for all $n \geq k$, then $\bigcup_{k=1}^{\infty} A_k = \limsup B_n \in \mathcal{A}$, and we have finished. ■

Next we consider the following question: Given a family $\mathcal{C}$ of subsets of X, what is the smallest family of subsets of X that contains the limits of all sequences of sets in $\mathcal{C}$? Or equivalently, which is the smallest σ-algebra of subsets of X that contains $\mathcal{C}$?

If $\mathcal{C}$ is a σ-algebra, then the answer is $\mathcal{C}$. Otherwise, let $\mathcal{F}$ be the family of all the σ-algebras of subsets of X which contain $\mathcal{C}$. Since $\mathcal{P}(X) \in \mathcal{F}$,

$\mathcal{F} \neq \emptyset$. As observed above, the intersection of an arbitrary family of σ-algebras is again a σ-algebra, and the intersection of all the σ-algebras in $\mathcal{F}$ is the smallest σ-algebra that contains $\mathcal{C}$. This σ-algebra is called the σ-algebra generated by $\mathcal{C}$ and it is denoted by $\mathcal{S}(\mathcal{C})$.

For instance, if $\mathcal{C}$ is the family of all the singletons $\{x\}$ of X, then $\mathcal{S}(\mathcal{C}) = \{E \subseteq X :$ either E is at most countable, or else $X \setminus E$ is at most countable$\}$. Of course, if X is at most countable, then $\mathcal{S}(\mathcal{C}) = \mathcal{P}(X)$.

There are two other examples which are useful in applications and we discuss them next.

Example 1.4. Let $\mathcal{A}_1$ be a σ-algebra of subsets of X_1 and $\mathcal{A}_2$ a σ-algebra of subsets of X_2. We are interested in constructing a σ-algebra of subsets of $X_1 \times X_2$ in terms of $\mathcal{A}_1$ and $\mathcal{A}_2$. Our first candidate is the family

$$\mathcal{C} = \{E_1 \times E_2 : E_1 \in \mathcal{A}_1 \text{ and } E_2 \in \mathcal{A}_2\},$$

but it is not hard to see that $\mathcal{C}$ is not necessarily closed under complementation, and consequently, $\mathcal{C}$ is not even an algebra.

Therefore we define the product σ-algebra $\mathcal{A}_1 \times \mathcal{A}_2 = \mathcal{S}(\mathcal{C})$. This is the natural thing to do, since $\mathcal{S}(\mathcal{C})$ is the smallest σ-algebra containing all the "rectangles" $E_1 \times E_2$. In some cases it is possible to give a concrete description of the product σ-algebra. For instance, if $\mathcal{A}_1 = \mathcal{S}(\mathcal{C}_1)$ and $\mathcal{A}_2 = \mathcal{S}(\mathcal{C}_2)$, then $\mathcal{A}_1 \times \mathcal{A}_2 = \mathcal{S}(\mathcal{C}_1 \times \mathcal{C}_2)$, cf. 4.22 below.

Example 1.5. Suppose $X = R^n$ is the Euclidean n-dimensional space endowed with the usual topology, and let $\mathcal{O}$ denote the family of open sets of R^n. Then $\mathcal{S}(\mathcal{O})$ is called the σ-algebra of Borel subsets of R^n and it is denoted by $\mathcal{B}_n$. There is yet a simpler way to generate $\mathcal{B}_n$. When $n = 1$, let $\mathcal{I}$ denote the collection of all open intervals of R. Since every open set is a disjoint, at most countable union of intervals in $\mathcal{I}$, it is clear that also $\mathcal{S}(\mathcal{I}) = \mathcal{B}_1$. In fact, since each interval $(a,b]$ of $\mathcal{E}$ can be expressed as the intersection

$$(a,b] = \bigcap_{n=1}^{\infty} (a, b + 1/n),$$

it also follows that $\mathcal{S}(\mathcal{E}) = \mathcal{B}_1$. When $n > 1$ we use the fact that every open set can be written as the countable union of nonoverlapping closed cubes, i.e., if the cubes intersect it is only along the faces, cf. 4.18 below. Thus, if $\mathcal{C}$ denotes now the family of all the closed cubes in R^n, we also have $\mathcal{S}(\mathcal{C}) = \mathcal{B}_n$. Finally, by 4.21 below, $\mathcal{B}_n \times \mathcal{B}_m = \mathcal{B}_{n+m}$.

2. ADDITIVE SET FUNCTIONS AND MEASURES

A measure on R^n is a natural generalization of such elementary notions as the length of a line segment, the area of a rectangle and the volume of a parallelepiped. In 1898, Borel formulated the following four postulates for defining measures of sets:

(i) A measure is always nonnegative.

(ii) The measure of the union of a finite number of pairwise disjoint sets is equal to the sum of their measures.

(iii) The measure of the complement of a set relative to another equals the difference of their measures.

(iv) Every set whose measure is not 0 is uncountable.

Based on these postulates we introduce the notion of an additive set function. Given a set X and an algebra $\mathcal{A}$ of subsets of X, a set function ψ on $\mathcal{A}$ is a function which assigns to each set of $\mathcal{A}$ a real value, or $\pm\infty$. To avoid technical difficulties we assume that if ψ takes infinite values, they are all of the same sign.

A set function ψ is said to be additive provided the following property holds: If $\{E_k\}_{k=1}^n \subseteq \mathcal{A}$ and the E_k's are pairwise disjoint, then

$$\psi\left(\bigcup_{k=1}^n E_k\right) = \sum_{k=1}^n \psi(E_k). \tag{2.1}$$

This property corresponds to Borel's postulate (ii).

Because ψ only takes infinite values of one sign, the right-hand side of (2.1) always makes sense under the usual arithmetic rules: $r + \infty = \infty$, $r + (-\infty) = -\infty$, $\infty + \infty = \infty$, $(-\infty) + (-\infty) = (-\infty)$.

These are two examples of additive set functions.

Example 2.1. Let X be an infinite set and let $\mathcal{A} = \{E \subseteq X :$ either E is finite or $X \setminus E$ is finite$\}$. Then $\mathcal{A}$ is an algebra and the mapping $\psi: \mathcal{A} \to [0,\infty]$ given by $\psi(E) = 0$ if E is finite and $\psi(E) = \infty$ if $X \setminus E$ is finite, is an additive set function.

Example 2.2. Let $X = (0,1]$, put $\mathcal{E} = \{E \subseteq X : E$ can be written as a finite disjoint union of half open intervals $(a,b]\}$, and let $f: X \to R$ be nondecreasing. Then $\mathcal{E}$ is an algebra of sets and $\psi: \mathcal{E} \to [f(0^+), f(1)]$ given by $\psi((a,b]) = f(b) - f(a)$ and extended additively otherwise, is an additive set function; one must check, of course, that the value $\psi(E)$ does not depend on the way in which E is represented as a disjoint union of half-open intervals. There is more to this example than meets the eye, and we will return to it in Chapter IX.

Additive set functions have the following property: If $E_1, E_2 \in \mathcal{A}$, $E_1 \subseteq E_2$ and $\psi(E_1)$ is a finite value, then Borel's postulate (iii) holds, to wit

$$\psi(E_2 \setminus E_1) = \psi(E_2) - \psi(E_1). \tag{2.2}$$

Indeed, by (2.1) we have that $\psi(E_2) = \psi(E_1) + \psi(E_2 \setminus E_1)$. Now, $\psi(E_2)$ is either a finite value or not. In the former case we subtract $\psi(E_1)$ from both sides of the above equality and (2.2) follows; in the latter case we observe that $\psi(E_2 \setminus E_1)$ is infinite of the same sign as $\psi(E_2)$ and (2.2) still holds.

An immediate consequence of (2.2) is that $\psi(\emptyset) = 0$. As a matter of fact, an additive set function ψ is either identically infinite, or else $\psi(\emptyset) = 0$.

From this point on we consider nonnegative set functions (Borel's postulate (i)), but we will have more to say concerning "signed" set functions, cf. 4.8 - 4.12 below and Chapter XI.

How do additive set functions behave with respect to limits? For instance, if in Example 2.1 the set $X = \{x_1, x_2, \ldots\}$ is countable and we put $E_n = \{x_1, \ldots, x_n\}$, $n \geq 1$, then it readily follows that $\psi(\lim E_n) = \psi(X) = \infty \neq \lim \psi(E_n) = 0$.

To deal with this inconvenience we restrict the domain of an additive set function to a σ-algebra and require an additional compatibility condition, the σ-additivity.

More precisely, given a set X and a σ-algebra $\mathcal{M}$ of subsets of X, we say that a set function μ defined on $\mathcal{M}$ is a measure provided the following three properties hold:

(i) $\mu\colon \mathcal{M} \to [0,\infty]$.

(ii) $\mu(\emptyset) = 0$.

(iii) If $\{E_k\}_{k=1}^{\infty} \subseteq \mathcal{M}$ is a sequence of pairwise disjoint sets, then

$$\mu\left(\bigcup_{k=1}^{\infty} E_k\right) = \sum_{k=1}^{\infty} \mu(E_k).$$

Condition (ii) is assumed in order to exclude the possibility that μ is identically ∞.

To emphasize the interrelation among these objects we say that μ is a measure on $(X, \mathcal{M})$, or that the triplet $(X, \mathcal{M}, \mu)$ is a measure space. The σ-algebra $\mathcal{M}$ is called the family of μ-measurable, or plainly measurable, sets.

A measure μ is said to be finite if $\mu(X) < \infty$. In this case, by (2.2), it follows that for every measurable set E, $\mu(E) \leq \mu(X)$, and μ only takes finite values. When μ is a finite measure we may rescale, i.e., consider $\mu_1(E) = \mu(E)/\mu(X), E \in \mathcal{M}$, instead, and assume that $\mu(X) = 1$; these measures are called probability measures.

The measure space $(X, \mathcal{M}, \mu)$ is said to be σ-finite if X is the countable union of measurable sets, each of finite μ measure. Informally, we also say that μ is σ-finite.

Some examples will clarify these concepts. Let $\mathcal{M}$ be the σ-algebra of subsets of an uncountable set X generated by the singletons of X. Then the set function $\psi(E) = 0$ when E is finite, and $\psi(E) = \infty$ otherwise, is an additive set function which is not a measure. On the other hand, the set functions $\mu(E) =$ number of elements of E when E is finite and $\mu(E) = \infty$ when E is measurable and infinite, and $\nu(E) = 0$ when E is at most countable and $\nu(E) = 1$ when $X \setminus E$ is at most countable, are measures. μ is not σ-finite and ν is a probability measure.

Also, if X is countable, the measure μ on $(X, \mathcal{P}(X))$ given by $\mu(E) =$ number of elements of E if E is finite and $\mu(E) = \infty$ otherwise, is σ-finite.

Next suppose that X is a nonempty set and that $\mathcal{M} = \mathcal{P}(X)$. Let $f: X \to [0,\infty]$, and for $E \in \mathcal{M}$ put

$$\mu(E) = \sum_{x \in E} f(x).$$

As usual the sum is defined as

$$\sup \sum_{k=1}^{n} f(x_k),$$

where the sup is taken over all finite subsets $\{x_1, \ldots, x_n\}$ of E.

To verify that $(X, \mathcal{M}, \mu)$ is a measure space, the only step that offers any difficulty is the σ-additivity of μ; we do this next.

Let $\{E_k\}_{k=1}^{\infty}$ be a sequence of pairwise disjoint measurable sets, and let E denote its union. Given a finite subset $\{x_1, \ldots, x_n\}$ of E, suppose that $\{x_1, \ldots, x_n\} \subseteq \bigcup_{i=1}^{m} E_{k_i}$, $m \le n$, and note that

$$\sum_{k=1}^{n} f(x_k) \le \sum_{i=1}^{m} \mu(E_{k_i}) \le \sum_{k=1}^{\infty} \mu(E_k). \tag{2.3}$$

Thus, taking the supremum of the left-hand side of (2.3) over all finite subsets of E, we get that

$$\mu(E) \le \sum_{k=1}^{\infty} \mu(E_k). \tag{2.4}$$

We show the opposite inequality next. If $\mu(E_k) = \infty$ for some k, since $\mu(E) = \mu(E \setminus E_k) + \mu(E_k) \geq \mu(E_k)$, $\mu(E)$ is also infinite and there is nothing to prove. Otherwise, given $\varepsilon > 0$, let $\{x_{1,k}, \dots, x_{n(k),k}\} \subseteq E_k$ be such that

$$\mu(E_k) \leq \sum_{i=1}^{n(k)} f(x_{i,k}) + \varepsilon 2^{-k}, \quad k = 1, 2, \dots \tag{2.5}$$

For each integer m, $\{x_{1,1}, \dots, x_{n(m),m}\}$ is a finite subset of E, and, by (2.5),

$$\sum_{k=1}^{m} \mu(E_k) \leq \sum_{k=1}^{m} \sum_{i=1}^{n(k)} f(x_{i,k}) + \varepsilon \sum_{k=1}^{m} 2^{-k} \leq \mu(E) + \varepsilon . \tag{2.6}$$

Since the right-hand side of (2.6) is independent of m, we may let $m \to \infty$ in the left-hand side there, and thus obtain

$$\sum_{k=1}^{\infty} \mu(E_k) \leq \mu(E) + \varepsilon .$$

But since ε above is arbitrary, the inequality opposite to (2.4) holds, μ is σ-additive.

Three particular instances of this example are of interest. If

$$\sum_{x \in X} f(x) = 1 ,$$

μ is a probability measure. On the other hand, if $f(x) = 1$ for all $x \in X$, μ is called, for obvious reasons, the counting measure on X. The counting measure is finite if X itself is finite and it is σ-finite if X is countable. Finally, if $f(x_0) = 1$ for some fixed $x_0 \in X$ and $f(x) = 0$ for $x \neq x_0$, μ is called the Dirac measure supported at x_0 and is denoted by δ_{x_0}; clearly $\delta_{x_0}(E) = 1$ or 0, according as to whether x_0 belongs to E or not.

The interesting question of when, in general, μ is finite or σ-finite, is left for the reader to answer, cf. 4.24 below.

We close this section with a simple criterion that enables us to determine when an additive set function is a measure.

Theorem 2.3. Let ψ be an additive, finitely valued set function defined on a σ-algebra $\mathcal{A}$. Then ψ is a measure iff for any nonincreasing sequence $\{E_k\}_{k=1}^{\infty} \subseteq \mathcal{A}$ with $\bigcap_{k=1}^{\infty} E_k = \emptyset$, we have $\lim_{k\to\infty} \psi(E_k) = 0$.

Proof. Assume that ψ is a measure and let $\{E_k\}$ be a nondecreasing sequence of sets in $\mathcal{A}$. Then by the σ-additivity of the finite set function ψ we have

$$\begin{aligned}\psi(E_1) &= \sum_{k=1}^{\infty}(\psi(E_k) - \psi(E_{k+1})) \\ &= \lim_{n\to\infty}\sum_{k=1}^{n-1}(\psi(E_k) - \psi(E_{k+1})) = \psi(E_1) - \lim_{n\to\infty}\psi(E_n)\,.\end{aligned}$$

Since $\psi(E_1)$ is finite, $\lim_{n\to\infty}\psi(E_n) = 0$, and the necessity follows.

As for the sufficiency, let $\{E_k\}$ be a disjoint sequence of measurable sets with union E and let $A_n = E\setminus(E_1\cup\ldots\cup E_n), n = 1,2\ldots$ Then $\{A_n\}$ is a nonincreasing sequence of measurable sets with $\bigcap_{n=1}^{\infty} A_n = \emptyset$, and $\lim_{n\to\infty}\psi(A_n) = 0$.

Now, since ψ is additive we have

$$\psi(E) = \sum_{k=1}^{n}\psi(E_k) + \psi(A_n)\,, \qquad n = 1,2,\ldots$$

Whence taking the limit as $n \to \infty$ it follows that

$$\psi(E) = \lim_{n\to\infty}\sum_{k=1}^{n}\psi(E_k) + \lim_{n\to\infty}(A_n) = \sum_{k=1}^{\infty}\psi(E_k)\,,$$

and the σ-additivity of ψ has been established. ■

3. PROPERTIES OF MEASURES

How do measures behave with respect to the usual set operations, and with respect to the limiting operations? Some of the basic properties are given in

Proposition 3.1. Suppose $(X, \mathcal{M}, \mu)$ is a measure space. Then the following properties hold:

(i) (Monotonicity) If E, F are measurable and $E \subseteq F$, then $\mu(E) \leq \mu(F)$. Moreover, if $\mu(E)$ is finite, then

$$\mu(F \setminus E) = \mu(F) - \mu(E)\,. \qquad (3.1)$$

(ii) (σ-subadditivity) If $\{E_k\}$ is a sequence of measurable sets, then

$$\mu\left(\bigcup E_k\right) \leq \sum \mu(E_k)\,. \tag{3.2}$$

(iii) (Continuity from below) If $E_1 \subseteq E_2 \subseteq \ldots$ is a nondecreasing sequence of measurable sets, then

$$\mu\left(\bigcup E_k\right) = \lim_{k\to\infty} \mu(E_k)\,. \tag{3.3}$$

(iv) (Continuity from above) If $E_1 \supseteq E_2 \supseteq \ldots$ is a nonincreasing sequence of measurable sets and for some $k\,, \mu(E_k) < \infty$, then

$$\mu\left(\bigcap E_k\right) = \lim_{k\to\infty} \mu(E_k)\,. \tag{3.4}$$

Proof. The monotonicity was essentially established in (2.2) above, so we say no more.

As for the σ-subadditivity, first note that on account of 5.2 in Chapter I and the properties of measurable sets, we may rewrite $\bigcup E_k = \bigcup F_k$, where $\{F_k\}$ is a sequence of pairwise disjoint measurable sets with $F_k \subseteq E_k$ for all k. Consequently, by the σ-additivity and the monotonicity of μ it follows that

$$\mu\left(\bigcup E_k\right) = \mu\left(\bigcup F_k\right) = \sum \mu(F_k) \leq \sum \mu(E_k)\,,$$

which is precisely (3.2).

Next note that if $\{E_k\}$ is nondecreasing, then $\lim E_k = \bigcup E_k$, so that in this particular instance the measure of the limit is the limit of the measures.

Now, if $\mu(E_1) = \infty$, by the monotonicity of μ, $\mu(E_k) = \infty$ for all k, and also $\mu(\bigcup E_k) = \infty$; in this case there is nothing to prove. Otherwise, since the sequence in question is nondecreasing, put $E_0 = \emptyset$, and note that $\bigcup_{k=1}^{\infty} E_k = \bigcup_{k=1}^{\infty}(E_k \setminus E_{k-1})$, where the sequence $\{E_k \setminus E_{k-1}\}$ is pairwise disjoint. Whence

$$\mu\left(\bigcup_{k=1}^{\infty} E_k\right) = \sum_{k=1}^{\infty} \mu(E_k \setminus E_{k-1}) = \lim_{n\to\infty} \sum_{k=1}^{n} \mu(E_k \setminus E_{k-1}).$$

Thus, by (3.1), we obtain that

$$\begin{aligned}\sum_{k=1}^{n} \mu(E_k \setminus E_{k-1}) &= \sum_{k=1}^{n} (\mu(E_k) - \mu(E_{k-1})) \\ &= \mu(E_n) - \mu(E_0) = \mu(E_n)\,,\end{aligned}$$

and (3.3) follows.

Finally, the idea to prove the continuity from above is to reduce the problem to one of continuity from below and to invoke (3.3). Replacing E_k by $E_k \bigcap E_{k_0}$ if necessary, where $\mu(E_{k_0}) < \infty$, we may assume that $\mu(E_1) < \infty$. Since $\{E_k\}$ is nonincreasing, the sequence $\{E_1 \setminus E_k\}$ is nondecreasing, and, by (3.3),

$$\mu\left(\bigcup_{k=1}^{\infty}(E_1 \setminus E_k)\right) = \lim_{k\to\infty} \mu(E_1 \setminus E_k). \tag{3.5}$$

Since $\bigcup_{k=1}^{\infty}(E_1 \setminus E_k) = E_1 \setminus \bigcap_{k=1}^{\infty} E_k$, and since by (3.1) it follows that $\mu\left(\bigcup_{k=1}^{\infty}(E_1 \setminus E_k)\right) = \mu(E_1) - \mu\left(\bigcap_{k=1}^{\infty} E_k\right)$, and that $\mu(E_1 \setminus E_k) = \mu(E_1) - \mu(E_k)$, by (3.5) we get that

$$\mu(E_1) - \mu\left(\bigcap_{k=1}^{\infty} E_k\right) = \mu(E_1) - \lim_{k\to\infty} \mu(E_k).$$

Moreover, $\mu(E_1) < \infty$, and this quantity may be cancelled in the above inequality to give (3.4), and to complete the proof. ∎

The restriction $\mu(E_1) < \infty$ is necessary for (3.4) to hold. Indeed, let μ be the measure on $(N, \mathcal{P}(N))$ given by $\mu(E) =$ number of elements of E if E is finite and $\mu(E) = \infty$ otherwise, and let $A_k = \{k, k+1, \ldots\}$. Then $\mu(A_k) = \infty$ for all k, but $\mu(\bigcap A_k) = \mu(\emptyset) = 0$.

In working with measures a useful result is the following

Theorem 3.2 (Borel-Cantelli Lemma). Suppose $(X, \mathcal{M}, \mu)$ is a measure space and let $\{E_n\}$ be a sequence of measurable sets with the property that $\sum_{n=1}^{\infty} \mu(E_n) < \infty$. Then

$$\mu(\limsup E_n) = 0. \tag{3.6}$$

Proof. First observe that by the σ-subadditivity of μ,

$$\mu\left(\bigcup_{n=1}^{\infty} E_n\right) \le \sum_{1}^{\infty} \mu(E_n) < \infty.$$

Now, the sequence consisting of $A_m = \bigcup_{n=m}^{\infty} E_n$, $m \ge 1$, is nonincreasing and $\mu(A_1) < \infty$. Whence, by (3.4)

$$\mu\left(\bigcap_{m=1}^{\infty} A_m\right) = \lim_{m\to\infty} \mu(A_m). \tag{3.7}$$

The set on the left-hand side of (3.7) is precisely $\limsup E_n$. As for the right-hand side, note that the measure of each set there does not exceed

$\sum_{n=m}^{\infty} \mu(E_n)$, and since these are the tails of a convergent series, we have $\lim_{m\to\infty} \mu(A_m) = 0$. Consequently, (3.7) gives at once (3.6). ■

Measures, or additive set functions for that matter, can be restricted or extended. More precisely, if $\mathcal{M}_1 \subseteq \mathcal{M}_2$ are σ-algebras of subsets of X, we say that a measure μ_1 on $(X, \mathcal{M}_1)$ is the restriction to $\mathcal{M}_1$ of the measure μ_2 on $(X, \mathcal{M}_2)$, and we write $\mu_1 = \mu_2|\mathcal{M}_1$, if $\mu_1(E) = \mu_2(E)$ for every $E \in \mathcal{M}_1$. In this case we also say that μ_2 is an extension of μ_1 to $\mathcal{M}_2$. For instance, given a measure space $(X, \mathcal{M}, \mu)$ and $A \in \mathcal{M}$, $\mu|\mathcal{M}_A$ can intuitively be thought of as the restriction of μ to A.

Sets of measure 0 play a special role in many questions of interest to us. Given a measure space $(X, \mathcal{M}, \mu)$, any measurable set of measure 0 is called a null set. Null sets are often denoted by N. If $\{N_k\}$ is a sequence of null sets, then by the σ-subaddivity of μ it readily follows that $\bigcup N_k$ is also a null set.

Also if N is a null set and $A \subseteq N$, then by the monotonicity of μ it follows that $\mu(A) = 0$, provided that A is measurable. This, of course, is not true of all measures. Consider, for instance, the following simple example: Let $X = \{a,b,c\}, \mathcal{M} = \{\emptyset, \{a\}, \{b,c\}, X\}$, and $\mu(\{a\}) = 1$, and $\mu(\{b,c\}) = 0$. Then μ is a probability measure and $N = \{b,c\}$ is a null set, but $\{b\}, \{c\} \subset \{b,c\}$ are not measurable. This motivates our next definition.

A measure space $(X, \mathcal{M}, \mu)$ is said to be complete if whenever $N \in \mathcal{M}$ is a null set and $A \subseteq N$, then A is also a measurable null set. In this case we also say, plainly, that μ is complete.

Since it is quite inconvenient to work with measure spaces which are not complete, we consider next whether a measure which is not complete can be extended, in a natural way, to complete measure.

Let, then, $(X, \mathcal{M}, \mu)$ be a measure space and put $\mathcal{N} = \{N \in \mathcal{M} : \mu(N) = 0\}$. If μ is not complete, then there is a set in $\mathcal{P}(\mathcal{N})$ which is not measurable, and the first step in constructing an extension of μ which is complete is to find a σ-algebra $\mathcal{M}_1$ which contains both $\mathcal{M}$ and $\mathcal{P}(\mathcal{N})$. The natural choice for $\mathcal{M}_1$ is $\mathcal{S}(\mathcal{M} \cup \mathcal{P}(\mathcal{N}))$; fortunately there is a simpler way to characterize $\mathcal{M}_1$. Indeed, if

$$\mathcal{A} = \{E \cup F : E \in \mathcal{M}, F \in \mathcal{P}(\mathcal{N})\},$$

then we claim that $\mathcal{M}_1 = \mathcal{A}$. Clearly $\mathcal{A} \subseteq \mathcal{M}_1$. If we can show that $\mathcal{A}$ is a σ-algebra of sets, since $\mathcal{A}$ contains $\mathcal{M}$ and $\mathcal{P}(\mathcal{N})$, it also contains the σ-algebra generated by $\mathcal{M} \cup \mathcal{P}(\mathcal{N})$, that is $\mathcal{M}_1$, and we are done.

First, $\mathcal{A} \neq \emptyset$. Next we show that if $A \in \mathcal{A}$, then also $X \setminus A \in \mathcal{A}$. Let $A = E \cup F$, where E is measurable and $F \subseteq N$, $N \in \mathcal{N}$. Since $N \in \mathcal{N}$

and $E \cup F = E \cup (F \setminus E)$, replacing N by $N \setminus E$ if necessary, we may assume that $E \cap F = E \cap N = \emptyset$. Now, since E and N are disjoint, we have that $E \cup F = (E \cup N) \cap (F \cup (X \setminus N))$, and consequently,

$$\begin{aligned} X \setminus (E \cup F) &= (X \setminus (E \cup N)) \cup (X \setminus (F \cup (X \setminus N))) \\ &= (X \setminus (E \cup N)) \cup ((X \setminus F) \cap N) = E_1 \cup F_1 , \end{aligned}$$

say. Since $E \cup N$ is measurable, $E_1 \in \mathcal{M}$, and since $F_1 \subseteq N$, $F_1 \in \mathcal{P}(\mathcal{N})$; in other words $X \setminus (E \cup F) \in \mathcal{A}$ as we wanted to show.

Finally we check that if $\{A_k\}$ is a sequence of subsets of $\mathcal{A}$, then also $\bigcup_{k=1}^{\infty} A_k$ belongs to $\mathcal{A}$. This is not hard: Since $A_k = E_k \cup F_k$, $E_k \in \mathcal{M}$, $F_k \in \mathcal{P}(\mathcal{N})$ for all k, it readily follows that $\bigcup_{k=1}^{\infty} A_k = (\bigcup_{k=1}^{\infty} E_k) \cup (\bigcup_{k=1}^{\infty} F_k) = E \cup F$, say. But as it is evident that $E \in \mathcal{M}$ and $F \in \mathcal{P}(\mathcal{N})$, our verification that $\mathcal{A}$ is a σ-algebra is finished.

We are now ready to prove

Theorem 3.3. Given a measure space $(X, \mathcal{M}, \mu)$, consider $\mathcal{N} = \{N \in \mathcal{M} : \mu(N) = 0\}$ and $\mathcal{M}_1 = \{E \cup F : E \in \mathcal{M}, F \in \mathcal{P}(\mathcal{N})\}$. Then there is a unique extension μ_1 of μ to $(X, \mathcal{M}_1)$ so that $(X, \mathcal{M}_1, \mu_1)$ is complete.

Proof. If μ is complete, $\mathcal{M}_1 = \mathcal{M}$, and by putting $\mu_1 = \mu$ we are done. Otherwise, if μ is not complete, define μ_1 on $(X, \mathcal{M}_1)$ as follows: If $A = E \cup F \in \mathcal{M}_1$, let

$$\mu_1(A) = \mu(E) . \tag{3.8}$$

First we show that μ_1 is a well-defined set function on $(X, \mathcal{M}_1)$, i.e., if $A = E_1 \cup F_1 = E_2 \cup F_2$, then we have $\mu(E_1) = \mu(E_2)$. This is not hard; indeed if $F_2 \subseteq N_2 \in \mathcal{N}$, note that

$$E_1 \subseteq E_1 \cup F_1 = E_2 \cup F_2 \subseteq E_2 \cup N_2 \in \mathcal{M} ,$$

and, by monotonicity, $\mu(E_1) \le \mu(E_2 \cup N_2) \le \mu(E_2) + \mu(N_2) = \mu(E_2)$. Reversing the roles of E_1 and E_2 we get that $\mu(E_2) \le \mu(E_1)$, and μ_1 is well-defined.

Next we check that μ_1 is a measure on $(X, \mathcal{M}_1)$. The only property that is not obvious is the σ-additivity. Let $\{A_k\}$ be a sequence of pairwise disjoint sets in $\mathcal{M}_1$, $A_k = E_k \bigcup F_k, F_k \subseteq N_k \in \mathcal{N}$ for all k. Then $F = \bigcup_k F_k \subseteq \bigcup_k N_k = N \in \mathcal{N}$, and since the E_k's are pairwise disjoint we have

$$\begin{aligned} \mu_1\left(\bigcup_k A_k\right) &= \mu_1\left(\left(\bigcup_k E_k\right) \cup F\right) \\ &= \mu\left(\bigcup_k E_k\right) = \sum \mu(E_k) = \sum \mu_1(A_k) , \end{aligned}$$

and the σ-additivity follows.

Finally we verify that $\mu_1|\mathcal{M} = \mu$ and that μ_1 is the only complete measure on $(X, \mathcal{M}_1)$ with this property. Since for $E \in \mathcal{M}$ we have $E = E \cup \emptyset$ and $\emptyset \in \mathcal{N}$, it is clear that $\mu_1(E) = \mu(E)$ and μ_1 is an extension of μ. To check that μ_1 is complete, let $N_1 = E_1 \cup F_1 \in \mathcal{M}_1$, $\mu_1(N_1) = 0$, and let $M \subseteq N_1$. Since N_1 is a null set we get that $\mu(E_1) = 0$ and $E_1 \cup F_1 \subseteq N \in \mathcal{N}$. Thus $M \subseteq N, M \in \mathcal{P}(\mathcal{N})$, $\mu_1(M) = 0$ and μ_1 is complete.

Further, to show that μ_1 is unique, suppose that μ_2 is another extension of μ to $\mathcal{M}_1$ and note that for $F \subseteq N \in \mathcal{N}$ we have $\mu_2(F) \leq \mu_2(N) = \mu(N) = 0$. Thus, if $A = E \cup F \in \mathcal{M}_1$, it readily follows that

$$\mu_2(A) \leq \mu_2(E) + \mu_2(F) = \mu(E) = \mu_1(A).$$

But since for $F \in \mathcal{P}(\mathcal{N})$ also $\mu_1(F) = 0$, we may essentially reverse the roles of μ_1 and μ_2 above and obtain that $\mu_1(A) \leq \mu_2(A)$ as well. In other words, $\mu_1(A) = \mu_2(A)$ for every $A \in \mathcal{M}_1$, and μ_1 is unique. ■

4. PROBLEMS AND QUESTIONS

4.1 Prove that $(\limsup A_n) \cap (\limsup B_n) \supseteq \limsup(A_n \cap B_n)$, and that $(\limsup A_n) \cup (\limsup B_n) = \limsup(A_n \cup B_n)$. What are the corresponding statements for the lim inf?

4.2 Let $\mathcal{A}$ be an algebra of subsets of X and let ψ be a finite additive set function defined on $\mathcal{A}$. Given $\{A_k\}_{k=1}^n \subseteq \mathcal{A}$, no two of the A_k's being the same, let $C_m = \{x : x$ belongs to exactly m of the A_k's$\}$, $m = 1, 2, \ldots, n$. Show that $\sum_{k=1}^n \psi(A_k) = \sum_{m=1}^n m\psi(C_m)$.

4.3 In the notation of problem 4.2, let $B_m = \{x : x$ belongs to at most m of the A_k's$\}$, $m = 1, 2, \ldots, n$. Show that $\sum_{k=1}^n \psi(A_k) = \sum_{m=1}^n \psi(B_m)$.

4.4 Let ψ be a finite additive set function defined on an algebra $\mathcal{A}$ of subsets of X, and let $A_1, A_2 \in \mathcal{A}$. Show that

$$\psi(A_1 \cap A_2) + \psi(A_1 \cup A_2) = \psi(A_1) + \psi(A_2).$$

4.5 It is possible to extend 4.4 to include more than two sets. Let $\{A_k\}_{k=1}^n \subseteq \mathcal{A}$ and for an integer $m \leq n$ put

$$r_m = \sum_{k_1 < \ldots < k_m} \psi(A_{k_1} \cap \ldots \cap A_{k_m}).$$

Show that
$$\psi\left(\bigcup_{k=1}^{n} A_k\right) = \sum_{m=1}^{n}(-1)^{m-1} r_m .$$

4.6 Assume ψ is an additive set function defined on an algebra $\mathcal{A}$ of subsets of X. Show that for arbitrary subsets $A_1, \ldots, A_n \subseteq \mathcal{A}$ we have

$$\sum_{i=1}^{n} \psi(A_i) - \sum_{i<j} \psi(A_i \cap A_j) \leq \psi\left(\bigcup_{i=1}^{n} A_i\right) .$$

4.7 An extended real-valued set function ψ defined on an algebra $\mathcal{A}$ of subsets of X is said to be bounded if there exists a constant M such that $|\psi(A)| \leq M$ for all $A \in \mathcal{A}$.
Show that any nonnegative additive set function which only assumes finite values is necessarily bounded.

4.8 Let ψ be an extended real-valued set function defined on an algebra $\mathcal{A}$ of subsets of X with the property that $\psi(\emptyset) = 0$. Given $A \in \mathcal{A}$, put

$$\psi_+(A) = \sup_{E \subseteq A, E \in \mathcal{A}} \psi(E), \quad \psi_-(A) = \sup_{E \subseteq A, E \in \mathcal{A}} (-\psi(E)),$$

and
$$|\psi|(A) = \psi_+(A) + \psi_-(A) .$$

ψ_+ is called the positive variation, ψ_- the negative variation and $|\psi|$ the total variation of ψ, respectively. Show that if ψ is additive, then all the variations are nonnegative additive set functions on $\mathcal{A}$.

4.9 In the setting of 4.8, if for $A \in \mathcal{A}$, $\psi(A)$ is finite, show that

$$|\psi|(A) = \sup(\psi(E_1) - \psi(E_2)) ,$$

where the sup is taken over those subsets E_1, E_2 of A which belong to $\mathcal{A}$.

4.10 In the setting of 4.8, show that if ψ is bounded above, i.e., if there exists a constant M such that $\psi(A) \leq M$ for all $A \in \mathcal{A}$, then the positive variation ψ_+ is a finite additive set function. Similarly, if ψ is bounded below, i.e., if there exists a constant m such that $\psi(A) \geq m$ for all $A \in \mathcal{A}$, then the negative variation ψ_- is a finite additive set function.

4.11 Assume ψ is an additive set function defined on an algebra $\mathcal{A}$ of subsets of X which is either bounded above or bounded below. Show that ψ can be represented as the differences of two nonnegative additive set functions. Specifically,

$$\psi(A) = \psi_+(A) - \psi_-(A), \quad A \in \mathcal{A}.$$

This relation is referred to as the Jordan decomposition of ψ, cf. Theorem 1.3 in Chapter III.

4.12 The total variation $|\psi|$ of the additive set function ψ defined on an algebra $\mathcal{A}$ of subsets of X can also be determined by the formula

$$|\psi|(A) = \sup \sum_{k=1}^{n} |\psi(E_k)|,$$

where the sup is taken over all finite partitions $\{E_k\}$ of A into disjoint sets of $\mathcal{A}$. Prove it.

4.13 Suppose ψ is a bounded additive set function defined on a σ-algebra $\mathcal{A}$ of subsets of X and suppose that ϕ is a finite measure defined on $\mathcal{A}$ with the property that for any sequence $\{E_k\} \subseteq \mathcal{A}$ with $\phi(E_k) \to 0$, also $\psi(E_k) \to 0$. Prove that ψ is also σ-additive.

4.14 In the notation of 4.8, if ψ is σ-additive, show that all the variations of ψ are also σ-additive.

4.15 Find $\mathcal{S}(\{\emptyset\})$. If $\emptyset \subsetneq A \subsetneq X$, what is $\mathcal{S}(\{A\})$? If A_1, A_2 are distinct subsets of X, show that $\mathcal{S}(\{A_1, A_2\})$ consists of, at most, sixteen sets.

4.16 Find $\mathcal{S}(\{A_1, \ldots, A_n\})$, where the A_i's are nonempty, pairwise disjoint, and $\bigcup_{i=1}^{n} A_i = X$.

4.17 Let $\mathcal{A}$ be an infinite σ-algebra of subsets of X. Show that $\mathcal{A}$ contains an infinite sequence of pairwise disjoint nonempty sets, and consequently, card $\mathcal{A} \geq c$.

4.18 Let $\mathcal{O}$ be an open set of R^n. Show that there is a sequence of nonoverlapping closed n-dimensional cubes $\{I_k\}$ such that $\mathcal{O} = \bigcup_k I_k$.

4.19 Let $\mathcal{A}$ be an algebra of subsets of X. Then the σ-algebra $\mathcal{S}(\mathcal{A})$ generated by A is the smallest family $\mathcal{F}$ of subsets of X that contains A and satisfies the following two conditions: (i) If $E_n \in \mathcal{F}$ and $E_n \subseteq E_{n+1}$ for $n = 1, 2, \ldots$ then $\bigcup_{n=1}^{\infty} E_n \in \mathcal{F}$, and, (ii) If $E_n \in \mathcal{F}$ and $E_n \supseteq E_{n+1}$ for $n = 1, 2, \ldots$, then $\bigcap_{n=1}^{\infty} E_n \in \mathcal{F}$.
Families that satisfy (i) and (ii) are said to be monotone. Prove that if the algebra $\mathcal{A}$ is a monotone family, then $\mathcal{A}$ is a σ-algebra.

4.20 If $\mathcal{A}_1$ is a σ-algebra of subsets of X_1 and $\mathcal{A}_2$ is a σ-algebra of subsets of X_2, show that $\mathcal{A}_1 \times \mathcal{A}_2$ is the smallest monotone family of subsets of $X_1 \times X_2$ that contains all finite disjoint unions of "rectangles" $E_1 \times E_2$ with $E_1 \in \mathcal{A}_1$ and $E_2 \in \mathcal{A}_2$.

4.21 Prove that $\mathcal{B}_n \times \mathcal{B}_m = \mathcal{B}_{n+m}$.

4.22 Prove that $\mathcal{S}(\mathcal{E}_1) \times \mathcal{S}(\mathcal{E}_2) = \mathcal{S}(\mathcal{E}_1 \times \mathcal{E}_2)$.

4.23 Let $\mathcal{A}$ be a σ-algebra of subsets of X and $E \subset X$. If $\mathcal{A}_1 = \mathcal{A} \cup \{E\}$, show that $\mathcal{S}(\mathcal{A}_1)$ consists of those subsets of X of the form $(A_1 \cap E) \cup (A_2 \cap (X \setminus E))$, where $A_1, A_2 \in \mathcal{A}$.

4.24 Let f be an extended real-valued function defined on X, and let μ be the measure on $(X, \mathcal{P}(X))$ given by $\mu(E) = \sum_{x \in E} f(x)$. Find necessary and sufficient conditions, in terms of f, for μ to be finite or σ-finite.

4.25 Let $(X, \mathcal{M}, \mu)$ be a measure space, and let $\{E_k\} \subseteq \mathcal{M}$. Show that if $\mu\left(\bigcup E_k\right) < \infty$, and $\mu(E_k) \geq \eta > 0$ for infinitely many k's, then $\mu\left(\limsup E_k\right) > 0$. By means of an example show that the condition $\mu\left(\bigcup E_k\right) < \infty$ cannot be removed.

4.26 Let $(X, \mathcal{M}, \mu)$ be a measure space, and $\{E_k\} \subseteq \mathcal{M}$. Show that $\mu\left(\liminf E_k\right) \leq \liminf \mu\left(E_k\right)$, and that, provided that $\mu\left(\bigcup E_k\right) < \infty$, $\limsup \mu\left(E_k\right) \leq \mu\left(\limsup E_k\right)$.
By means of examples show that we may have strict inequalities above.

4.27 Let $(X, \mathcal{M}, \mu)$ be a probability measure space. If $\mu(\limsup A_n) = 1$ and $\mu(\liminf B_n) = 1$, prove that $\mu(\limsup(A_n \cap B_n)) = 1$. What happens if we assume instead that $\mu(\limsup B_n) = 1$?

4.28 Let μ be a measure on $(R, \mathcal{B}_1)$ with the property that $\mu(I) < \infty$ for every finite interval I, $y \in R$, and put

$$F_y(x) = \begin{cases} \mu((y, x]) & \text{if } x > y \\ 0 & \text{if } x = y \\ -\mu((x, y]) & \text{if } x < y. \end{cases}$$

Show that F_y is a nondecreasing right-continuous function; F_y is called a distribution function induced by μ.

4.29 Let $(X, \mathcal{M}, \mu)$ be a measure space, τ be a mapping of X onto Y and set $\mathcal{N} = \{A \subseteq Y : \tau^{-1}(A) \in \mathcal{M}\}$. Furthermore, let ν be the set function defined on $\mathcal{N}$ by $\nu(A) = \mu(\tau^{-1}(A))$. Prove that $\mathcal{N}$ is a σ-algebra of subsets of Y and that $(Y, \mathcal{N}, \nu)$ is a measure space.

4.30 Let μ_1, μ_2 be measures on $(X, \mathcal{M})$, $\mu_2(X) < \infty$, and suppose that $\mu_1(E) \geq \mu_2(E)$ for all $E \in \mathcal{M}$. Show that there exists a (unique) measure μ_3 on $(X, \mathcal{M})$ such that

$$\mu_1(E) = \mu_2(E) + \mu_3(E), \quad \text{all } E \in \mathcal{M}.$$

Is the restriction $\mu_2(X) < \infty$ necessary?

4.31 Let $\{\mu_k\}$ be a sequence of measures on $(X, \mathcal{M})$ with the property that $\mu_k(E) \leq \mu_{k+1}(E)$ for all $E \in \mathcal{M}$. Is the set function μ defined on $(X, \mathcal{M})$ by

$$\mu(E) = \lim_{k \to \infty} \mu_k(E),$$

necessarily a measure?

4.32 A useful concept in measure theory is that of semifiniteness, a notion weaker than that of σ-finiteness. A measure μ on $(X, \mathcal{M})$ is said to be semifinite if for each $A \in \mathcal{M}$ with $\mu(A) = \infty$, there is a measurable set $E \subset A$, with $0 < \mu(E) < \infty$. Show that every σ-finite measure is semifinite, as is the counting measure, and give an example of a measure that is not semifinite.
Also prove that if μ is semifinite and $\mathcal{M}$ does not contain an uncountable, pairwise disjoint collection of measurable sets of positive measure, then μ is σ-finite. Show by means of an example that this does not hold without the assumption that there are no "infinite atoms."

4.33 Let μ be a semifinite measure defined on $(X, \mathcal{M})$. Show that every measurable set with infinite measure contains a measurable subset with arbitrarily large, finite, measure.

4.34 Let μ be a measure on $(X, \mathcal{M})$. Show that μ can be written as the sum $\mu(A) = \mu_1(A) + \mu_2(A)$, $A \in \mathcal{M}$, where μ_1 is a semifinite measure and the measure μ_2 only assumes the values 0 and ∞; the decomposition need not be unique.

4.35 Assume that $(X, \mathcal{M}, \mu)$ is a measure space, and for $E_1, E_2 \in \mathcal{M}$ put

$$d(E_1, E_2) = \mu(E_1 \,\triangle\, E_2).$$

Discuss under what conditions endowed with this distance, $(\mathcal{M}, d)$ becomes a complete metric space.

4.36 Assume that $(X_k, \mathcal{M}_k, \mu_k)$ are measure spaces for $k = 1, 2, \ldots$, and that the X_k's are pairwise disjoint. Put $X = \bigcup_k X_k$, and let $\mathcal{M}$ be the class of subsets A of X of the form $A = \bigcup_k A_k$, where $A_k \in \mathcal{M}_k$ for all k. Show that $\mathcal{M}$ is a σ-algebra of subsets of X and that the set function μ given on $\mathcal{M}$ by $\mu(A) = \sum_k \mu_k(A_k)$ is a measure on $(X, \mathcal{M})$. When is μ finite? σ-finite?

CHAPTER V

The Lebesgue Measure

In this chapter we introduce the most important example of a measure on R^n, the Lebesgue measure. I learned this construction from A. Zygmund.

1. LEGESGUE MEASURE ON R^n

In defining a measure on R^n we must first decide on the σ-algebra of measurable sets; in the case at hand geometric considerations are also of importance.

We call a closed parallelepiped, i.e., a closed bounded set of the form $\{(x_1, \ldots x_n) : a_k \leq x_k \leq b_k, 1 \leq k \leq n\}$, a closed interval. An open interval is the interior of a closed interval, i.e., a set of the form $\{(x_1, \ldots, x_n) : a_k < x_k < b_k, 1 \leq k \leq n\}$. Intervals I, open and closed, have volume $v(I)$ equal to $\prod_{k=1}^{n}(b_k - a_k)$.

We require that the σ-algebra of Lebesgue measurable sets contain all open and closed intervals and that the Lebesgue measure agree with the volume for intervals. Since each open set in R^n is the countable union of nonoverlapping closed intervals, Lebesgue sets include all open sets, and consequently, also all closed and Borel sets. It is not intuitively clear what the Lebesgue measure of these general sets is; however, we expect the Lebesgue measure to satisfy two additional properties: It should be complete and translation invariant, i.e., if A is Lebesgue measurable and $A_y = y + A = \{y + x : x \in A\}$, then A_y is also measurable and A and A_y have the same measure.

Translation invariance plays an essential role in the determination of the nature of the σ-algebra of Lebesgue measurable sets. In the early 1900's Vitali showed that if we accept the Axiom of choice, then not all subsets of R are Lebesgue measurable. In the early 1970's Solovay proved that it is consistent with the usual axioms of Set Theory, excluding the Axiom of choice, for every subset of R to be Lebesgue measurable. One cannot conclude from Solovay's result, however, that the Axiom of choice and the existence of a subset of R which is not Lebesgue measurable are equivalent.

Since the construction of a subset of R which is not Lebesgue measurable depends on general, rather than specific, properties of the measure, including translation invariance, we describe such a set at this point.

Let $I = [-1/2, 1/2]$ and let $\sim$ be the relation defined for numbers x, y in I by $x \sim y$ iff $x - y$ is a rational number (between -1 and 1); it is not hard to check that $\sim$ is an equivalence relation. Let $\mathcal{R}$ be the collection of all the distinct equivalence classes of $\sim$; clearly $\mathcal{R}$ can be indexed by a subset of I. Now, by the Axiom of choice, there is a set $A \subset I$ which contains exactly one element from each equivalence class of $\sim$; it is the set A we propose to show is not Lebesgue measurable.

Suppose, to the contrary, that A is Lebesgue measurable, enumerate the rational numbers in $[-1,1], r_0 = 0, r_1, \ldots$, and let $A_k = r_k + A, k \geq 0$, be the rational translates of A. Since we assume that A is Lebesgue measurable, the A_k's are also Lebesgue measurable and they all have the same measure. Note that translates corresponding to different r_k's are disjoint: Indeed, if for $j \neq k$ there is $y \in A_j \cap A_k$, then there are $x, x' \in A$ such that $y = x + r_j = x' + r_k$. Now, since $r_j \neq r_k$, then also $x \neq x'$, and x and x' correspond to two distinct equivalence classes of $\sim$. But we also have that $x - x' = r_k - r_j$ is a rational number (between -1 and 1), and consequently, $x \sim x'$. However, since A contains one element from each equivalence class this cannot happen, and the A_k's are pairwise disjoint.

Next observe that

$$I \subseteq \bigcup_k A_k \subseteq [-3/2, 3/2] . \tag{1.1}$$

Suppose we have shown (1.1) to be true. Then the left-hand side inclusion implies that the Lebesgue measure of A cannot be 0, for otherwise $\bigcup_k A_k$ would be a null set which contains a subset of measure 1, and the right-hand side inclusion implies that the Lebesgue measure of A cannot be positive, for otherwise $\bigcup_k A_k$ would be a measurable set of infinite measure contained in a set of measure 3. In short, (1.1) implies that A cannot be Lebesgue measurable.

To prove that (1.1) holds is not hard. For $x \in I$, let x' be the representative of the equivalence class of x which belongs to A. Then there is a rational number r_k in $[-1,1]$, such that $x = x' + r_k$. Whence $x \in A_k$, and the left-hand side inclusion in (1.1) holds. Moreover, since each $A_k \subseteq [-3/2,3/2]$, the same is true of the union, and the right-hand side inclusion in (1.1) also holds.

Now that we are satisfied that not every subset of R^n can be Lebesgue measurable, how do we go about constructing the σ-algebra of Lebesgue measurable sets? The idea is to introduce the Lebesgue measure as we go along. First we define a nonnegative σ-subadditive set function on $\mathcal{P}(R^n)$, the Lebesgue outer measure, and then select as Lebesgue measurable sets those subsets of R^n which can be approximated, in a sense made precise by the outer measure, by open sets.

Since intervals are the only sets we know how to measure at this point, it is natural that the definition of Lebesgue outer measure involve coverings with intervals. Given a subset A of R^n, we define the Lebesgue outer measure $|A|_e$ of A as the quantity

$$\inf\left\{\sum\nolimits_k v(I_k): A \subseteq \bigcup\nolimits_k I_k\right\}, \tag{1.2}$$

where the infimum in (1.2) is taken over the family of all at most countable coverings of A by closed intervals. Now, a countable mesh of nonoverlapping unit intervals (actually n-dimensional cubes of sidelength one) covers R^n, and consequently it also covers every subset of R^n. Thus the inf in (1.2) is a well-defined quantity.

By (1.2) it is clear that if $A \subseteq B$, then $|A|_e \le |B|_e$ (monotonicity). Moreover, as expected, the outer measure of an interval coincides with its volume.

Proposition 1.1. Let I be a closed interval in R^n. Then $|I|_e = v(I)$.

Proof. Since I is a finite covering of itself, $|I|_e \le v(I) < \infty$. To prove the opposite inequality note that since $|I|_e < \infty$, given $\varepsilon > 0$, there is a family of closed intervals $\{I_k\}$ such that $I \subseteq \bigcup I_k$, and $\sum v(I_k) \le |I|_e + \varepsilon$. Furthermore, by extending the sidelengths of the I_k's and then deleting the edges, it is readily seen that given $\eta > 0$, there are open intervals $I'_k \supseteq I_k$, such that $v(I'_k) \le (1+\eta)v(I_k)$, all k. Since $I \subseteq \bigcup I'_k$, and since I is compact and the I'_k's are open, by the Heine-Borel theorem there is a finite subcovering, which for simplicity we also denote by $\{I'_k\}$, such that $I \subseteq \bigcup_{k=1}^N I'_k$. In this case it is intuitively clear, although involved to prove, that $v(I) \le \sum_{k=1}^N v(I'_k)$, and consequently we also have

$$v(I) \le (1+\eta)\sum\nolimits_{k=1}^N v(I_k) \le (1+\eta)\sum v(I_k) \le (1+\eta)(|I|_e + \varepsilon). \tag{1.3}$$

Since ε and η are both arbitrary, from (1.3) it readily follows that $v(I) \le |I|_e$, and we have finished. ∎

Along the same lines it is not hard to see that the following result holds: Let $\{I_k\}_{k=1}^m$ be a finite collection of nonoverlapping closed intervals in R^n, then

$$\left|\bigcup_{k=1}^m I_k\right|_e = \sum_{k=1}^m v(I_k).$$

Corollary 1.2. Let I be an open interval. Then $|I|_e = v(I)$.

Proof. Let $\bar{I}$ denote the closure of I, then we have $|I|_e \le v(\bar{I}) = v(I)$, and one inequality holds. Also, if J is any closed interval contained in I, monotonicity implies that $v(J) = |J| \le |I|_e$. But since $v(I) = \sup v(J)$, where the sup is taken over the family of the closed subintervals J of I, we get that $v(I) \le |I|_e$, and the opposite inequality holds. ■

Next we show that the outer measure is σ-subadditive.

Proposition 1.3. Let $\{E_k\}_{k=1}^\infty$ be any sequence of subsets of R^n. Then

$$\left|\bigcup_k E_k\right|_e \le \sum_k |E_k|_e . \tag{1.4}$$

Proof. If $|E_k|_e = \infty$ for some k, then by monotonicity also the union has infinite outer measure, and we have equality in (1.4). Otherwise, suppose that $|E_k|_e < \infty$ for all k, and let $\varepsilon > 0$ be arbitrary. For each k let $\{I_{k,j}\}$ be a covering of E_k by closed intervals with the property that

$$\sum_j v(I_{k,j}) \le |E_k|_e + \varepsilon/2^k . \tag{1.5}$$

Clearly

$$\bigcup_k E_k \subseteq \bigcup_{j,k} I_{j,k} , \tag{1.6}$$

and consequently, the closed intervals on the right-hand side of (1.6) form a covering of $\bigcup_k E_k$. Whence, by definition, we have

$$\left|\bigcup_k E_k\right|_e \le \sum_{j,k} v(I_{j,k}) , \tag{1.7}$$

and since the summands on the right-hand side of (1.7) are nonnegative we may interchange freely the order of summation and estimate that expression by

$$\sum_k \left(\sum_j v(I_{j,k})\right) \le \sum_k \left(|I_k|_e + \varepsilon/2^k\right) = \sum_k |E_k|_e + \varepsilon .$$

But since $\varepsilon > 0$ is arbitrary, we also have $|\bigcup_k E_k|_e \le \sum_k |E_k|_e$. ■

In contrast to the case of measures, it is interesting to point out that strict inequality may occur in (1.4), even if the E_k's are pairwise disjoint. To see this observe that the Lebesgue outer measure is translation invariant, and that with the notation of (1.1) above, $|A|_e > 0$. Then, again by (1.1), $|\bigcup_k A_k|_e \le 3 < \sum_k |A_k|_e = \infty$.

Since open sets can be expressed as the union of closed intervals, it is reasonable to attempt to compute the outer measure of subsets of R^n in terms of the outer measure of open sets. Specifically, we have

Proposition 1.4. Let E be any subset of R^n. Then

$$|E|_e = \inf \{|\mathcal{O}|_e : \mathcal{O} \text{ is open, and } \mathcal{O} \supseteq E\} . \tag{1.8}$$

Proof. If $|E|_e = \infty$, by monotonicity every open set which contains E (and this class is nonempty since R^n is one such set) also has infinite outer measure and (1.8) holds. On the other hand, if $|E|_e$ is finite, given $\varepsilon > 0$, let $\{I_k\}_{k=1}^{\infty}$ be a covering of E by closed intervals such that

$$\sum_k v(I_k) \le |E|_e + \varepsilon/2 .$$

Furthermore, for each k, let I'_k be an open interval containing I_k such that $v(I'_k) \le v(I_k) + \varepsilon/2^{k+1}$, and put $\mathcal{O} = \bigcup_k I'_k$. By construction $\mathcal{O}$ is open and it contains E. Moreover, by Proposition 1.3 and Corollary 1.2 it readily follows that

$$\begin{aligned} |E|_e \le |\mathcal{O}|_e \le \sum_k |I'_k|_e &= \sum_k v(I'_k) \\ \le \sum_k \left(v(I_k) + \varepsilon/2^{k+1}\right) &= \sum_k v(I_k) + \varepsilon/2 \le |E|_e + \varepsilon . \end{aligned} \tag{1.9}$$

Thus, for any $\varepsilon > 0$, there is an open set $\mathcal{O} \supseteq E$ such that $|E|_e \le |\mathcal{O}|_e \le |E|_e + \varepsilon$, and (1.8) holds. ■

Proposition 1.4 is important in applications; to state some we need a definition.

We say that a subset H of R^n is a G_δ set if H is the intersection of an at most countable family of open sets. The complement of a G_δ set is an F_σ set, i.e., an at most countable union of closed sets.

Corollary 1.5. Let E be an arbitrary subset of R^n. Then there is a G_δ set H which contains E and such that $|H|_e = |E|_e$.

Proof. If $|E|_e = \infty$, put $H = R^n$. If $|E|_e < \infty$, by (1.9) there is a sequence $\{\mathcal{O}_k\}_{k=1}^\infty$ of open sets containing E such that $|\mathcal{O}|_e \le |E|_e + 1/k$. Let now $H = \bigcap_k \mathcal{O}_k$; by construction H is a G_δ set, and $E \subseteq H \subseteq \mathcal{O}_k$ for all k. Thus by monotonicity it follows that

$$|E|_e \le |H|_e \le |\mathcal{O}_k|_e \le |E|_e + 1/k\,, \quad \text{all } k\,. \tag{1.10}$$

Since k is arbitrary we get that $|E|_e = |H|_e$. ■

A closer look at inequality (1.10) for sets E with finite outer measure indicates that the following property is true: Given $\varepsilon > 0$, there is an open set $\mathcal{O} \supseteq E$ such that $|\mathcal{O}|_e - |E|_e < \varepsilon$. This estimate does not make sense when $|E|_e = \infty$, but a closely related result holds for any subset of R^n, to wit, there exists an open set $\mathcal{O} \supseteq E$ such that $|\mathcal{O}|_e \le |E|_e + |\mathcal{O} \setminus E|_e$. This inequality is a simple consequence of the monotonicity and it hints that rather to seek to control $|\mathcal{O}|_e - |E|_e$, which is meaningless when $|E|_e = \infty$, we may try to control $|\mathcal{O} \setminus E|_e$. This control is, indeed, all that is needed.

We say that $E \subseteq R^n$ is Lebesgue measurable if for any $\varepsilon > 0$, there exists an open set $\mathcal{O} \supseteq E$ such that

$$|\mathcal{O} \setminus E|_e < \varepsilon\,.$$

The class of Lebesgue measurable sets is denoted by $\mathcal{L}_n$, or plainly by $\mathcal{L}$. Notice that open sets, as well as sets with outer measure equal to 0, belong to $\mathcal{L}$. In the case of open sets this is obvious, and for sets E of outer measure 0 observe that by (1.10) there are open sets $\mathcal{O}_k \supseteq E$ with $|\mathcal{O}_k| \le 1/k$ for all k. Then, given $\varepsilon > 0$, let $k \ge 1/\varepsilon$, and note that

$$|\mathcal{O}_k \setminus E|_e \le |\mathcal{O}_k|_e \le 1/k \le \varepsilon\,.$$

In order to show that $\mathcal{L}$ is a σ-algebra we must verify that $\mathcal{L}$ is closed under countable unions, which is easy, and under complementation, which requires some work. We begin with the easier part.

Proposition 1.6. Let $\{E_k\}_{k=1}^\infty$ be a sequence of subsets in $\mathcal{L}$, then $\bigcup_k E_k \in \mathcal{L}$.

Proof. For a given $\varepsilon > 0$, we must find an open set $\mathcal{O} \supseteq \bigcup_k E_k$ so that $|\mathcal{O} \setminus \bigcup_k E_k|_e < \varepsilon$. Since each E_k belongs to $\mathcal{L}$, there are open sets $\mathcal{O}_k \supseteq E_k$ such that $|\mathcal{O}_k \setminus E_k|_e \le \varepsilon/2^k$ for all k. Hence, $\mathcal{O} = \bigcup_k \mathcal{O}_k$ is an open set which contains $\bigcup_k E_k$, and since as is readily seen $(\bigcup_k \mathcal{O}_k) \setminus (\bigcup_k E_k) \subseteq \bigcup_k (\mathcal{O}_k \setminus E_k)$, by Proposition 1.3 we get that

$$|\mathcal{O} \setminus \textstyle\bigcup_k E_k|_e \le \sum_k |\mathcal{O}_k \setminus E_k|_e \le \sum_k \varepsilon/2^k = \varepsilon\,.$$

Thus, the union of the E_k's belongs to $\mathcal{L}$. ∎

To complete the verification that $\mathcal{L}$ is a σ-algebra, we begin by showing that closed sets are indeed Lebesgue measurable. This requires some preliminary results.

Lemma 1.7. Let E_1, E_2 be subsets of R^n with the property that

$$d(E_1, E_2) = \inf\{|x - x'| : x \in E_1, x' \in E_2\} > 0\,.$$

Then, $|E_1 \cup E_2|_e = |E_1|_e + |E_2|_e$. In particular, this is true if E_1, E_2 are compact and disjoint.

Proof. If either $|E_1|_e$ or $|E_2|_e$ is infinite, then the same is true of $|E_1 \cup E_2|_e$, and we are done. Now, if both are finite, since the outer measure is subadditive, it suffices to show that $|E_1|_e + |E_2|_e \le |E_1 \cup E_2|_e$. But in this case we also have that $|E_1 \cup E_2|_e < \infty$, and consequently, given $\varepsilon > 0$, there is a covering $\{I_k\}$ of $E_1 \cup E_2$ consisting of closed intervals such that

$$\sum\nolimits_k v(I_k) \le |E_1 \cup E_2|_e + \varepsilon\,.$$

There are three relevant kinds of I_k's, to wit:

(i) Those I_k's such that $I_k \cap E_1 \ne \emptyset$, $I_k \cap E_2 = \emptyset$, call them I_k^1's;

(ii) Those I_k's such that $I_k \cap E_1 = \emptyset$, $I_k \cap E_2 \ne \emptyset$, call them I_k^2's;

(iii) Those I_k's which intersect both E_1 and E_2.

The intervals in the third class above may be subdivided into nonoverlapping closed subintervals of diameter less than or equal to $d(E_1, E_2)$. Each subinterval thus obtained either belongs to the first family (i), or to the second family (ii), or it does not intersect $E_1 \cup E_2$ and it can be discarded. Therefore, we divide the I_k's into a covering of E_1, a covering of E_2, and throw away the rest. By definition we have

$$|E_1|_e + |E_2|_e \le \sum\nolimits_k v(I_k^1) + \sum\nolimits_k v(I_k^2) \le \sum\nolimits_k v(I_k) \le |E_1 \cup E_2|_e + \varepsilon\,,$$

which implies, since ε is arbitrary, that, as asserted, $|E_1|_e + |E_2|_e \le |E_1 \cup E_2|_e$. ∎

We are now ready to prove

Theorem 1.8. Closed subsets of R^n are Lebesgue measurable.

Proof. Suppose first that the closed set F in question is bounded, and hence compact. Then, given $\varepsilon > 0$, by inequality (1.10) there is an open set $\mathcal{O} \supset F$ such that $|\mathcal{O}|_e \le |F|_e + \varepsilon$; we would like to show that $|\mathcal{O} \setminus F|_e \le \varepsilon$ as well. Now, $\mathcal{O} \setminus F$ is also open, and consequently it can be expressed as the countable union of nonoverlapping closed intervals, $\bigcup_k I_k$, say. By Proposition 1.3, it follows that $|\mathcal{O} \setminus F|_e \le \sum_k v(I_k)$. On the other hand, since

$$\mathcal{O} = F \cup (\bigcup_k I_k) \supset F \cup \left(\bigcup_{k=1}^N I_k\right), \quad \text{all } N,$$

by monotonicity we get that

$$\left| F \cup \left(\bigcup_{k=1}^N I_k\right)\right|_e \le |\mathcal{O}|_e, \quad \text{all } N.$$

Furthermore, since F and $\bigcup_{k=1}^N I_k$ are both compact and disjoint and the I_k's are nonoverlapping, by Lemma 1.7 it readily follows that

$$|F|_e + \left|\bigcup_{k=1}^N I_k\right|_e = |F|_e + \sum_{k=1}^N v(I_k) \le |\mathcal{O}|_e.$$

In particular

$$\sum_{k=1}^N v(I_k) \le |\mathcal{O}|_e - |F|_e \le \varepsilon, \quad \text{all } N.$$

But since this inequality holds with a bound independent of N, we have $\sum_k v(I_k) \le \varepsilon$, and consequently also $|\mathcal{O} \setminus F|_e \le \varepsilon$. Thus, in this case, F is Lebesgue measurable.

For general closed subsets F of R^n, let $F_k = \{x \in F : |x| \le k\}$, and observe that each F_k is closed and bounded, and that $F = \bigcup_k F_k$. By the above argument each F_k is measurable, and by Proposition 1.6 so is their union, F. ■

Theorem 1.9. Suppose $E \in \mathcal{L}$, then $R^n \setminus E \in \mathcal{L}$.

Proof. Let $\mathcal{O}_k \supset E$ be a family of open sets such that $|\mathcal{O}_k \setminus E|_e \le 1/k, k = 1, 2, \ldots$; each $R^n \setminus \mathcal{O}_k$ is closed, and hence measurable. Furthermore, since $R^n \setminus \mathcal{O}_k \subseteq R^n \setminus E$ for all k, $H = \bigcup_k (R^n \setminus \mathcal{O}_k)$ is an F_σ measurable subset of $R^n \setminus E$.

Let $A = (R^n \setminus E) \setminus H$; since $R^n \setminus E = H \cup A$, we will be done once we check that A is measurable. To do this we show that $|A|_e = 0$. Indeed, since for each k, $A \subseteq (R^n \setminus E) \setminus (R^n \setminus \mathcal{O}_k) = \mathcal{O}_k \setminus E$, it readily follows that

$$|A|_e \le |\mathcal{O}_k \setminus E|_e \le 1/k, \quad \text{all } k.$$

Thus $|A|_e = 0$, and we are done. ■

Theorem 1.9 completes the verification that $\mathcal{L}$ is a σ-algebra and, at the same time, it provides a description of the Lebesgue measurable sets. Indeed, the argument of Theorem 1.9 applied to the (measurable) complement $R^n \setminus E$ of a measurable set E gives that $E = R^n \setminus (R^n \setminus E) = H \cup N$, where H is an F_σ set and $|N|_e = 0$. It also gives that the Lebesgue measurable sets are precisely those subsets E of R^n which satisfy the following property: Given $\varepsilon > 0$, there is a closed set $F \subseteq E$ such that $|E \setminus F|_e \leq \varepsilon$.

Next we construct the Lebesgue measure on $(R^n, \mathcal{L})$; it is the restriction of the Lebesgue outer measure to $\mathcal{L}$. More precisely, we have

Theorem 1.10. The set function $|\cdot|_e$ restricted to $\mathcal{L}$ is a measure. We call this measure the Lebesgue measure on R^n and denote it by $|\cdot|$.

Proof. The proof amounts to showing that the set function $|\cdot|_e$ is σ-additive on $\mathcal{L}$. For this purpose, let $\{E_k\}_{k=1}^{\infty}$ be a sequence of pairwise disjoint measure sets, and let E denote its union. Note that, by Proposition 1.3, $|E| \leq \sum_k |E_k|$. As for the opposite inequality, assume first that the E_k's are bounded. Given $\varepsilon > 0$, let $F_k \subseteq E_k$ be a sequence of closed sets such that $|F_k \setminus E_k| \leq \varepsilon/2^k$ for all k. Since $E_k = F_k \cup (E_k \setminus F_k)$ we also have $|E_k| \leq |F_k| + \varepsilon/2^k$. Furthermore, since the E_k's are pairwise disjoint, the sequence of F_k's is composed of pairwise disjoint compact subsets of R^n. Fix N, and note that by (a simple extension of) Lemma 1.7, $|\bigcup_{k=1}^N F_k| = \sum_{k=1}^N |F_k|$. Thus, since $\bigcup_{k=1}^N F_k \subseteq E$ for all N, it follows that $\sum_{k=1}^N |F_k| \leq |E|$, all N, and consequently also $\sum_{k=1}^\infty |F_k| \leq |E|$. Whence

$$\sum_{k=1}^{\infty} |E_k| \leq \sum_{k=1}^{\infty} \left(|F_k| + \varepsilon/2^k\right) \leq |E| + \varepsilon\,,$$

and, since ε is arbitrary, we are done in this case.

In the general case, fix an increasing sequence $\{I_j\}$ of bounded intervals so that $\bigcup_j I_j = R^n$, $I_0 = \emptyset$, and put $S_j = I_j \setminus I_{j-1}, j = 1, 2, \ldots$ Then, the sets $E_{k,j} = E_k \cap S_j$ are measurable, pairwise disjoint and bounded, and for each k we have $\bigcup_j E_{k,j} = E_k$.

Thus, on the one hand, $\left|\bigcup_{j,k} E_{k,j}\right| = |\bigcup_k E_k|$, and, on the other hand,

$$\left|\bigcup_{j,k} E_{k,j}\right| = \sum_{k,j} |E_{k,j}| = \sum_k \sum_j |E_{k,j}| = \sum_k |E_k|\,.$$

In other words, $|\cdot|$ is σ-additive on $\mathcal{L}$, and we have finished. ■

The following characterization of $\mathcal{L}$, due to Carathéodory (1873–1950), highlights the interplay between the Lebesgue measurable sets and the Lebesgue measure, and it is very interesting since it can be used to define $\mathcal{L}$, and more general σ-algebras of sets, cf. 3.40 below.

Theorem 1.11 (Carathéodory). A subset E of R^n is Lebesgue measurable iff for every subset A of R^n we have

$$|A|_e = |A \cap E|_e + |A \setminus E|_e . \tag{1.11}$$

Proof. We begin by assuming that E is measurable and A is any subset of R^n. Since $A = (A \cap E) \cup (A \setminus E)$, by the subadditivity of the Lebesgue outer measure it follows that

$$|A|_e \leq |A \cup E|_e + |A \setminus E|_e .$$

To prove the opposite inequality, note that by Corollary 1.5 there is a G_δ measurable set $H \supseteq A$ such that $|A|_e = |H|$. Now, since H is also measurable, we have $|H| = |H \cap E| + |H \setminus E|$. Whence, by monotonicity

$$|A|_e = |H| \geq |A \cap E|_e + |A \setminus E|_e ,$$

and (1.11) holds.

Next assume that (1.11) is true for every subset A of R^n; we distinguish the cases $|E|_e < \infty$ and $|E|_e = \infty$. In the former case, by Corollary 1.5 there is a G_δ set $H \supseteq E$ such that $|H| = |E|_e$. By (1.11) we have

$$|H| = |H \cap E|_e + |H \setminus E|_e = |E|_e + |H \setminus E|_e . \tag{1.12}$$

(1.12) gives at once that $H \setminus E$ is a measurable set of measure 0, and that $E = H \setminus (H \setminus E)$ is also measurable.

As for the latter case, let $E_k = \{x \in E : |x| \leq k\}$ be so that $|E_k| < \infty$, and let H_k be a G_δ set containing E_k such that $|H_k| = |E_k|_e$ for all k. By (1.11), with $A = H_k$ there, we get

$$|H_k| = |H_k \cap E|_e + |H_k \setminus E|_e \geq |E_k|_e + |H_k \setminus E|_e .$$

Since $|H_k| = |E_k|_e$, we have $|H_k \setminus E|_e = 0$ for all k. Whence, setting $H = \bigcup_k H_k$, it readily follows that H is a measurable set which contains E, and that $H \setminus E = \bigcup_k (H_k \setminus E)$ is a measurable set of measure 0. Thus, $E = H \setminus (H \setminus E)$ is also measurable. ■

2. THE CANTOR SET

It is easy to see that there are uncountable sets of measure 0 in $R^n, n \geq 2$; indeed, the boundary of any interval is such a set. How about R? The Cantor set is such an example, and we construct it next.

Consider the closed interval $C_0 = [0,1]$. The first stage of the construction is to trisect C_0 and to remove the interior of the middle interval, $(1/3,2/3)$. Each successive step is essentially the same. Let $C_1 = [0,1/3] \cup [2/3,1]$; C_1 is the union of $2^1 = 2$ closed disjoint intervals. At the second stage we subdivide each of the closed intervals of C_1 into thirds and remove from each one the middle open thirds, $(1/9,2/9)$ and $(7/9,8/9)$. Suppose that C_n has been constructed and that it consists of 2^n closed disjoint intervals, each of length 3^{-n}. Subdivide each of the closed intervals of C_n into thirds and remove from each one of them the interior of the middle intervals. What is left from C_n is C_{n+1}; note that C_{n+1} is the union of 2^{n+1} closed intervals, each of length $3^{-(n+1)}$.

The Cantor set C is now defined as $C = \bigcap_{n=0}^{\infty} C_n$. Some of the elementary properties of C are the following: It is closed, it contains the endpoints of all intervals in C_n, and any point of C is the limit of a nondecreasing (and a nonincreasing) sequence of endpoints of the intervals of the C_n's.

It is not hard to give an analytical description of the elements of C. Let $x = \sum_{n=1}^{\infty} a_n 3^{-n}$ be the tryadic expansion of an arbitrary $x \in C$. We observe that since $x \notin (1/3,2/3), a_1 \neq 1$; similarly, since $x \notin (1/9,2/9) \cup (7/9,8/9), a_2 \neq 1$, and so on. In other words, by induction we see that $a_n \neq 1$ for all n, and C consists precisely of those points with $a_n = 0,2$ in their tryadic expansion. For example, the number $1/4 = \sum_{n=1}^{\infty} 2 \cdot 3^{-2n}$ is in C, but is not an endpoint of any of the intervals of the C_n's.

0 1/3 2/3 1

0 1/9 2/9 1/3 2/3 7/9 8/9 1

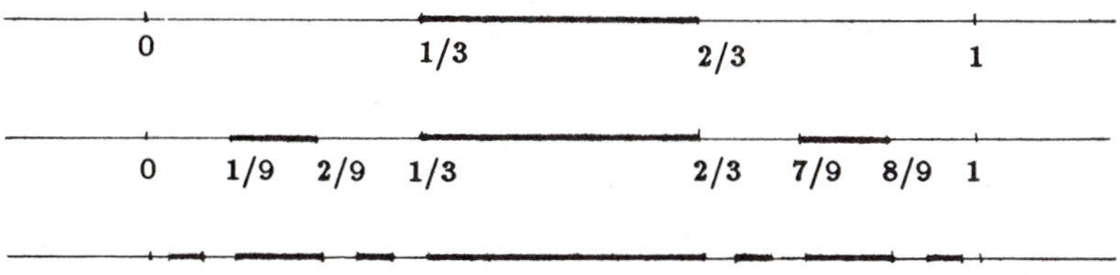

Figure 2

As for the cardinality of C, we have

Proposition 2.1. C is uncountable.

Proof. The idea is to show that $C \sim 2^N, 2 = \{0,1\}$. If $(x_n) \in 2^N$, let $y_n = 2x_n$, and put $f((x_n)) = \sum_{n=1}^{\infty} y_n 3^{-n}$. Since $y_n \neq 1$ for all n, f maps 2^N into C; we want to show that f is one-to-one and onto.

Suppose that $(x_n) \neq (x'_n)$, and let $m = \min\{n : x_n \neq x'_n\}$; we may assume that $x_m = 0$ and $x'_m = 1$. Since $2\sum_{n=m+1}^{\infty} 3^{-n} = 3^{-m}$, it follows that

$$f((x'_n)) = 2\sum_{n=1}^{\infty} x'_n 3^{-n} \geq 2\sum_{n=1}^{m-1} x_n 3^{-n} + 2 \cdot 3^{-m}$$

$$> 2\sum_{n=1}^{\infty} x_n 3^{-n} = f((x_n)),$$

and f is one-to-one.

Since given $x = \sum_{n=1}^{\infty} a_n 3^{-n}$ in C, we have $f((a_n/2)) = x$, f is also onto. ■

Is C measurable, and if so, what is its measure? Since C is covered by the intervals in any C_n we have $|C|_e \leq 2^n 3^{-n}$ for all n, and consequently, $|C| = 0$. Thus C is an example of an uncountable set of measure 0 in the line.

3. PROBLEMS AND QUESTIONS

3.1 Suppose A, B are not Lebesgue measurable, is the same true of $A \cup B$?

3.2 Suppose $|A|_e = 0$ and show that for every subset of B of R^n we have $|B \cup A|_e = |B \setminus A|_e = |B|_e$.

3.3 Let $A, B \subseteq R^n$. Show that

$$\big||A|_e - |B|_e\big| \leq |A \,\triangle\, B|_e .$$

3.4 Suppose $\{E_k\}$ is a nondecreasing sequence of subsets of R^n and let $E = \bigcup_k E_k$. Is it true that $\lim_{k\to\infty} |E_k|_e = |E|_e$?

3.5 Does the notion of outer measure change if we replace the coverings by intervals by coverings with balls? How about parallelepipeds with a fixed orientation?

3.6 Show that if $\sum |E_k|_e < \infty$, then $|\limsup E_k|_e = 0$.

3.7 Suppose $A, B \subseteq R^n$, $A \in \mathcal{L}$ and $|A \triangle B|_e = 0$. Show that $B \in \mathcal{L}$ and $|A| = |B|$.

3.8 Assume $\{E_k\}$ is a sequence of pairwise disjoint Lebesgue measurable sets and let A be any set. Is it true that

$$\left|A \cap \left(\bigcup_{n=1}^{\infty} E_k\right)\right|_e = \sum_{k=1}^{\infty} |A \cap E_k|_e \ ?$$

3.9 Consider the transformation $\phi(x) = \eta x + \delta$ from R into itself, where $\eta \neq 0$ and δ are real numbers. Show that: (i) For any set E, $|\phi(E)|_e = |\eta||E|_e$. (ii) E is Lebesgue measurable iff $\phi(E)$ is Lebesgue measurable, and in this case $|\phi(E)| = |\eta||E|$.
Can you think of extensions of this result to R^n?

3.10 A mapping ϕ from R into itself is said to be an isometry if for any x, x' in R we have $|\phi(x) - \phi(x')| = |x - x'|$. Show that if ϕ is an isometry and $E \in \mathcal{L}$, then $\phi(E) \in \mathcal{L}$ and $|\phi(E)| = |E|$.

3.11 Assume that $|N| = 0$ and show that $\{x^3 : x \in N\}$ is a null Lebesgue set.

3.12 Suppose $|E|_e < \infty$ and show that $E \in \mathcal{L}$ iff for any $\varepsilon > 0$, we can write $E = (A \cup A_1) \setminus A_2$, where A is the union of a finite collection of nonoverlapping intervals and $|A_1|_e, |A_2|_e < \varepsilon$.

3.13 Is the set of irrational numbers in the line a G_δ set?

3.14 Show that $E \in \mathcal{L}$ iff $E = H \setminus N$, where H is a G_δ set and $|N| = 0$.

3.15 Does there exist a function $f: R \to [0,1]$ such that the set D of its discontinuities has $|D| = 0$ and $D \cap I$ is uncountable for every interval I of R?

3.16 Assume A is a Lebesgue measurable subset of R of finite measure and put $\phi(x) = |A \cap (-\infty, x]|$. Show that ϕ is continuous at each x of R.

3.17 Let A be a Lebesgue measurable subset of R and let $0 < \eta < |A|$. Show that there exists a Lebesgue measurable set B so that $B \subseteq A$ and $|B| = \eta$.

3.18 Given $\varepsilon > 0$, show that there exists a dense open subset $\mathcal{O}$ of $[0,1]$ with $|\mathcal{O}| < \varepsilon$ so that its boundary $\partial\mathcal{O}$ satisfies $|\partial\mathcal{O}| \geq 1 - \varepsilon$.

3.19 Let $A = \{x \in [0,1] : x = .a_1 a_2 \ldots, a_n \neq 7, \text{all } n\}$. Prove that $|A| = 0$. Generalize this result to different configurations of a_n's and to dyadic, tryadic expansions.

3.20 Let $A = \{x \in [0,1] : x = .a_1a_2\ldots, a_n = 2 \text{ or } 3, \text{ all } n\}$. Show that A is measurable and compute $|A|$.

3.21 Let $A = \{x \in R$: there exist infinitely many pairs of integers p, q such that $|x - p/q| \le 1/q^3\}$. Show that $|A| = 0$.

3.22 Suppose $I_1, \ldots, I_n$ are open intervals in R, so that if $Q_1 = Q \cap [0,1]$ denotes the rational numbers in $[0,1]$, then $Q_1 \subset \bigcup_{j=1}^n I_j$. Prove that $\sum_{j=1}^n |I_j| \ge 1$.
Is the conclusion true if the I_j's are measurable sets rather than intervals?
What if we allow the collection of intervals to be infinite rather than finite?

3.23 Let $r_1, r_2, \ldots$ be an enumeration of Q. Show that

$$R \setminus \bigcup_n (r_n - 1/n^2, r_n + 1/n^2) \ne \emptyset.$$

On the other hand, also show that

$$R \setminus \bigcup_n (r_n - 1/n, r_n + 1/n)$$

may, or may not, be empty.

3.24 Suppose E is bounded measurable subset of R, $|E| > 0$. Prove that there exist $x_1, x_2 \in E$ so that $x_1 - x_2 \in Q$.

3.25 Show that if B is a Hamel basis for R, then $|B| = 0$.

3.26 Construct Lebesgue null subsets B_1, B_2 of R such that

$$B_1 + B_2 = \{x : x = b_1 + b_2, b_1 \in B_1, b_2 \in B\} = R.$$

3.27 Construct a Cantor-type subset C_η of $[0,1]$ by removing at the nth stage a "middle" interval of length $(1-\eta)3^{-n}$, $0 < \eta < 1$. Show that C_η enjoys all the properties of C, but it has Lebesgue measure $|C_\eta| = \eta$.

3.28 If $-1 \le r \le 1$, show there exist $x, y \in C$ such that $y - x = r$.

3.29 Does the Cantor set contain a Hamel basis for R?

3.30 Construct a Cantor-like subset of $[0,1]$ which consists entirely of irrational numbers.

3.31 Let (a_n) be a fixed decreasing sequence of real numbers such that $a_0 = 1$, and $0 < 2a_n < a_{n-1}$, and define the sequence (d_n) by $d_n = a_{n-1} - 2a_n$, $n \ge 1$.

Now let $I_{0,1} = [0,1]$, $I_{1,1} = [0,a_1]$, $I_{1,2} = [1 - a_1,1]$, $I_{2,2} = [a_1 - a_2,a_1]$, $I_{3,3} = [a_1 - a_2,a_1 - a_2 + a_3]$, $I_{3,4} = [a_1 - a_3,a_1]$, and so on; this definition can be made precise by induction. Now put

$$F_n = \bigcup_{k=1}^{2^n} I_{n,k} \quad \text{and} \quad P = \bigcap_{n=1}^{\infty} F_n \,.$$

Show that $P \in \mathcal{L}$ and $|P| = \lim_{n\to\infty} 2^n a_n$. Moreover, if $0 \le \eta < 1$, the a_n's can be chosen so that $|P| = \eta$.
Also, if $r_n = a_{n-1} - a_n$, then the elements of P are precisely those real numbers of the form $\sum_{k=1}^{n} \varepsilon_k r_k, \varepsilon_k = 0$ or 1.

3.32 Show that the class of Lebesgue measurable subsets of R^n has cardinality 2^c.
Consider now the following relation on $\mathcal{L}$: Given E_1, E_2 in $\mathcal{L}$, we say that $E_1 \sim E_2$ if $|E_1 \,\Delta\, E_2| = 0$. Show that $\sim$ is an equivalence relation on $\mathcal{L}$ and that the family of the equivalence classes has cardinality c.

3.33 Prove that there is no Lebesgue measurable subset A of R such that $a|I| \le |A \cap I| \le b|I|$ for all open intervals I of the line. More precisely, prove the following two assertions: (a) If $|A \cap I| \le b|I|$ for all open intervals $I \subset R$ and $b < 1$, then $|A| = 0$, and, (b) If $a|I| \le |A \cap I|$ for all open intervals $I \subset R$ and $a > 0$, then $|A| = 1$.

3.34 Prove there exists a Lebesgue measurable set $E \subset R$ such that

$$0 < |E \cap I| < |I| \,, \quad \text{all bounded intervals } I \subset R \,.$$

3.35 Does there exist a measurable subset E of R such that

$$0 < |E \cap I| \,, \quad 0 < |I \setminus E| \,, \quad \text{all intervals } I \subset R?$$

3.36 A measurable subset A of R is said to have a well-defined density, if the limit

$$D(A) = \lim_{\lambda\to\infty} \frac{|A \cap (-\lambda,\lambda)|}{2\lambda}$$

exists. In this case $D(A)$ is called the density of A. Give an example of a measurable set whose density is defined, and one whose density is not defined. Further, prove that if A_1 and A_2 are disjoint and have well-defined density, then $A_1 \cup A_2$ also has a well-defined density, and $D(A_1 \cup A_2) = D(A_1) + D(A_2)$.

3.37 Prove that the Lebesgue measure enjoys the following property, known as regularity: Given a measurable set A we have

$$|A| = \sup\{|K| : K \text{ is compact, and } K \subseteq A\} \,.$$

3.38 It is difficult to approximate sets that are not Lebesgue measurable with measurable ones. Specifically, suppose $E \subset R^n$ is not Lebesgue measurable. Show there is $\eta > 0$ such that if $E \subseteq A$ and $R^n \setminus E \subseteq B$, and if A and B are Lebesgue measurable, then $|A \cap B| \geq \eta$.

3.39 Decide whether the following statement is true: $A \subseteq R^n$ is Lebesgue measurable iff for every open subset G of R^n we have

$$|G| = |G \cap A|_e + |G \setminus A|_e .$$

3.40 Suppose μ^* is a nonnegative σ-subadditive monotone set function defined for all the subsets of a set X such that $\mu^*(\emptyset) = 0$. We say that $E \subseteq X$ is measurable with respect to μ^* if for every subset $A \subseteq X$ we have

$$\mu^*(A) = \mu^*(A \cap E) + \mu^*(A \setminus E) .$$

Let $\mathcal{M}$ be the class of subsets of X which are measurable with respect to μ^*. Show that $\mathcal{M}$ is a σ-algebra of subsets of X and that the restriction of μ^* to $\mathcal{M}$ defines a measure on $(X, \mathcal{M})$. This construction is known as the Carathéodory extension of an outer measure.

CHAPTER VI

Measurable Functions

In this chapter we introduce the class of measurable functions, for which the integral will be defined, and discuss some of its basic properties.

1. ELEMENTARY PROPERTIES OF MEASURABLE FUNCTIONS

Let $\mathcal{M}$ be a σ-algebra of (measurable) subsets of X and suppose f is an extended real-valued function defined on X; by this we mean that, in addition to real values, f may also assume the values $\pm\infty$. We say that f is measurable if for any real number λ, $\{x \in X : f(x) > \lambda\} = \{f > \lambda\} \in \mathcal{M}$; that is to say, all the level sets of f are measurable.

For instance, for any $\mathcal{M}$, $f = \chi_A$ is measurable iff $A \in \mathcal{M}$. If $\mathcal{M} = \{\emptyset, X\}$, only constant functions are measurable, and if $\mathcal{M} = \mathcal{P}(X)$, all functions are measurable.

We begin by exploring some simple properties of measurable functions.

Proposition 1.1. Suppose $\mathcal{M}$ is a σ-algebra of subsets of X and let f be an extended real-valued function defined on X. Then, the following statements are equivalent:

(i) f is measurable.

(ii) For any real $\lambda, \{f \geq \lambda\} \in \mathcal{M}$.

(iii) For any real $\lambda, \{f < \lambda\} \in \mathcal{M}$.

(iv) For any real $\lambda, \{f \leq \lambda\} \in \mathcal{M}$.

Proof. (i) implies (ii). Fix λ, and for $n \geq 1$ let $A_n = \{f > \lambda - 1/n\}$; by assumption $A_n \in \mathcal{M}$, all n. Now, since $\{f \geq \lambda\}$ is the intersection of the A_n's, it also belongs to $\mathcal{M}$, and (ii) holds.

(ii) implies (iii). $\{f < \lambda\} = X \setminus \{f \geq \lambda\}$.
(iii) implies (iv). $\{f \leq \lambda\} = \bigcap_{n=1}^{\infty}\{f < \lambda + 1/n\}$.
(iv) implies (i). $\{f > \lambda\} = X \setminus \{f \leq \lambda\}$. ■

In working with measurable functions it is essential to know whether certain sets are measurable. Since these sets are readily obtained from those introduced in Proposition 1.1 we merely indicate how their measurability is established.

$$\{f = \infty\} = \bigcap_{n=1}^{\infty}\{f > n\}, \quad \{f = -\infty\} = \bigcap_{n=1}^{\infty}\{f < -n\}. \tag{1.1}$$

$$\{f < \infty\} = \bigcup_{n=1}^{\infty}\{f < n\}, \quad \{f > -\infty\} = \bigcup_{n=1}^{\infty}\{f > -n\}. \tag{1.2}$$

$$\{-\infty < f < \infty\} = \{f > -\infty\} \cap \{f < \infty\}. \tag{1.3}$$

Also, for any real numbers λ, η, we will have the occasion to deal with the measurable sets

$$\begin{aligned}\{\lambda < f < \infty\} &= \{f > \lambda\} \cap \{f < \infty\}, \\ \{-\infty < f < \lambda\} &= \{f < \lambda\} \cap \{f > -\infty\}.\end{aligned} \tag{1.4}$$

$$\begin{aligned}\{\lambda < f < \mu\} &= \{f > \lambda\} \cap \{f < \mu\}, \\ \{\lambda < f \leq \mu\} &= \{f > \lambda\} \cap \{f \leq \mu\}.\end{aligned} \tag{1.5}$$

$$\begin{aligned}\{\lambda \leq f < \mu\} &= \{f \geq \lambda\} \cap \{f < \mu\}, \\ \{\lambda \leq f \leq \mu\} &= \{f \geq \lambda\} \cap \{f \leq \mu\}.\end{aligned} \tag{1.6}$$

$$\begin{aligned}\{f = \lambda\} &= \{f \leq \lambda\} \cap \{f \geq \lambda\}, \\ \{f \neq \lambda\} &= X \setminus \{f = \lambda\}.\end{aligned} \tag{1.7}$$

Our next result indicates how to handle the infinite values of a measurable function.

Proposition 1.2. Let $\mathcal{M}$ be a σ-algebra of subsets of X and let f be an extended real-valued function defined on X. Then, f is measurable iff $\{f = -\infty\} \in \mathcal{M}$ and for each real λ, $\{\lambda < f < \infty\} \in \mathcal{M}$.

Proof. The necessity has been established in (1.1) and (1.4). As for the sufficiency, first observe that

$$\{f < \infty\} = \{f = -\infty\} \cup \left(\bigcup_{n=-\infty}^{\infty}\{n < f < \infty\}\right)$$

belongs to $\mathcal{M}$ by assumption. Whence $\{f = \infty\} = X \setminus \{f < \infty\}$ is also in $\mathcal{M}$, and since for each real λ we have

$$\{f > \lambda\} = \{\lambda < f < \infty\} \cup \{f = \infty\} \in \mathcal{M},$$

the level sets of f are measurable and f is measurable. ■

In fact, a more general statement is true.

Proposition 1.3. Let $\mathcal{M}$ be a σ-algebra of subsets of X and suppose f is an extended real-valued function defined on X. Then, f is measurable iff $\{f = -\infty\} \in \mathcal{M}$ and for each open subset $\mathcal{O} \subseteq R$, $f^{-1}(\mathcal{O}) \in \mathcal{M}$.

Proof. Since for each real λ, (λ, ∞) is open, the sufficiency follows from Proposition 1.2. As for the necessity, suppose $\mathcal{O}$ is an open subset of R and write $\mathcal{O} = \bigcup I_k$, where the I_k's are an at most countable collection of pairwise disjoint open intervals, one or two of which are possibly unbounded. By 5.8 in Chapter I

$$f^{-1}(\mathcal{O}) = \bigcup_k f^{-1}(I_k). \tag{1.8}$$

Now, by (1.3), (1.4) and (1.5), the sets in the union on the right-hand side of (1.8) all belong to $\mathcal{M}$ and $f^{-1}(\mathcal{O})$ is measurable. Since $\{f = -\infty\} \in \mathcal{M}$ whenever f is measurable, we have finished. ■

So far the role of measures on $(X, \mathcal{M})$ is not apparent, but in dealing with measurable functions sets of measure 0 are important and the following concept essential.

Given a measure space $(X, \mathcal{M}, \mu)$, we say that a property $P(x)$ is true μ-almost everywhere on a measurable subset E of X, and denote this by μ-a.e. on E, if $\mu(\{x \in E : P(x) \text{ is not true }\}) = 0$.

For instance, we say that a measurable function f is finite μ-a.e. on E if $\mu(\{x \in E : f(x) = \pm\infty\}) = 0$.

It is natural to expect that measurable functions that coincide μ-a.e. on X be, in some sense, equivalent. A more precise statement is

Theorem 1.4. Let μ be a complete measure on $(X, \mathcal{M})$, and let f, g be extended real-valued functions defined on X. If f is measurable and $g = f$ μ-a.e., then g is also measurable and

$$\mu(\{g > \lambda\}) = \mu(\{f > \lambda\}), \quad \text{all real } \lambda. \tag{1.9}$$

Proof. Let $N = \{g \neq f\}$; by assumption N is a null, measurable, set. Now, for each real λ we have

$$\{g > \lambda\} \cup N = \{f > \lambda\} \cup N\,. \tag{1.10}$$

Since f is measurable, the set on the right-hand side of (1.10) is measurable, and so is the set on the left-hand side there. Moreover, since μ is complete and $N_0 = \{x \in N : g(x) \leq \lambda\} \subseteq N$, then N_0 is also a null, measurable set and consequently,

$$\{g > \lambda\} = (\{g > \lambda\} \cup N) \setminus N_0 \tag{1.11}$$

is also measurable.

Next observe that since N is null, we have $\mu(\{f > \lambda\} \cup N) = \mu(\{f > \lambda\})$ for all real λ. Whence, by (1.11) and (an argument similar to) 3.2 in Chapter V, we get

$$\begin{aligned}\mu(\{g > \lambda\}) &= \mu(\{g > \lambda\} \cup N) - \mu(N_0)\\ &= \mu(\{f > \lambda\} \cup N) = \mu(\{f > \lambda\})\,. \quad \blacksquare\end{aligned}$$

Theorem 1.4 states that functions that coincide μ-a.e. are roughly interchangeable; this property is essential in operating with extended real-valued functions. Consider, for instance, addition: $f(x)+g(x)$ is undefined for those x's where f and g assume infinite values of opposite sign. The idea is to work with functions $\tilde{f}$ and $\tilde{g}$ which are closely related to f and g, and for which the sum makes sense. We proceed as follows: Let the (bad) set

$$B = \{x \in X : f(x) = \infty, g(x) = -\infty\} \bigcup \{x \in X : f(x) = -\infty, g(x) = \infty\}\,.$$

Since B is measurable, $\mathcal{M}_{X \setminus B} = \{E \cap (X \setminus B) : E \in \mathcal{M}\}$ is a σ-algebra of subsets of $X \setminus B$. Observe that $\tilde{f} = f|(X \setminus B)$ and $\tilde{g} = g|(X \setminus B)$ are also measurable on $(X \setminus B, \mathcal{M}_{X \setminus B})$, and that $\tilde{f}(x) + \tilde{g}(x)$ is defined for any $x \in X \setminus B$. In fact, as we shall prove below, $\tilde{f} + \tilde{g}$ is also measurable.

To avoid having to go through various technical considerations each time we discuss an operation involving measurable functions, we sort the functions out into equivalence classes and operate at the level of classes.

Let $(X, \mathcal{M}, \mu)$ be a measure space. We consider the collection $\mathcal{F}$ consisting of those functions f that satisfy the following properties:

(i) f is an extended real-valued function defined on $X \setminus N$, where N is a null subset of X.

(ii) f is measurable, as a function on $(X \setminus N, \mathcal{M}_{X\setminus N})$.
Note that we only require functions in $\mathcal{F}$ to be defined μ-a.e. on X.

Next we identify those measurable functions which coincide μ-a.e.; more precisely, given $f, g \in \mathcal{F}$, we say that $f \sim g$ iff there is a null subset N of X such that $f(x) = g(x)$ for $x \in X \setminus N$. It is clear that $\sim$ is an equivalence relation on $\mathcal{F}$; the only property that offers any difficulty is the transitivity, and this follows at once from the fact that the union of null sets is null.

We return to the addition: Given equivalence classes $\tilde{f}, \tilde{g} \in \mathcal{F}$ corresponding to the finite μ-a.e. functions f, g, by removing the bad set B we readily see that $h(x) = f(x) + g(x)$ is a finite quantity for $x \in X \setminus N$, N null, and we put $\tilde{f} + \tilde{g} = \tilde{h}$. It is straightforward to verify that $\tilde{h}$ is well-defined, i.e., it is independent of the representatives of the classes $\tilde{f}, \tilde{g}$, and this completes our discussion.

Now that we know how things should be done we agree to denote the equivalence class $\tilde{f}$ of a function f once again by f, and to operate with the classes as if they were functions. This should cause no undue stress, and the reader should keep in mind that a statement such as "a function f defined on X" actually means "an equivalence class $\tilde{f}$ of a function f defined μ-a.e. on X."

To deal with the usual arithmetic operations we need a preliminary result.

Lemma 1.5. Let $(X, \mathcal{M}, \mu)$ be a measure space, and suppose f, g are extended real-valued measurable functions defined on X. Then $\{f > g\} \in \mathcal{M}$.

Proof. Let $\{r_k\}$ be an enumeration of the rational numbers and observe that by Proposition 1.1

$$E_k = \{f > r_k\} \cap \{g < r_k\} \in \mathcal{M}\,, \quad k = 1, 2, \ldots$$

Thus $\{f > g\} = \bigcup_k E_k$ is also measurable. ■

We are now ready to prove

Theorem 1.6. Let $(X, \mathcal{M}, \mu)$ be a measure space and f, g be extended real-valued measurable functions defined on X. Then $f \pm g$ is also measurable.

Proof. We only do the addition. Observe that for any real $\lambda, \lambda - g$ is measurable. Since

$$\{f + g > \lambda\} = \{f > \lambda - g\}\,, \quad \text{real } \lambda\,,$$

the conclusion follows at once from Lemma 1.5. ■

The other operations of interest to us are covered by the following result.

Theorem 1.7. Let $(X, \mathcal{M}, \mu)$ be a measure space, assume f is a measurable, finite μ-a.e. function defined on X and let ϕ be a real-valued continuous function defined on R. Then the composition $\phi \circ f$ is measurable.

Proof. Since $\{f = \pm\infty\}$ is a null set we may assume that $\phi \circ f$ is well-defined and that $\{\phi \circ f = -\infty\} = \emptyset$. By Proposition 1.2 the measurability of $\phi \circ f$ will be established once we show that

$$(\phi \circ f)^{-1}((\lambda, \infty)) = f^{-1}\left(\phi^{-1}((\lambda, \infty))\right) \in \mathcal{M}\,, \quad \text{all real } \lambda\,. \tag{1.12}$$

But this is not hard; indeed, since ϕ is continuous, $\phi^{-1}((\lambda, \infty)) = \mathcal{O}$ is an open subset of R, and, by Proposition 1.3, $f^{-1}(\mathcal{O}) \in \mathcal{M}$. Thus (1.12) holds and we have finished. ■

Theorem 1.7 shows that the composition $\phi \circ f$ of a measurable function f with a continuous function ϕ is measurable; it is not intuitively apparent that the composition $f \circ \phi$ should also be measurable. In fact, it is not, as the following example shows.

Let $\{K_n\}$ be a sequence of Cantor-like sets, $|K_n| = 1 - 1/n, n = 2, 3, \ldots$, and let $A = \bigcup_n K_n$. Since

$$|[0,1] \setminus A| \le |[0,1] \setminus K_n| \le 1/n\,, \quad n = 2, 3 \ldots$$

it readily follows that $[0,1] = \bigcup_n K_n \cup Z, |Z| = 0$, and consequently for any subset B of $[0,1]$ we have

$$B = \bigcup_n (B \cap K_n) \cup (B \cap Z)\,, \quad |B \cap Z| = 0\,.$$

In particular, if B is not Lebesgue measurable, there is an index N so that $B \cap K_N$ is not Lebesgue measurable.

Referring to the construction of the Cantor set, let $D_n = [0,1] \setminus C_n$, where as usual C_n denotes the union of the intervals remaining after n steps. D_n consists of $2^n - 1$ open intervals, I_n^k say, ordered from left to right by k, removed in the first n steps of the construction of C. Since K_n is a Cantor-like set, there also is a sequence of open intervals, J_n^k say, $1 \le k \le 2^n - 1$, ordered from left to right by k, removed in the first n steps of the construction of K_N.

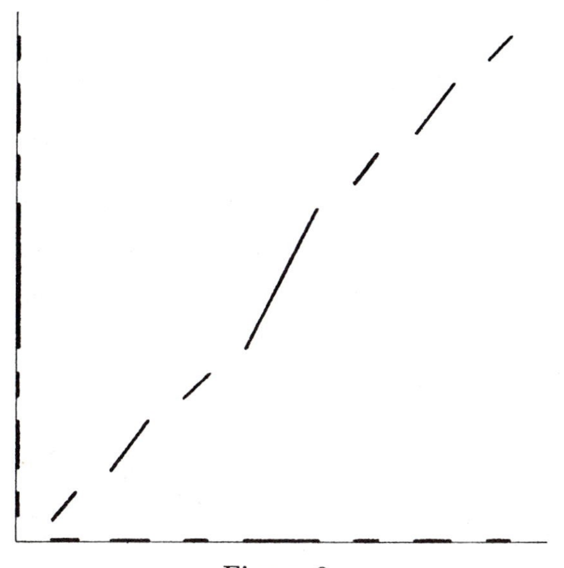

Figure 3

We define now a function h from [0,1] onto [0,1] as follows: Construct K_N in the interval [0,1] corresponding to the domain of h, and C in the interval [0,1] that corresponds to the range of h. Then h is the function that maps the left-end point of J_n^k into the left-end point of I_n^k, the right-end point of J_n^k into the right-end point of I_n^k, and is extended to $[0,1]\setminus K_N$ by continuity. It is not hard to check that h is well-defined, one-to-one (if this were not the case K_N would contain an interval, and this is not possible) and onto [0,1]. Let $B \cap K_N$ be a non-Lebesgue measurable subset of K_N and put $A = h(B \cap K_N) \subseteq C$. Then A is null, and consequently measurable; in other words the image of this non-Lebesgue measurable set by a continuous function is Lebesgue measurable. Another way to express this situation is the following: If $\phi = h^{-1}$ is the continuous inverse of h, then $\phi(A) = B \cap K_N$, and the image of a Lebesgue measurable set by a continuous function is not necessarily Lebesgue measurable. Also $\phi(C) = K_N$, and ϕ takes a null set onto a set of positive Lebesgue measure.

Returning to the question at hand, let $f = \chi_A$; since A is null f is Lebesgue measurable. Consider now the composition $f \circ \phi$. The inverse image $(f \circ \psi)^{-1}((1/2, 3/2))$ is readily seen to equal

$$\phi^{-1}\left(f^{-1}((1/2,3/2))\right) = \phi^{-1}(A) = B \cap K_N\,,$$

which is not measurable. Thus, as asserted, $f \circ \phi$ is not measurable.

The measurability of several expressions involving f follow at once from Theorem 1.7, with an appropriate choice of ϕ. Some important instances are:

$$\begin{aligned}
f(x) \pm \lambda\,, \lambda \text{ real} \qquad & \phi(t) = t \pm \lambda \\
\lambda f(x), \lambda \text{ real} \qquad & \phi(t) = \lambda t \\
f(x)^p, p \geq 1 \qquad & \phi(t) = t^p \\
|f(x)|^p, p \geq 0 \qquad & \phi(t) = |\,t|^p \\
f^+(x) = \vee(0, f(x)) \qquad & \phi(t) = t^+ \\
f^-(x) = \vee(0, -f(x)) \qquad & \phi(t) = t^-
\end{aligned}$$

and when $f \neq 0$ μ-a.e.,

$$\begin{aligned}
1/f(x) \qquad & \phi(t) = 1/t \\
|f(x)|^\eta, \eta \text{ real} \qquad & \phi(t) = |\,t|^\eta\,.
\end{aligned}$$

We also have

Theorem 1.8. Let $(X, \mathcal{M}, \mu)$ be a measure space and assume f, g are measurable, finite μ-a.e. functions defined on X. Then fg is measurable, and, if $g \neq 0$ μ-a.e., also f/g is measurable.

Proof. By Theorem 1.6, $f \pm g$ are measurable, and by the above remarks so are $(f \pm g)^2$. Whence, once again by Theorem 1.6,

$$4fg = (f+g)^2 - (f-g)^2$$

is also measurable, and so is fg.

Since $1/g$ is measurable, it readily follows that $f/g = f \cdot \dfrac{1}{g}$ is also measurable. ■

Corollary 1.9. Let $(X, \mathcal{M}, \mu)$ be a measure space, and suppose f is a "simple" function defined on X, i.e., $f(x) = \sum_{k=1}^{n} c_k \chi_{E_k}(x)$, where the c_k's are real constants and the E_k's form a measurable partition of X. Then f is measurable.

Next we consider whether measurability is preserved under limiting operations. We begin by discussing the inf and sup of a sequence $\{f_n(x)\}$ of measurable functions, which exist for any $x \in X$.

Lemma 1.10. Let $(X, \mathcal{M}, \mu)$ be a measure space, and let $\{f_n\}$ be a sequence of extended real-valued measurable functions defined on X. Then the functions

$$\inf f_n(x), \quad \text{and} \quad \sup f_n(x), \quad x \in X$$

are measurable.

Proof. Since for arbitrary sequences $\{g_n\}$ we have

$$\inf g_n(x) = -\sup(-g_n(x)), \quad \text{all } x \in X,$$

we only need prove one statement, the one with the sup, say. But this is not hard; indeed given any real λ, note that

$$\{\sup f_n > \lambda\} = \bigcup_n \{f_n > \lambda\}. \tag{1.13}$$

By assumption the set on the right-hand side of (1.13) is the countable union of measurable sets, and so it is measurable. Thus the set on the left-hand side is measurable, and the sup of the f_n's is measurable. ∎

Next we consider the lim sup and lim inf of a sequence of functions; these limits exist for every $x \in X$, even if the sequence does not converge.

Theorem 1.11. Assume $(X, \mathcal{M}, \mu)$ is a measure space and let $\{f_n\}$ be a sequence of extended real-valued measurable functions defined on X. Then $\liminf f_n$ and $\limsup f_n$ are measurable. In particular, if $f(x) = \lim f_n(x)$ exists for all $x \in X$, then f is measurable.

Proof. Since

$$\liminf f_n(x) = \sup_{k \geq 1} \left(\inf_{m \geq k} f_m(x) \right), \quad x \in X,$$

the measurability of the lim inf follows at once from Lemma 1.10. Similarly, since

$$\limsup f_n(x) = \inf_{k \geq 1} \left(\sup_{m \geq k} f_m(x) \right), \quad x \in X,$$

also the lim sup is measurable. Finally, if the sequence converges, $f(x)$ coincides with both the $\limsup f_n$ and $\liminf f_n$, and it is measurable as well. ∎

We close this section with an interesting and important result; it shows how to approximate arbitrary functions by functions that assume finitely many values.

Theorem 1.12. Let $(X, \mathcal{M}, \mu)$ be a measure space, and f be an extended real-valued function defined on X. Then there is a sequence $\{f_n\}$ of simple real-valued functions defined on X, i.e.,

$$f_n(x) = \sum_{i=1}^{k_n} c_i \chi_{E_i}(x), \quad c_i \text{ real}, \quad E_i \text{ pairwise disjoint},$$

so that $\lim_{n\to\infty} f_n(x) = f(x), x \in X$.

Furthermore,

(i) If f is measurable, so are the f_n's.

(ii) If f is nonnegative, then the sequence $\{f_n\}$ is nondecreasing, and

$$0 \le f_n(x) \le f(x), \quad \text{all } x \in X, \quad n = 1, 2, \ldots$$

(iii) If f is bounded, i.e., $|f(x)| \le M$ for all $x \in X$, then the f_n's converge uniformly to f.

Proof. The f_n's are defined by looking closely at the level sets of f. Suppose first that f is nonnegative, fix an integer $n \ge 1$, and consider the $n2^n$ pairwise disjoint subintervals of $[0, n)$ given by

$$[(k-1)2^{-n}, k2^{-n}), \quad 1 \le k \le n2^n.$$

Put now

$$f_n(x) = \begin{cases} (k-1)2^{-n} & \text{if } (k-1)2^{-n} \le f(x) < k2^{-n}, \, 1 \le k \le n2^n \\ n & \text{otherwise.} \end{cases}$$

Clearly $f_n(x) \le f(x)$ for all $x \in X$ and $n = 1, 2, \ldots$ Also, each f_n assumes a finite number of values; more precisely, if

$$A_{k,n} = \{(k-1)2^{-n} \le f < k2^{-n}\} \quad \text{and} \quad A_n = \{f \ge n\},$$

we have

$$f_n(x) = \sum_{k=1}^{n2^n} (n-1)\, 2^{-n} \chi_{A_{k,n}}(x) + n\chi_{A_n}(x). \tag{1.14}$$

Observe that $f_{n+1}(x)$ is obtained from $f_n(x)$ by dividing each interval $[(k-1)2^{-n}, k2^{-n})$ in half, and then only increasing $f_n(x)$ to $f_{n+1}(x)$ at those x's where $f_n(x)$ is changed; this proves the remarks in (ii).

We claim that

$$\lim_{n\to\infty} f_n(x) = f(x), \quad x \in X. \tag{1.15}$$

Now, if $f(x) = \infty$, then $f_n(x) = n$ for all n, and $f_n(x) \to \infty$. On the other hand, if $f(x)$ is finite, by the definition of $f_n(x)$ it readily follows that

$$f(x) - f_n(x) \le 2^{-n}\,, \quad \text{all } n > f(x)\,, \tag{1.16}$$

and (1.15) holds. Moreover, if f is measurable, by (1.6) the $A_{k,n}$'s are measurable, and, by (ii) in Proposition 1.1, the A_n's are measurable. Whence f_n is a simple measurable function and (i) is proved.

Furthermore, if f is bounded, (1.16) is true for $n > M$, uniformly for x in X, the convergence is uniform and (iii) holds.

In short, we have obtained the desired conclusion for nonnegative functions. To complete the proof recall that each function f is the difference of two nonnegative functions, $f = f^+ - f^-$, and apply the first part of the proof to f^+ and f^- separately. Note that in this case, the f_n's also have the property that

$$|f_n(x)| \le |f(x)|\,, \quad \text{all } x \in X\,, \quad n = 1, 2, \ldots \quad \blacksquare$$

2. STRUCTURE OF MEASURABLE FUNCTIONS

What does a measurable function f look like? In case there is a topology defined on X, how far is f from being continuous?

First an example: Let f be the characteristic function of the set of irrational numbers I in [0,1]. f is Lebesgue measurable, yet discontinuous at every point of [0,1]. There is, however, another way to interpret this situation: Since f is constant on I, it is continuous there in the relative topology in I with respect to the usual topology of R. Now, $|[0,1] \setminus I| = 0$ and consequently, we have that f is continuous on a subset I of [0,1] of full measure in the relative topology in I.

This observation points to a general fact concerning measurable functions. In order to avoid technical difficulties, and since the nature of the question is already apparent there, we restrict our attention to the Lebesgue measure on $X = [0,1]$.

We begin by considering the simpler question of the boundedness of measurable functions.

Proposition 2.1. Assume f is an extended real-valued Lebesgue measurable function defined on X such that $|\{|f| = \infty\}| = 0$. Then, for any $\varepsilon > 0$, there exist a measurable subset B of X and a constant M with the following properties:

(i) $|B| < \varepsilon$.

(ii) $|f(x)| \le M$ for $x \in X \setminus B$.

Proof. Let $B_n = \{|f| > n\}, n = 1, 2, \ldots$; $\{B_n\}$ is a nonincreasing sequence of measurable subsets of X. Suppose that for all n, $|B_n| > \varepsilon$. Since $\bigcap B_n = \{|f| = \infty\}$ and $|X| = 1$, by (3.4) in Chapter IV we get

$$\lim_{n\to\infty} |B_n| = |\{|f| = \infty\}| \ge \varepsilon\,,$$

which is impossible. Thus, there exists m such that $|B_m| \le \varepsilon$, and the desired conclusion obtains with $B = B_m$ and $M = m$. ■

As for the continuity of f, the following result of Lusin (1883–1950) provides the answer.

Theorem 2.2 (Lusin). Suppose f is an extended real-valued measurable function defined on X with the property that $|\{|f| = \infty\}| = 0$. Then, given $\varepsilon > 0$, there is a closed subset F of X such that

(i) $|X \setminus F| < \varepsilon$.

(ii) $f|F$ is continuous on F, in the relative topology in F.

Proof. Since $f = f^+ - f^-$ is the difference of two nonnegative measurable functions, we may assume that $f \ge 0$. Now, given $\varepsilon > 0$, let B be a measurable subset of X corresponding to the choice $\varepsilon/2$ in Proposition 2.1. The restriction $f|(X \setminus B)$ of f to $X \setminus B$ is measurable and bounded, and with no fear of confusion we also denote it by f. By Theorem 1.12 there is a sequence $\{f_n\}$ of simple functions defined on $X \setminus B$ that converges uniformly to f there. Let

$$f_n(x) = \sum\nolimits_{i=1}^{i_n} c_{i,n}\chi_{E_{i,n}}(x)\,, \quad n = 1, 2, \ldots$$

where the $c_{i,n}$'s are real numbers and the $E_{i,n}$'s form a pairwise disjoint partition of $X \setminus B$.

By the regularity properties of the Lebesgue measure, given $\eta_n > 0$, there exist closed subsets $F_{i,n}$ of $E_{i,n}$ with the property that

$$|E_{i,n} \setminus F_{i,n}| \le \eta_n \quad i = 1, \ldots, i_n,\ n = 1, 2, \ldots$$

Put now $F_n = \bigcup_{i=1}^{i_n} F_{i,n}$; since each $F_{i,n}$ is closed, F_n is a closed subset of X. Moreover, since the $F_{i,n}$'s are pairwise disjoint and compact, they can be separated by pairwise disjoint open subsets of X. Whence, the restriction of f_n to F_n, or any of its subsets for that matter, is continuous in the relative topology in F_n.

Next observe that

$$
\begin{aligned}
|(X \setminus B) \setminus F_n| &= |\bigcup_{i=1}^{i_n} E_{i,n} \setminus \bigcup_{i=1}^{i_n} F_{i,n}| \\
&\leq \sum_{i=1}^{i_n} |E_{i,n} \setminus F_{i,n}| \leq i_n \eta_n \,. \qquad (2.1)
\end{aligned}
$$

A good choice for η_n is $\varepsilon / i_n 2^{n+1}$ for then, by (2.1), we have

$$|(X \setminus B) \setminus F_n| \leq \varepsilon / 2^{n+1}, \quad \text{all } n\,.$$

Put now $F = \bigcap_{n=1}^{\infty} F_n$; F is a compact subset of $X \setminus B$ and $\{f_n | F\}$ is a sequence of continuous functions in the relative topology in F which converges uniformly to f on F. Since the uniform limit of a sequence of continuous functions on a compact set is continuous, $f|F$ is continuous in the relative topology in F.

It only remains to estimate $|X \setminus F|$. First observe that $|(X \setminus B) \setminus F|$ equals

$$
\begin{aligned}
|(X \setminus B) \setminus \textstyle\bigcap_n F_n| &= |\textstyle\bigcup_n ((X \setminus B) \setminus F_n)| \\
&\leq \sum_n |(X \setminus B) \setminus F_n| \leq \sum_n \varepsilon / 2^{n+1} = \varepsilon / 2\,. \qquad (2.2)
\end{aligned}
$$

Now, since $X \setminus F = ((X \setminus B) \setminus F) \cup B$, by (2.2) and the choice of B it readily follows that

$$|X \setminus F| \leq \varepsilon/2 + \varepsilon/2 = \varepsilon\,. \quad \blacksquare$$

3. SEQUENCES OF MEASURABLE FUNCTIONS

Let $(X, \mathcal{M}, \mu)$ be a measure space. The pointwise convergence of a sequence $\{f_n\}$ of real-valued functions defined on X is a well-defined notion. Here, and unless otherwise specified, the term "convergence" means that the convergence is to a finite limit.

Actually, a more relevant concept is that of μ-a.e. convergence; sets of measure 0 play a special role in this setting. For instance, on $X = [0,1]$, the sequence $f_n(x) = 0$ if $x \neq 1/2$ and $f_n(1/2) = 1 + 1/n$ tends to

$f(x) = 0$, δ_a-a.e., $a \neq 1/2$, but $f_n \not\to f$ $\delta_{1/2}$-a.e. Other important notions of convergence involve uniformity, and we shall discuss them shortly.

How can we express the notion of μ-a.e. convergence in terms of familiar concepts? First an observation: Since the limiting function f is finite μ-a.e., the statements $f_n(x) \to f(x)$ μ-a.e. and $|f_n(x) - f(x)| \to 0$ μ-a.e. are equivalent. So, we may restrict our attention to sequences of nonnegative functions which converge to 0 μ-a.e. For such sequences we have

Proposition 3.1. Let $(X, \mathcal{M}, \mu)$ be a measure space and let $\{f_n\}$ be a sequence of measurable nonnegative functions defined on X. Then, $f_n \to 0$ μ-a.e. on a subset of $M \in \mathcal{M}$ iff for each $\eta > 0$, the sets $B_{n,\eta} = \{x \in M : f_n(x) > \eta\}$ satisfy $\mu(\limsup B_{n,\eta}) = 0$.

Proof. Let $f_n \to 0$ μ-a.e. on M and suppose that for some $\eta > 0$ we have $\mu(\limsup B_{n,\eta}) > 0$. Then each $x \in \limsup B_{n,\eta}$ belongs to infinitely many of the $B_{n,\eta}$'s, and consequently, there is a sequence $n_k \to \infty$ such that $f_{n_k}(x) > \eta$. Whence, $\limsup f_n(x) \geq \eta > 0$, and the f_n's do not converge to 0 on a subset of M of positive measure; this is a contradiction.

Conversely, given $\varepsilon > 0$, pick $0 < \eta < \varepsilon$, and consider a point x in $M \setminus (\limsup B_{n,\eta})$. Since x belongs to at most finitely many of the $B_{n,\eta}$'s there is a n_0 such that $f_n(x) \leq \eta < \varepsilon$ for all $n \geq n_0$. In other words, $f_n(x) \to 0$ for $x \in M \setminus N, \mu(N) = 0$, which is precisely what we wanted to show. ■

Since $\limsup B_{n,\eta} = \bigcap_{k=1}^{\infty}(\bigcup_{n=k}^{\infty} B_{n,\eta})$, if $\mu(\bigcup_{n=k}^{\infty} B_{n,\eta}) < \infty$ for some k, by (3.4) in Chapter IV the conditions of Proposition 3.1 are satisfied iff

$$\lim_{k\to\infty} \mu\left(\bigcup_{n=k}^{\infty} B_{n,\eta}\right) = 0\,, \quad \text{all } \eta > 0\,. \tag{3.1}$$

In particular, if $\mu(X) < \infty$, (3.1) describes convergence μ-a.e.

The relation (3.1) points to a possible limitation of the concept of μ-a.e. convergence, namely, we require the control of all the $B_{n,\eta}$'s, from one index on. To illustrate this point, consider the sequence of (dyadic) subintervals of $I = [0,1]$ defined as follows: $I_0 = I$, $I_1 = [0,1/2]$, $I_2 = [1/2,1]$, $I_3 = [0,1/4]$, and so on. In other words, the sequence consists of successive blocks of 2^n nonoverlapping intervals, each of length 2^{-n}, and the union of the intervals in each block is I. Let $\{f_n\}$ be the sequence consisting of the characteristic functions of the I_n's. Clearly $\{f_n\}$ does not converge to 0 anywhere on I, yet in some sense the f_n's are getting close to 0. Specifically,

$$\lim_{n\to\infty} |\{f_n > \eta\}| = 0\,, \quad \text{all real } \eta > 0\,. \tag{3.2}$$

Notice that in contrast to (3.1), we are dealing with one $B_{n,\eta}$ at a time.

Motivated by this remark we introduce the following definition. Given a measure space $(X, \mathcal{M}, \mu)$ and a sequence $\{f_n\}$ of measurable nonnegative extended real-valued finite μ-a.e. functions defined on X, we say that f_n converges to 0 in μ-measure iff

$$\lim_{n\to\infty} \mu(\{f_n > \eta\}) = 0\,, \quad \text{all real } \eta > 0\,. \tag{3.3}$$

If $\mu(X) < \infty$ we refer to convergence in μ-measure as convergence in the sense of probability, or convergence in probability. Thus, μ-a.e. convergence implies convergence in probability, but the opposite is not true. Also, μ-a.e. convergence does not, in general, imply convergence in μ-measure. To see this consider $(R, \mathcal{L}, |\cdot|)$, and observe that the sequence $f_n(x) = \chi_{[n,\infty)}(x)$ tends to 0 everywhere, but $|\{f_n > 1/2\}| = \infty$ for all n.

Nevertheless, a closer look at the first example indicates that there is a subsequence $\{f_{n_k}\}$ of the f_n's, specifically that consisting of the characteristic functions of the intervals $[0,1/2^n]$, $n \ge 1$, with the property that $\lim_{k\to\infty} f_{n_k}(x) = 0$ for all $x \in (0,1]$. The remarkable fact is that this property is true for arbitrary sequences; before we prove this we need a bit of information concerning convergence in μ-measure.

Proposition 3.2. Let $(X, \mathcal{M}, \mu)$ be a measure space and suppose $\{f_n\}$ is a sequence of measurable nonnegative extended real-valued finite μ-a.e. functions defined on X. Then, $f_n \to 0$ in μ-measure iff for any $\varepsilon, \delta > 0$, there exists a constant $N = N_{\varepsilon,\delta}$ such that

$$\mu(\{f_n > \delta\}) < \varepsilon\,, \quad \text{all } n \ge N\,. \tag{3.4}$$

Proof. The necessity of the condition is obvious. As for the sufficiency, if $f_n \not\to 0$ in μ-measure, then by (3.3) there exist $\eta > 0$ and a sequence $n_k \to \infty$ such that

$$L = \limsup \mu\left(\{f_{n_k} > \eta\}\right) > 0\,.$$

Thus, (3.4) cannot hold for $\varepsilon = L$ and $\delta = \eta$, and this contradiction completes the proof. ∎

Next we show that convergence in probability implies μ-a.e. convergence along a subsequence.

Proposition 3.3. Let $(X, \mathcal{M}, \mu)$ be a finite measure space, and assume $\{f_n\}$ is a sequence of measurable, nonnegative, extended real-valued, finite μ-a.e. functions defined on X. If $f_n \to 0$ in probability, then there is an increasing sequence $n_k \to \infty$ such that

$$\lim_{k \to \infty} f_{n_k} = 0 \ \mu\text{-a.e. on } X\,.$$

Proof. By Proposition 3.2 it follows that for each n we may find an index $n_k < n_{k+1}$ with the property that

$$\mu(\{f_n > 1/2^k\}) < 1/2^k\,, \quad \text{all } n \geq n_k\,. \tag{3.5}$$

Let $B_k = \{f_{n_k} > 1/2^k\}$ and consider the (bad) set $B = \limsup B_k$. Since by (3.5) $\sum_{k=1}^{\infty} \mu(B_k) < \infty$, by the Borel-Cantelli Lemma we have $\mu(B) = 0$. Now, it is not hard to see that

$$\lim_{k \to \infty} f_{n_k}(x) = 0\,, \quad x \in X \setminus B\,. \tag{3.6}$$

Indeed, if $x \notin B$, then x belongs to at most finitely many of the B_k's and (3.6) holds. ■

Sometimes we have to deal with questions of convergence when no limit is in evidence. For μ-a.e. convergence this can be reduced to the numerical case, where the Cauchy criterion is available. Specifically, let $(X, \mathcal{M}, \mu)$ be a measure space, and let $\{f_n\}$ be a sequence of measurable extended real-valued finite μ-a.e. functions defined on X. We say that $\{f_n\}$ is Cauchy μ-a.e. if for μ almost every $x \in X$, given $\varepsilon > 0$, there is an integer $n_0 = n_0(x)$ such that

$$|f_n(x) - f_{n'}(x)| \leq \varepsilon, \quad \text{all } n, n' \geq n_0\,.$$

By the Cauchy criterion of convergence of numerical sequences, if $\{f_n\}$ is Cauchy, then $\lim_{n \to \infty} f_n(x) = f(x)$ exists μ-a.e.

The same is true for convergence in probability. Let $(X, \mathcal{M}, \mu)$ be a finite measure space, and let $\{f_n\}$ be a sequence of measurable, extended real-valued, finite μ-a.e. functions defined on X. We say that $\{f_n\}$ is Cauchy in probability if given $\varepsilon, \delta > 0$, there is an integer n_0 such that

$$\mu(\{|f_n - f_{n'}| > \delta\}) < \varepsilon\,, \quad \text{all } n, n' \geq n_0\,.$$

Sequences which are Cauchy in probability converge in the sense of probability, cf. 4.30 below, and convergence in probability corresponds to a notion of "metric" convergence, cf. 4.31 below.

Next we discuss the concept of uniform convergence. Let $(X, \mathcal{M}, \mu)$ be a measure space and assume $\{f_n\}$ is a sequence of measurable nonnegative extended real-valued finite μ-a.e. functions defined on X. We say that $f_n \to 0$ almost uniformly if given $\varepsilon > 0$, we can find a measurable subset B of X such that $\mu(B) < \varepsilon$ and

$$\lim_{n\to\infty} f_n(x) = 0\,, \quad \text{uniformly for } x \in X \setminus B\,.$$

Proposition 3.4. Let $(X, \mathcal{M}, \mu)$ be a measure space, let $\{f_n\}$ be a sequence of measurable nonnegative extended real-valued finite μ-a.e. functions defined on X, and suppose that $f_n \to 0$ almost uniformly. Then $f_n \to 0$ μ-a.e.

Proof. For every positive integer k there is a measurable subset B_k of X such that $\mu(B_k) < 1/k$ and $f_n(x) \to 0$, uniformly for $x \in X \setminus B_k$. Clearly $f_n \to 0$ pointwise on the (good) set $G = \bigcup_{k=1}^{\infty}(X \setminus B_k)$. It only remains to check that $\mu(X \setminus G) = 0$; this is not hard. Since $X \setminus G$ equals

$$X \setminus (\textstyle\bigcup_{k=1}^{\infty}(X \setminus B_k)) = \bigcap_{k=1}^{\infty}(X \setminus (X \setminus B_k)) = \bigcap_{k=1}^{\infty} B_k\,,$$

and $\mu(B_1) < \infty$, it readily follows that

$$\mu(X \setminus G) = \lim_{k\to\infty} \mu(B_k) = 0\,. \quad \blacksquare$$

How about the converse to Proposition 3.4? To decide whether it is true we investigate the rate at which arbitrary sequences converge pointwise to 0. First we show that the convergence occurs at a fairly rapid rate.

Theorem 3.5. Let $(X, \mathcal{M}, \mu)$ be a finite measure space, and suppose $\{f_n\}$ is a sequence of measurable nonnegative extended real-valued finite μ-a.e. functions defined on X so that $f_n \to 0$ μ-a.e. Then there exists a nondecreasing sequence of integers $\lambda_n \to \infty$ with the property that

$$\lim_{n\to\infty} \lambda_n f_n = 0 \ \mu\text{-a.e. on } X\,. \tag{3.7}$$

Proof. Redefining the f_n's if necessary on a set of measure 0, we may assume that the f_n's are finite everywhere and that $f_n(x) \to 0$ for every $x \in X$. Let

$$g_n(x) = \sup_{k\geq n} f_k(x)\,, \quad x \in X\,.$$

Clearly the g_n's are measurable, $g_n(x) \geq f_n(x)$, all $n, x \in X$, and

$$g_n(x) \geq g_{n+1}(x), \quad \lim_{n \to \infty} g_n(x) = 0.$$

In other words, working with the g_n's instead, we may also assume that, in fact, the f_n's decrease to 0 everywhere.

The first step is to construct the λ_n's. Let $n_1 = 1$, and note that since $f_n \to 0$ in probability, for each integer $k = 2, 3, \ldots$, there exist integers $n_k > n_{k-1}$ such that

$$\mu(\{f_{n_k} > 1/k^2\}) \leq 1/2^k, \quad k = 2, 3, \ldots \tag{3.8}$$

Put now

$$\lambda_n = k, \quad \text{for } n_k \leq n < n_{k+1}, \; k = 1, 2, \ldots \tag{3.9}$$

In other words, the sequence of λ_n's is defined in blocks: The first $(n_2 - n_1)$ entries are 1's, the next $(n_3 - n_2)$ entries are 2's, and so on. Furthermore, since $n_k \to \infty$ as $k \to \infty$, also $\lambda_n \to \infty$ as $n \to \infty$.

Next we deal with the convergence of the sequence $\{\lambda_n f_n\}$. Let

$$B_m = \bigcup_{n \geq n_m} \{\lambda_n f_n > 1/m\}, \quad m = 1, 2, \ldots \tag{3.10}$$

It is not hard to estimate $\mu(B_m)$. First observe that since the λ_n's are constant on blocks we have

$$\begin{aligned} B_m &= \bigcup_{k=m}^{\infty} \bigcup_{n=n_k}^{n_{k+1}-1} \{\lambda_n f_n > 1/m\} \\ &= \bigcup_{k=m}^{\infty} \bigcup_{n=n_k}^{n_{k+1}-1} \{k f_n > 1/m\}. \end{aligned} \tag{3.11}$$

Furthermore, since the sequence $\{f_n\}$ is nonincreasing and since $k \geq m$, the innermost union in (3.11) is contained in $\{f_{n_k} > 1/k^2\}$, and

$$B_m \subseteq \bigcup_{k=m}^{\infty} \{f_{n_k} > 1/k^2\}. \tag{3.12}$$

Whence, by (3.12) and (3.8)

$$\mu(B_m) \leq \sum_{k=m}^{\infty} \mu(\{f_{n_k} > 1/k^2\}) \leq 2^{-m+1}.$$

Let the (bad) set $B = \limsup B_m$. Since $\sum \mu(B_m) < \infty$, by the Borel-Cantelli Lemma we have $\mu(B) = 0$. It only remains to check that for any $x \in X \setminus B$, we have $\lim_{n\to\infty} \lambda_n f_n(x) = 0$. But this is not hard: Given $\varepsilon > 0$, let m be so large that $1/m \le \varepsilon$ and $x \notin B_m$; such a choice is always possible since x belongs to at most finitely many of the B_m's. Then by (3.10) there exists n_m so that

$$\lambda_n f_n(x) \le 1/m \le \varepsilon\,, \quad \text{all } n \ge n_m\,,$$

and $\lambda_n f_n(x) \to 0$. ∎

Our next result is an interesting interpretation of Theorem 3.5.

Theorem 3.6. Let $(X, \mathcal{M}, \mu)$ be a finite measure space, and let $\{f_n\}$ be a sequence of measurable nonnegative extended real-valued finite μ-a.e. functions defined on X such that $f_n \to 0$ μ-a.e. Then there exist a measurable nonnegative finite μ-a.e. function f defined on X and a sequence of real numbers $\eta_n \to 0$ such that

$$f_n \le \eta_n f \ \mu\text{-a.e. on } X\,. \tag{3.13}$$

Proof. In the notation of Theorem 3.5, let

$$f(x) = \sup_n \{\lambda_n f_n(x)\}\,, \quad x \in X\,.$$

Clearly f is measurable and nonnegative, and since $\lambda_n f_n \to 0$ μ-a.e., f is also finite μ-a.e. Put now $\eta_n = 1/\lambda_n$, and note that

$$f_n \le \eta_n f \ \mu\text{-a.e.,} \quad \text{with } \eta_n \to 0\,,$$

as asserted. ∎

We are now ready to show that convergence μ-a.e. implies almost uniform convergence; this result is due to Egorov (1869–1931).

Theorem 3.7 (Egorov). Let $(X, \mathcal{M}, \mu)$ be a finite measure space, and let $\{f_n\}$ be a sequence of measurable nonnegative finite μ-a.e. functions defined on X such that $f_n \to 0$ μ-a.e. Then $f_n \to 0$ almost uniformly.

Proof. We must show that given $\varepsilon > 0$, there exists $B \in \mathcal{M}$ such that $\mu(B) < \varepsilon$ and $f_n(x) \to 0$ uniformly on $X \setminus B$. Let f be the μ-a.e. finite function corresponding to the sequence $\{f_n\}$ constructed in Theorem 3.6. By Proposition 2.1 there is a constant M such that $f(x) \leq M$ for $x \in X \setminus B$, $\mu(B) < \varepsilon$. By Theorem 3.6

$$f_n(x) \leq \eta_n M\,, \quad x \in X \setminus B\,, \mu(B) < \varepsilon\,,$$

and $f_n(x) \to 0$ uniformly for $x \in X \setminus B$. ■

The measurability of the f_n's is essential to the validity of Egorov's theorem, cf. 4.38 below, as is the assumption $\mu(X) < \infty$. Indeed, in the measure space $(R, \mathcal{L}, |\cdot|)$ the sequence $f_n = \chi_{[n,\infty)}$ tends to 0 everywhere on R, but not uniformly on any unbounded subset of R.

4. PROBLEMS AND QUESTIONS

The setting of the first thirteen problems and questions is the following: $\mathcal{M}$ is a σ-algebra of (measurable) subsets of X and f is an extended real-valued function defined on X.

4.1 Suppose $\{f > \lambda\} \in \mathcal{M}$ for each rational number λ; is f measurable?

4.2 Suppose f is a measurable real-valued function defined on X, and put $g(x) = 0$ if $f(x)$ is rational and $g(x) = 1$ if $f(x)$ is irrational; is g measurable?

4.3 Suppose f is measurable and $B \in \mathcal{B}_1$ is a Borel subset of R; does it follow that $f^{-1}(B) \in \mathcal{M}$?

4.4 Suppose f is a measurable real-valued function defined on X, and let ϕ be a real-valued Borel measurable function defined on R. Show that the composition $\phi \circ f$ is measurable.

4.5 Suppose f is measurable and show that for each real $r, s > 0$, the truncations

$$f_{r,s}(x) = \begin{cases} r & \text{if } f(x) > r \\ f(x) & \text{if } -s \leq f(x) \leq r \\ -s & \text{if } f(x) < -s\,, \end{cases}$$

are measurable.

4.6 If f, g are measurable real-valued functions defined on X and ϕ is a continuous real-valued function defined on $R \times R$, show that $\phi(f, g)$ is measurable.

4.7 Suppose $X = R$, and show that $\mathcal{B}_1$ is the smallest σ-field with respect to which all continuous functions are measurable. More precisely, show that every continuous function is measurable iff $\mathcal{B}_1 \subseteq \mathcal{M}$.

4.8 Let $\{f_n\}$ be a sequence of measurable functions defined on X, and let $A = \{x \in X : \lim_{n\to\infty} f_n(x) \text{ exists}\}$. Show that $A \in \mathcal{M}$.

4.9 An elementary function is one which assumes at most countably many values. Show that if f is an everywhere finite real-valued measurable function, then it is the uniform limit of a sequence of elementary functions. Also, unless f is bounded, it is not the uniform limit of a sequence of simple functions.

4.10 Let μ be a measure on $(X, \mathcal{M})$. Show that the following assertions are equivalent: (a) $(X, \mathcal{M}, \mu)$ is a complete measure space, and, (b) If $\{f_n\}$ is a sequence of measurable functions defined on X which converges μ-a.e., and if f is any extended real-valued function defined on X such that $f = \lim_{n\to\infty} f_n$ μ-a.e., then f is measurable.

4.11 Suppose $(X, \mathcal{M}, \mu)$ is a σ-finite measure space and let f be a measurable function defined on X. Show that the function

$$\mu(\{|f| > \lambda\}), \quad \lambda > 0$$

is nonincreasing and right-continuous. Furthermore, if f, f_1 and f_2 are nonnegative and measurable and η_1, η_2 are nonnegative real numbers so that $f \le \eta_1 f_1 + \eta_2 f_2$ μ-a.e., then for any $\lambda > 0$

$$\mu(\{f > (\eta_1 + \eta_2)\lambda\}) \le \mu(\{f_1 > \lambda\}) + \mu(\{f_2 > \lambda\}).$$

4.12 With the notation of 4.11, if $\{f_n\}$ is a nondecreasing sequence of measurable functions and $f = \lim_{n\to\infty} f_n$, then for any $\lambda > 0$,

$$\lim_{n\to\infty} \mu(\{|f_n| > \lambda\}) = \mu(\{|f| > \lambda\}).$$

4.13 With the notation of 4.11, suppose that $\mu(\{|f| > \lambda\}) \to 0$ as $\lambda \to \infty$, and let

$$f^*(t) = \inf\{\lambda : \mu(\{|f| > \lambda\} \le t\}, \quad t \ge 0.$$

Show that f^* satisfies the following properties:

(i) It is nonincreasing and right-continuous.

(ii) If $0 < \eta \le \mu(\{|f| > \lambda\}) < \infty$, then

$$f^*(\mu(\{|f| > \lambda\})) \le \lambda < f^*(\mu(\{|f| > \lambda\}) - \eta)\,.$$

(iii) If f^* is continuous at $t = \mu(\{|f| > \lambda\})$, then $f^*(\mu(\{|f| > \lambda\})) = \lambda$.

(iv) f and f^* are equimeasurable, i.e., for $\lambda > 0$, $\mu(\{|f| > \lambda\}) = |\{f^* > \lambda\}|$.

Because of property (iv), f^* is called the nonincreasing equimeasurable rearrangement of f.

The next six problems and questions deal with the Lebesgue measure on R.

4.14 Suppose f is defined a.e. on [0,1] and it is continuous a.e. there. Is f Lebesgue measurable? What if f is right-continuous instead?

4.15 Suppose f is differentiable on [0,1]. Is f' Lebesgue measurable?

4.16 Show that f is Lebesgue measurable iff there is a Borel measurable function g such that $f = g$ Lebesgue-a.e.

4.17 Suppose $f(x, y)$ is a Lebesgue measurable real-valued function defined on R^2 with the property that $f(x, \cdot)$ is Lebesgue measurable as a function of $x \in R$, and $f(\cdot, y)$ is continuous as a function of $y \in R$. Let now $\phi(x) = \max_{c \le y \le d} f(x, y)$. Prove that ϕ is Lebesgue measurable.

4.18 Suppose that f is Lebesgue measurable and ϕ is real-valued, continuous, and has the following property: For any null set N, $\phi^{-1}(N) \in \mathcal{L}$. Show that $f \circ \phi$ is Lebesgue measurable.

4.19 Suppose $\{A_n\}$ is a sequence of Lebesgue measurable subsets of R, and let $f_n = \chi_{A_n}, n = 1, 2, \ldots$ Find necessary and sufficient conditions for the sequence $\{f_n\}$ to: (a) converge Lebesgue-a.e., and, (6) converge uniformly.

In the next three problems and questions we assume that $(X, \mathcal{M}, \mu)$ is a probability measure space.

4.20 An extended real-valued measurable function f defined on X is said to be bounded in probability if given $\varepsilon > 0$, there exists a finite real number M_ε such that $\mu(\{|f| \le M_\varepsilon\}) \ge 1 - \varepsilon$. Prove that f is bounded in probability iff f is finite μ-a.e.

4.21 A sequence of extended real-valued measurable functions $\{f_n\}$ is said to be bounded in probability iff $\sup |f_n|$ is bounded in probability. The sequence $\{f_n\}$ is said to diverge to ∞ in probability iff for each $M > 0$ and $\varepsilon > 0$, there exists a finite integer $n_0(M, \varepsilon)$ such that if $n > n_0$, then $\mu(\{|f_n| > M\}) > 1 - \varepsilon$. Prove that if $\{f_n\}$ diverges to ∞ in probability and $\{g_n\}$ is bounded in probability, then the sequence $\{f_n + g_n\}$ diverges to ∞ in probability.

4.22 Suppose that $\sup f_n = \infty$ μ-a.e. Does there necessarily exist a subsequence $\{f_{n_k}\}$ that diverges to ∞ in probability?

In the next nine problems and questions we assume that $(X, \mathcal{M}, \mu)$ is a measure space.

4.23 This result concerns the approximation Theorem 1.12. Suppose the function f there enjoys the property that $\{x \in X : f(x) \neq 0\}$ is σ-finite. Then prove that the approximating sequence $\{f_n\}$ may be constructed with the additional property that $\mu(\{f_n \neq 0\}) < \infty$ for each n. Conversely, if each of the simple functions f_n satisfies $\mu(\{f_n \neq 0\}) < \infty$ and for each $x \in X$ we have $\lim_{n\to\infty} f_n(x) = f(x)$, then the set $\{f \neq 0\}$ is σ-finite.

4.24 Let $\mu(X) < \infty$, and suppose $\{f_n\}$ is a sequence of extended real-valued measurable functions defined on X that satisfies

$$\sum_{n=1}^{\infty} \mu(\{|f_n|/\lambda_n > 1\}) < \infty .$$

Prove that $\limsup(|f_n|/\lambda_n) \leq 1$ μ-a.e.

4.25 Suppose $\mu(X) = 1$. Show that the sequence of extended real-valued measurable functions $\{f_n\}$ converges μ-a.e. to the measurable function f iff for any $\varepsilon > 0$ and N we have

$$\mu(\{|f_n - f| \leq \varepsilon\,, \text{ for all } n \geq N\}) = 1\,.$$

4.26 Suppose $\mu(X) < \infty$ and let $\{f_n\}$ be a sequence of extended real-valued measurable functions defined on X which converges to f in probability. Show that f is measurable and that $f_n^+ \to f^+$, $f_n^- \to f^-$ and $|f_n| \to |f|$ in the sense of probability. Are the analogous statements for convergence in measure valid?

4.27 In the setting of 4.26, suppose that $\{g_n\}$ is a sequence of extended real-valued functions defined on X which converges to g in probability, and $\{\lambda_n\}$ a sequence of real numbers, $\lambda_n \to \lambda$. Does it follow that $f_n + \lambda_n g_n \to f + \lambda g$ in probability? Is it also true that $f_n/g_n \to f/g$ in probability? How about the corresponding statements for convergence in measure?

Consider further the following statement: If $\phi: R^2 \to R$ is continuous, then $\phi(f_n, g_n) \to \phi(f, g)$ in probability, or in measure. Is it true?

4.28 In the setting of 4.26, assume that the sequence $\{f_n\}$ is nondecreasing. Show that in this case we have $\lim_{n\to\infty} f_n = f$ μ-a.e.

4.29 In the setting of 4.26, suppose that μ is a probability measure and, for $t \neq 0$ real, let $D(t) = 1$ if $t < 0$ and $D(t) = 0$ if $t > 0$. Show that $f_n \to 0$ in probability iff

$$\lim_{n\to\infty} \mu(\{f_n > t\}) = D(t)\,, \quad \text{all } t \neq 0\,.$$

4.30 Suppose that $\mu(X) < \infty$, and let $\{f_n\}$ be a Cauchy sequence in the sense of probability, i.e., for any $\varepsilon > 0$ there exists an integer $N = N_\varepsilon$ such that

$$\mu(\{|f_n - f_m| > \varepsilon\}) < \varepsilon, \quad \text{all } n, m \geq N\,.$$

Show that there exists a measurable function f such that $\lim_{n\to\infty} f_n = f$ in probability.

4.31 Suppose that $\mu(X) < \infty$, and for extended real-valued measurable functions f, g defined on X put

$$d(f,g) = \inf\{\varepsilon > 0 : \mu(\{|f - g| > \varepsilon\}) \leq \varepsilon\}\,.$$

Observe that $d(f,g) = 0$ iff $f = g$ μ-a.e.; whence (at the level of classes) upon identifying functions which agree μ-a.e. we have $d(f,g) = 0$ iff $f = g$.
Show that $d(f,g)$ is a distance function and that $f_n \to f$ in probability iff $d(f_n, f) \to 0$. Furthermore, show that endowed with the metric d, the space of measurable functions is a complete metric space.
Is a similar result, involving now convergence in measure, true for arbitrary measure spaces ?

4.32 Show that if μ is the counting measure on the integers $X = Z$, then convergence in measure is equivalent to uniform convergence.

4.33 Suppose $(X, \mathcal{M}, \mu)$ is a finite measure space, and let $\{f_n\}$ be a sequence of measurable real-valued functions defined on X. Show that given $\varepsilon > 0$, there exist a (bad) set B with $\mu(B) < \varepsilon$ and a constant M such that

$$|f_n(x)| \leq M\,, \quad \text{all } x \in X \setminus B\,, \quad \text{all } n\,.$$

Is the conclusion true if we assume instead that the f_n's are finite μ-a.e.?

4.34 Suppose $(X, \mathcal{M}, \mu)$ is a finite measure space, and let $\{f_n\}$ be a sequence of extended real-valued measurable functions defined on X. Show that $\{f_n\}$ converges to a finite limit μ-a.e. iff given $\varepsilon > 0$, there exists a finite constant M_ε such that

$$\mu(\{\sup |f_n| \leq M_\varepsilon\}) \geq 1 - \varepsilon .$$

4.35 Suppose $(X, \mathcal{M}, \mu)$ is a finite measure space, and let $\{f_n\}$ be a sequence of measurable real-valued functions defined on X. Show that there exists a sequence (a_n) of positive real numbers with the property that

$$\lim_{n \to \infty} a_n f_n = 0 \ \mu\text{-a.e.}$$

4.36 Show that the proof of Lusin's theorem may be adjusted to give the following result: Suppose f is an extended real-valued finite Lebesgue-a.e. measurable function defined on $X = [a,b]$. Then, given $\varepsilon > 0$, there is a continuous function ϕ defined on X such that

$$|\{x \in X : f(x) \neq \phi(x)\}| < \varepsilon .$$

4.37 In the setting of 4.36 show that the following is true: There is a sequence of continuous functions $\{f_n\}$ defined on X such that $\lim_{n \to \infty} f_n = f$ Lebesgue-a.e.

4.38 By means of an example show that the conclusion of Egorov's theorem does not necessarily hold if the f_n's are not Lebesgue measurable.

4.39 Construct a sequence of Lebesgue measurable functions $\{f_n\}$ defined on $X = [a,b]$ with the following properties: The f_n's converge at every point of X, but they do not converge uniformly on any Lebesgue measurable subset E of X with $|X \setminus E| = 0$.

4.40 Let $\{f_n\}$ be a sequence of Lebesgue measurable functions defined on $X = [a,b]$ and suppose that $\lim_{n \to \infty} f_n = f$ exists Lebesgue-a.e. on X. If $f \neq 0$ Lebesgue-a.e. and $f_n \neq 0$ Lebesgue-a.e. on X, prove that given $\varepsilon > 0$, there exist $c > 0$ and a sequence $\{E_n\}$ of Lebesgue measurable subsets of X such that

$$|f_n(x)| \geq c\,,\ x \in E_n\,, \quad \text{and} \quad |X \setminus E_n| \leq \varepsilon\,, \quad n = 1, 2, \ldots$$

4.41 Prove the following variant of Egorov's theorem: Let $X = [a,b]$ and suppose $f, f_n, n = 1, 2, \ldots$, are Lebesgue measurable functions defined on X such that $\limsup f_n \le f$ on X. Then, given $\varepsilon > 0$, there is a Lebesgue measurable set $E \subseteq X$ with $|E| < \varepsilon$ such that for each $\eta > 0$ there is N with the property that

$$f_n(x) \le f(x) + \eta\,, \quad \text{all } x \in X \setminus E\,, n \ge N\,.$$

4.42 Prove the following extension of Egorov's theorem due to Lusin. Let $(X, \mathcal{M}, \mu)$ be a σ-finite measure space, and f, f_n be extended real-valued measurable functions defined μ-a.e. on X, $n = 1, 2, \ldots$, such that $\lim_{n\to\infty} f_n = f$ μ-a.e. Then there exists measurable sets $N, E_1, E_2, \ldots$ such that $X = N \cup \bigcup_k E_k$, $\mu(N) = 0$, and the sequence $\{f_n\}$ converges uniformly to f on each E_k.

4.43 Let $\mathcal{M}$ be a σ-algebra of (measurable) subsets of X, and let f be a complex-valued function defined on X. We say that $f = \Re f + i\Im f$ is measurable iff the (real-valued) functions $\Re f$ and $\Im f$ are measurable. This is an open ended question: Discuss the properties of complex-valued measurable functions. For instance, show that f is measurable iff for any open subset $\mathcal{U}$ in the complex plane, $f^{-1}(\mathcal{U}) \in \mathcal{M}$.

CHAPTER VII

Integration

In this chapter we introduce the integral and establish its basic properties, including taking limits under the integral sign. The relation between the Riemann and the Lebesgue integrals of a bounded function is elucidated.

1. THE INTEGRAL OF NONNEGATIVE FUNCTIONS

Suppose $(X, \mathcal{M}, \mu)$ is a measure space and let ϕ be a nonnegative simple function defined on X. Specifically,

$$\phi(x) = \sum_{k=1}^{n} a_k \chi_{A_k}(x), \quad a_k \text{ real},$$

where the A_k's form a measurable pairwise disjoint partition of X. The integral of ϕ over X with respect to μ is denoted by $\int_X \phi(x)\, d\mu(x)$, or plainly by $\int_X \phi\, d\mu$, and it is defined as the quantity

$$\int_X \phi\, d\mu = \sum_{k=1}^{n} a_k \mu(A_k). \tag{1.1}$$

The usual convention $0 \cdot \infty = 0$ is in effect, so that $\int_X \phi\, d\mu = 0$ whenever $\phi = 0$ μ-a.e. on X.

We begin by proving some elementary properties of the integral of simple functions.

Theorem 1.1. The integral defined by the expression (1.1) is a nonnegative real number or ∞, and it satisfies the following properties:

(i) It is well-defined, i.e., if also $\phi(x) = \sum_{h=1}^{m} b_h \chi_{B_h}(x)$, then

$$\sum_{k=1}^{n} a_k \mu(A_k) = \sum_{k=1}^{m} b_h \mu(B_h).$$

(ii) If ϕ and ψ are nonnegative simple functions and $\phi = \psi$ μ-a.e., then $\int_X \phi \, d\mu = \int_X \psi \, d\mu$.

(iii) It is positively linear, i.e., if ϕ, ψ are nonnegative simple functions and $\lambda > 0$, then

$$\int_X (\phi + \lambda\psi) \, d\mu = \int_X \phi \, d\mu + \lambda \int_X \psi \, d\mu. \tag{1.2}$$

(iv) It is monotone, i.e., if $0 \le \phi \le \psi$ are simple functions, then

$$\int_X \phi \, d\mu \le \int_X \psi \, d\mu. \tag{1.3}$$

(v) For each nonnegative simple function ϕ, the set function ν given by

$$\nu(E) = \int_X \phi \chi_E \, d\mu = \int_E \phi \, d\mu, \quad E \in \mathcal{M},$$

is a measure on $(X, \mathcal{M})$.

Proof. Since the proof of (i) follows along the lines to that of (iii), we only do (iii). Let, then, $\phi = \sum a_n \chi_{A_n}$ and $\psi = \sum b_m \chi_{B_m}$ be nonnegative simple functions; $\phi + \lambda\psi$ is then the simple function that takes the value $a_n + \lambda b_m$ on the set $A_n \cap B_m \in \mathcal{M}$. Note that the $a_n + \lambda b_m$'s are not necessarily distinct, but that the $A_n \cap B_m$'s are pairwise disjoint. Thus, by the definition of the integral, we have

$$\int_X (\phi + \lambda\psi) \, d\mu = \sum_{n,m} (a_n + \lambda b_m) \mu(A_n \cap B_m). \tag{1.4}$$

If the sum in (1.4) is infinite there are indices n, m such that $a_n + \lambda b_m \ne 0$, $\mu(A_n \cap B_m) = \infty$, and either $a_n \mu(A_n) = \infty$ or $b_m \mu(B_m) = \infty$. In the former case we have $\int_X \phi \, d\mu = \infty$, in the latter case it follows that $\int_X \psi \, d\mu = \infty$, and in either case (1.2) holds. On the other hand, if the sum in (1.4) is finite, since the summands there are nonnegative they may be rearranged freely and we obtain at once that the sum equals

$$\sum_n a_n \sum_m \mu(A_n \cap B_m) + \lambda \sum_m b_m \sum_n \mu(A_n \cap B_m).$$

By the additivity of μ this expression is

$$\sum_n a_n\mu(A_n) + \lambda \sum_m b_m\mu(B_m) = \int_X \phi\, d\mu + \lambda \int_X \psi\, d\mu\,,$$

and (iii) holds.

We prove (ii) next. Since $\phi = \psi$ μ-a.e., there are nonnegative simple functions ζ, ϕ' and ψ' and a null set A such that ϕ' and ψ' vanish off A and

$$\phi = \zeta + \phi', \quad \psi = \zeta + \psi'.$$

By (i) it readily follows that

$$\int_X \phi\, d\mu = \int_X \zeta\, d\mu = \int_X \psi\, d\mu\,,$$

and (ii) is true.

As for (iv), let $\phi = \sum a_n \chi_{A_n}$, $\psi = \sum b_m \chi_{B_m}$, and observe that if $\mu(A_n \cap B_m) \neq 0$, then $a_n \le b_m$. Whence, since $\bigcup_{n,m}(A_n \cap B_m) = X$, we have

$$\int_X \phi\, d\mu = \sum_{n,m} a_n\mu(A_n \cap B_m) \le \sum_{n,m} b_m\mu(A_n \cap B_m) = \int_X \psi\, d\mu\,,$$

and (iv) is true.

(v) is a useful property and, among other things, it gives new examples of measures on $(X, \mathcal{M})$. Clearly ν is a nonnegative set function and $\nu(\emptyset) = 0$; only the σ-additivity requires some work.

First observe that if ϕ is a nonnegative simple function and $E \in \mathcal{M}$, then $\phi\chi_E$ is also a nonnegative simple function and

$$\nu(E) = \int_X \phi\chi_E\, d\mu = \sum_n a_n\mu(A_n \cap E)\,.$$

Suppose, then, that $\{E_k\}$ is a sequence of pairwise disjoint, measurable subsets of X, and let E denote its union. Now, since μ is a measure, the right-hand side of the above expression equals

$$\sum_{k=1}^{\infty} \sum_n a_n\mu(A_n \cap E_k) = \sum_k \int_X \phi\chi_{E_k}\, d\mu = \sum \nu(E_k)\,.$$

Whence ν is a measure on $(X, \mathcal{M})$. ∎

How about the integral of arbitrary nonnegative measurable functions? By Theorem 1.12 in Chapter VI, these functions are limits of nondecreasing sequences of simple functions, and this fact suggests a way of defining the integral.

Let $(X, \mathcal{M}, \mu)$ be a measure space, f a nonnegative measurable function defined on X, and set

$$\mathcal{F}_f = \{\phi : \phi \text{ is simple, and } 0 \le \phi \le f\}.$$

The integral of f over X with respect to μ is denoted by $\int_X f(x)\, d\mu(x)$, or simply $\int_X f\, d\mu$, and it is defined as the quantity

$$\int_X f\, d\mu = \sup\left\{ \int_X \phi\, d\mu : \phi \in \mathcal{F}_f \right\}. \tag{1.5}$$

By Theorem 1.12 in Chapter VI, $\mathcal{F}_f \neq \emptyset$, and consequently, $\int_X f\, d\mu$ is a well-defined nonnegative real number or ∞.

(1.5) is similar to the definition of the lower Riemann integral of a nonnegative function f, but with a crucial difference: Rather than considering partitions of the "domain of integration" X, we work with partitions of the "range" of f, in a manner compatible with each individual function f. More precisely, $\mathcal{F}_f$ contains the ϕ's constructed in Theorem 1.12 in Chapter VI, and these simple functions are closely related to the level sets of f.

Before we go on we must check that if f is simple, then the definitions in (1.1) and (1.5) coincide. If $\int_X f\, d\mu$ denotes the expression given by (1.1), since $f \in \mathcal{F}_f$ it readily follows that

$$\int_X f\, d\mu \le \sup_{\mathcal{F}_f} \int_X \phi\, d\mu. \tag{1.6}$$

On the other hand, if $\phi \in \mathcal{F}_f$, then by (1.3),

$$\int_X \phi\, d\mu \le \int_X f\, d\mu, \quad \text{all } \phi \in \mathcal{F}_f,$$

the inequality opposite to (1.6) holds, and the definitions given by (1.1) and (1.5) coincide.

We are now ready to consider some elementary properties of the integral.

Theorem 1.2. Assume $(X, \mathcal{M}, \mu)$ is a measure space, and let f, g be nonnegative measurable functions defined on X. We then have

(i) If $f = g$ μ-a.e., then $\int_X f\,d\mu = \int_X g\,d\mu$.

(ii) $\int_X (f+g)\,d\mu \le \int_X f\,d\mu + \int_X g\,d\mu$. (1.7)

(iii) If $f \le g$, then $\int_X f\,d\mu \le \int_X g\,d\mu$.

(iv) If $A \subseteq B$ are measurable, then

$$\int_A f\,d\mu = \int_X f\chi_A\,d\mu \le \int_B f\,d\mu\,. \tag{1.8}$$

Proof. (i) follows at once since for any $\phi \in \mathcal{F}_f$ there exists a $\psi \in \mathcal{F}_g$ such that $\phi = \psi$ μ-a.e., and, by property (ii) in Theorem 1.1, the integrals of ϕ and ψ over X with respect to μ are equal.

As for (ii), observe that since for any $\phi \in \mathcal{F}_f$ and $\psi \in \mathcal{F}_g$ we have $\phi + \psi \in \mathcal{F}_{f+g}$, by (1.2) it follows that

$$\int_X \phi\,d\mu + \int_X \psi\,d\mu = \int_X (\phi+\psi)\,d\mu \le \int_X (f+g)\,d\mu\,.$$

Whence taking the sup of the left-hand side in the above inequality over $\phi \in \mathcal{F}_f$ and $\psi \in \mathcal{F}_g$ gives (1.7), and (ii) holds.

As for (iii), note that in this case we have $\mathcal{F}_f \subseteq \mathcal{F}_g$, and consequently,

$$\int_X \phi\,d\mu \le \int_X g\,d\mu\,, \quad \text{all } \phi \in \mathcal{F}_f\,.$$

Taking the sup over the ϕ's above gives (iii).

To verify (1.8) it suffices to note that in this case $\mathcal{F}_{f\chi_A} \subseteq \mathcal{F}_{f\chi_B}$. Thus (iv) holds, and the proof is complete. ■

It is of interest to determine whether equality holds in (1.7). To address this question, and to investigate the behavior of the integral with respect to limits, we consider the following result.

Theorem 1.3 (Beppo Levi). Let $(X, \mathcal{M}, \mu)$ be a measure space and assume $\{f_n\}$ is a nondecreasing sequence of nonnegative finite μ-a.e. measurable functions defined on X. Then, $\lim_{n\to\infty} f_n(x) = f(x)$ exists everywhere on X, $f(x)$ is nonnegative and measurable and

$$\int_X f\,d\mu = \lim_{n\to\infty} \int_X f_n\,d\mu \left(= \sup \int_X f_n\,d\mu\right)\,. \tag{1.9}$$

Proof. That f is nonnegative and measurable is clear. By monotonicity, the numerical sequence $\int_X f_n \, d\mu, n = 1, 2, \ldots$, is nondecreasing, and consequently, it has a limit L, say. Also, by monotonicity, $\int_X f_n \, d\mu \leq \int_X f \, d\mu$ for all n, and

$$L = \sup_n \int_X f_n \, d\mu \leq \int_X f \, d\mu \,. \tag{1.10}$$

If $L = \infty$, the right-hand side of (1.10) is also ∞, and (1.9) holds in this case. On the other hand, if L is finite we must show the inequality opposite to (1.10), and this requires some work. Given $0 < \eta < 1$ and $\phi \in \mathcal{F}_f$, let $E_n = \{f_n > \eta\phi\}$; $\{E_n\}$ is a sequence of measurable sets and since the f_n's are nondecreasing and $\phi \leq f$, it readily follows that $E_n \subseteq E_{n+1}$ for all n, and that $\lim E_n = X$.

Consider now the measure $\nu(E) = \int_E \phi \, d\mu, E \in \mathcal{M}$, and observe that by monotonicity we have

$$\int_X f_n \, d\mu \geq \int_{E_n} f_n \, d\mu \geq \int_{E_n} \eta\phi \, d\mu = \eta\nu(E_n) \,. \tag{1.11}$$

By (1.10) and (v) in Theorem 1.1 both sides of (1.11) have a finite limit as $n \to \infty$. Whence, taking limits there we obtain at once

$$L \geq \lim_{n\to\infty} \eta\nu(E_n) = \eta\nu(X) = \eta \int_X \phi \, d\mu \,. \tag{1.12}$$

Now, (1.12) holds for each $\phi \in \mathcal{F}_f$, and taking sup over $\mathcal{F}_f$ we get

$$\eta \int_X f \, d\mu \leq L \,. \tag{1.13}$$

Since $\eta < 1$ is arbitrary it is clear that (1.13) is also true with $\eta = 1$ and the inequality opposite to (1.10) holds. ■

This result of Beppo Levi (1875–1961), also known as the Monotone convergence theorem, or MCT, has many important applications; before we discuss them we present a simple extension of MCT, also useful in applications.

Corollary 1.4 (μ-a.e. version of MCT). Let $(X, \mathcal{M}, \mu)$ be a measure space and $\{f_n\}$ a μ-a.e. nondecreasing sequence of nonnegative finite μ-a.e. measurable functions defined on X. Then $\lim_{n\to\infty} f_n(x) = f(x)$ exists μ-a.e on X, $f(x)$ is nonnegative and measurable and

$$\lim_{n\to\infty} \int_X f_n \, d\mu = \int_X f \, d\mu \,.$$

Proof. Let N be the null set outside of which the f_n's increase to f, and put $g_n = f_n$ on $X \setminus N$, and $g_n = 0$ on N, and $g = f$ on $X \setminus N$, and $g = 0$ on N.

The point is that now the g_n's converge to g everywhere, and they coincide with the f_n's and f at the level of integrals. More precisely, by property (i) of Theorem 1.2,

$$\int_X f\,d\mu = \int_X g\,d\mu\,, \quad \text{and} \quad \int_X f_n\,d\mu = \int_X g_n\,d\mu\,, \quad \text{all } n\,,$$

and consequently

$$\lim_{n\to\infty} \int_X f_n\,d\mu = \lim_{n\to\infty} \int_X g_n\,d\mu = \int_X g\,d\mu = \int_X f\,d\mu\,. \quad \blacksquare$$

As for the applications, we do the additivity and σ-additivity of the integral first.

Proposition 1.5. Suppose $(X, \mathcal{M}, \mu)$ is a measure space and let f, g be nonnegative extended real-valued measurable functions defined on X. Then

$$\int_X (f+g)\,d\mu = \int_X f\,d\mu + \int_X g\,d\mu\,. \tag{1.14}$$

Proof. Let $\{\phi_n\} \subseteq \mathcal{F}_f$ and $\{\psi_n\} \subseteq \mathcal{F}_g$ be sequences of nonnegative simple functions that increase to f and g respectively; for instance, the sequences constructed in Theorem 1.12 of Chapter VI will do. Observe that $\{\phi_n + \psi_n\} \subseteq \mathcal{F}_{f+g}$, and that this sequence increases to $f + g$. Now, by (1.2) we have

$$\int_X (\phi_n + \psi_n)\,d\mu = \int_X \phi_n\,d\mu + \int_X \psi_n\,d\mu\,, \quad \text{all } n\,. \tag{1.15}$$

By MCT the left-hand side of (1.15) converges to $\int_X (f+g)\,d\mu$ as $n \to \infty$, and also by MCT the right-hand side converges to $\int_X f\,d\mu + \int_X g\,d\mu$ as $n \to \infty$. Thus (1.14) holds. $\blacksquare$

We are now ready to prove

Theorem 1.6. Assume $(X, \mathcal{M}, \mu)$ is a measure space, suppose $\{f_n\}$ is a sequence of nonnegative extended real-valued measurable functions defined on X and let $f = \sum_n f_n$. Then f is nonnegative, extended real-valued and measurable, and

$$\int_X f\,d\mu = \sum_n \int_X f_n\,d\mu\,. \tag{1.16}$$

Proof. Let $s_k = \sum_{n=1}^{k} f_n, k = 1, 2, \ldots$, and observe that the s_k's form a nondecreasing sequence of nonnegative, extended real-valued measurable functions defined on X; moreover, $\lim_{k\to\infty} s_k = f$, and f is nonnegative, extended real-valued and measurable. By MCT

$$\int_X f\,d\mu = \lim_{k\to\infty} \int_X s_k\,d\mu\,. \tag{1.17}$$

Now, by (a simple extension of) Proposition 1.5, we have

$$\int_X s_k\,d\mu = \sum_{n=1}^{k} \int_X f_n\,d\mu\,,$$

and consequently,

$$\lim_{k\to\infty} \int_X s_k\,d\mu = \lim_{k\to\infty} \sum_{n=1}^{k} \int_X f_n\,d\mu = \sum_{n=1}^{\infty} \int_X f_n\,d\mu\,.$$

Whence, replacing this expression in the right-hand side of (1.17), (1.16) follows. ■

An interesting consequence of Theorem 1.6 is

Proposition 1.7. Suppose $(X, \mathcal{M}, \mu)$ is a measure space and let f be a nonnegative extended real-valued measurable function defined on X. Then the set function

$$\nu(E) = \int_E f\,d\mu\,, \quad E \in \mathcal{M}\,,$$

is a measure on $(X, \mathcal{M})$.

Proof. That ν is nonnegative and $\nu(\emptyset) = 0$ is clear. As for the σ-additivity, let $\{E_n\}$ be a sequence of pairwise disjoint measurable sets, and put $f_n = f\chi_{E_n}, n = 1, 2, \ldots$ The sequence $\{f_n\}$ satisfies the hypothesis of Theorem 1.6, and consequently, by (1.16) we have

$$\begin{aligned}\sum_n \nu(E_n) = \sum_n \int_X f_n\,d\mu &= \int_X \sum_n f_n\,d\mu \\ &= \int_X f \sum \chi_{E_n}\,d\mu = \int_X f\chi_{\bigcup E_n}\,d\mu = \nu\left(\bigcup E_n\right).\end{aligned}$$

In other words, ν is σ-additive. ■

It is natural to consider whether Theorem 1.3 can be extended to include more general classes of functions. First we discuss a result due to Fatou (1878–1929); since its statement involves the lim inf it applies to arbitrary sequences.

Thereom 1.8 (Fatou's Lemma). Suppose $(X, \mathcal{M}, \mu)$ is a measure space and let $\{ f_n \}$ be a sequence of nonnegative extended real-valued measurable functions defined on X. Then

$$\int_X \liminf f_n \, d\mu \le \liminf \int_X f_n d\mu \,. \tag{1.18}$$

Proof. The idea of the proof is to invoke MCT at the appropriate spot. Let

$$g_k = \inf_{n \ge k} f_n \,, \quad k = 1, 2, \ldots$$

The sequence $\{g_k\}$ satisfies the following properties:

(i) The g_k are nonnegative extended real-valued measurable functions.

(ii) The sequence is nondecreasing.

(iii) $g_k \le f_n$, all $n \ge k \ge 1$.

(iv) $\lim_{k \to \infty} g_k = \sup_{k \ge 1} (\inf_{n \ge k} f_n) = \liminf f_n$.

By (iv) and MCT we have

$$\lim_{k \to \infty} \int_X g_k \, d\mu = \int_X \lim_{k \to \infty} g_k \, d\mu = \int_X \liminf f_n \, d\mu \,. \tag{1.19}$$

On the other hand, by (iii) and monotonicity we get

$$\int_X g_k \, d\mu \le \int_X f_n \, d\mu \,, \quad \text{all } n \ge k \,,$$

and consequently for each fixed k we have

$$\int_X g_k \, d\mu \le \liminf \int_X f_n \, d\mu \,. \tag{1.20}$$

Combining now (1.19) and (1.20) it readily follows that (1.18) holds, and we have finished. ■

It is not hard to see that strict inequality may occur in Fatou's Lemma. Indeed, for the Lebesgue measure on [0,1] and the sequence $\{f_n\}$ given by $f_n(x) = n\chi_{(0,1/n)}(x)$, $n = 1, 2, \ldots$, we have $\liminf f_n(x) = 0$ for all x, and consequently, the left-hand side of (1.18) is 0, but $\int_{[0,1]} f_n \, d\mu = 1$ for all n, and the right-hand side there is 1.

Fatou's Lemma is very important in applications; Fatou discovered it while investigating the convergence properties of the Poisson integral, a problem that lies at the heart of Harmonic Analysis. A related result will be discussed in Theorem 3.1 in Chapter XVII.

2. THE INTEGRAL OF ARBITRARY FUNCTIONS

As interesting a result as Theorem 1.8 is, it still does not address the question of interchanging limits with integrals. We discuss this general question in the context of integrals of functions of arbitrary sign.

Let $(X, \mathcal{M}, \mu)$ be a measure space, and let f be an extended real-valued measurable function defined on X; we can write $f = f^+ - f^-$ as the difference of two nonnegative functions. In particular, both of the integrals $\int_X f^+ \, d\mu$ and $\int_X f^- \, d\mu$ exist, and if either of these integrals is finite we define the integral $\int_X f \, d\mu$ of f over X with respect to μ as

$$\int_X f \, d\mu = \int_X f^+ \, d\mu - \int_X f^- d\mu \,. \tag{2.1}$$

Observe that if f and g are measurable and $f = g$ μ-a.e., then

$$f^+ = g^+ \, , \; f^- = g^- \quad \mu\text{-a.e.}$$

and consequently, by (i) in Theorem 1.2, the integral of f over X with respect to μ exists iff that of g exists, and in this case they are equal.

Some of the properties of the integral are readily obtained, for instance

$$-\infty \le \int_X f \, d\mu \le \infty \,, \tag{2.2}$$

and

$$\int_X \lambda f \, d\mu = \lambda \int_X f \, d\mu \,, \quad \lambda \text{ real} \,. \tag{2.3}$$

Indeed, (2.2) follows at once from (2.1), and (2.3) is a simple consequence of this observation: For $\lambda > 0$ we have $(\lambda f)^+ = \lambda f^+$ and $(\lambda f)^- = \lambda f^-$, and for $\lambda < 0$ we have $(\lambda f)^+ = -\lambda f^-$ and $(\lambda f)^- = -\lambda f^+$.

An interesting theory may be developed for those functions for which both of the integrals on the right-hand side of (2.1) are finite. This class of functions is denoted by $L^1(X, \mu)$, or plainly $L(X, \mu)$, $L(X)$, or $L(\mu)$, and it is called the Lebesgue class L^1; the functions f in L^1 are said to be integrable.

Note that since

$$f^+ \, , f^- \le |f| = f^+ + f^- \,, \tag{2.4}$$

and for functions f in $L(\mu)$ the expression $\int_X (f^+ + f^-) \, d\mu = \int_X |f| \, d\mu$ is finite, it is built into the definition of $L(\mu)$ that $f \in L(\mu)$ iff the integral

$\int_X f\,d\mu$ is defined and $\left|\int_X f\,d\mu\right| < \infty$. On the other hand, as simple examples show, it is possible for $|f|$ to be integrable, and yet for f not to be measurable. Thus the measurability of f is essential in the definition of $L(\mu)$.

The following estimate, in the spirit of (3.1) in Chapter III, is quite useful.

Proposition 2.1. Let $(X, \mathcal{M}, \mu)$ be a measure space, and let f be an extended real-valued measurable function defined on X for which the integral over X with respect to μ is defined. Then we have

$$\left|\int_X f\,d\mu\right| \leq \int_X |f|\,d\mu\,. \tag{2.5}$$

Proof. If the right-hand side of (2.5) is infinite there is nothing to prove. On the other hand, if $f \in L(\mu)$, then by (2.4) and Proposition 1.5 the integrals of f^+ and f^- over X with respect to μ are also finite and we have

$$\begin{aligned}\left|\int_X f\,d\mu\right| &= \left|\int_X f^+\,d\mu - \int_x f^-\,d\mu\right| \\ &\leq \int_X f^+\,d\mu + \int_X f^-\,d\mu = \int_X (f^+ + f^-)\,d\mu = \int_X |f|\,d\mu\,. \quad \blacksquare\end{aligned}$$

When f is integrable, (2.5) may be interpreted as a statement concerning its "size." A more precise estimate was proved by Chebychev (1821–1894); the result is known as Chebychev's inequality.

Theorem 2.2. Let $(X, \mathcal{M}, \mu)$ be a measure space and suppose f is an extended real-valued measurable function defined on X. Then for any real $\lambda > 0$ we have

$$\lambda\,\mu(\{|f| > \lambda\}) \leq \int_X |f|\,d\mu\,. \tag{2.6}$$

Proof. Let $A_\lambda = \{|f| > \lambda\}$; A_λ is measurable and $\lambda\chi_{A_\lambda} \leq |f|$ μ-a.e. Thus, by Theorem 1.2 (iii), it readily follows that

$$\int_X \lambda\chi_{A_\lambda}\,d\mu = \lambda\mu(A_\lambda) \leq \int_X |f|\,d\mu\,. \quad \blacksquare$$

Corollary 2.3. Let $(X, \mathcal{M}, \mu)$ be a measure space and let $f \in L(\mu)$. Then f is finite μ-a.e. Further, if f is nonnegative and $\int_X f \, d\mu = 0$, then $f = 0$ μ-a.e.

Proof. By Chebychev's inequality we have

$$\mu(\{|f| > n\}) \leq \frac{1}{n} \int_X |f| \, d\mu \to 0 \quad \text{as } n \to \infty .$$

Whence

$$\mu(\{|f| = \infty\}) = \lim_{n \to \infty} \mu(\{|f| > n\}) = 0 ,$$

and f is finite μ-a.e. Moreover, if f is nonnegative and its integral over X vanishes, then, also by Chebychev's inequality, $\mu(\{f > \lambda\}) = 0$ for all $\lambda > 0$, and f vanishes μ-a.e. ■

How does the integral behave with respect to addition?

Proposition 2.4. Let $(X, \mathcal{M}, \mu)$ be a measure space and suppose λ is a real number and $f, g \in L(\mu)$. Then $f + \lambda g$ is integrable and

$$\int_X (f + \lambda g) \, d\mu = \int_X f \, d\mu + \lambda \int_X g \, d\mu . \tag{2.7}$$

Proof. The integrability of $f + \lambda g$ follows at once from the estimate $|f + \lambda g| \leq |f| + |\lambda| \, |g|$. Now, since $h = f + \lambda g$ is integrable, the integral of h over X with respect to μ is a well-defined finite number. Furthermore, we have

$$h^+ - h^- = f^+ - f^- + (\lambda g)^+ - (\lambda g)^- ,$$

and consequently, also

$$h^+ + f^- + (\lambda g)^- = h^- + f^+ + (\lambda g)^+ . \tag{2.8}$$

All the summands in (2.8) are nonnegative, and by Corollary 2.3 finite μ-a.e. By Proposition 1.5 then, it readily follows that

$$\int_X h^+ \, d\mu + \int_X f^- \, d\mu + \int_X (\lambda g)^- \, d\mu = \int_X h^- \, d\mu + \int_X f^+ \, d\mu + \int_X (\lambda g)^+ \, d\mu ,$$

and since all the integrals are finite we may move them freely and obtain

$$\int_X h^+ \, d\mu - \int_X h^- \, d\mu = \int_X f^+ \, d\mu - \int_X f^- \, d\mu + \int_X (\lambda g)^+ \, d\mu - \int_X (\lambda g)^- \, d\mu .$$

Thus (2.7) holds. ■

The following variant of Proposition 2.4 is important in applications since it allows us to consider arbitrary functions for which the integral is defined.

Proposition 2.5. Let $(X, \mathcal{M}, \mu)$ be a measure space and suppose f, g are extended real-valued measurable functions defined on X which satisfy the following conditions: The integral of f over X with respect to μ is defined, and g is integrable. Then the integral of $f+g$ over X with respect to μ is defined and

$$\int_X (f+g)\,d\mu = \int_X f\,d\mu + \int_X g\,d\mu\,. \tag{2.9}$$

Proof. By Proposition 2.4, (2.9) is only novel when f is not integrable. If this is the case one of the quantities, $\int_X f^+\,d\mu$ or $\int_X f^-\,d\mu$, is infinite and the other one is finite. To fix ideas suppose that $\int_X f^-\,d\mu = \infty$, and observe that with the notation of Proposition 2.4, recalling that $h = f + g$ and setting $\lambda = 1$ in (2.8), we have

$$h^+ + f^- + g^- = h^- + f^+ + g^+\,. \tag{2.10}$$

Since $\int_X f^+\,d\mu, \int_X g^+\,d\mu < \infty$, from (2.10) it follows that $\int_X h^-\,d\mu = \infty$. Furthermore, since $h^- = 0$ when $h^+ \neq 0$, by (2.10) we also get that $h^+ \leq f^+ + g^+$, and consequently $\int_X h^+\,d\mu < \infty$. Thus the integral of h over X with respect to μ exists and it equals $-\infty$. Whence the left-hand side of (2.9) equals $-\infty$, and so does the right-hand side. Thus (2.9) holds, and we are done. ■

We are now in a position to explore what Fatou's Lemma says in the general context of functions of arbitrary sign.

Theorem 2.6 (Fatou's Lemma). Let $(X, \mathcal{M}, \mu)$ be a measure space and suppose $\{f_n\}$ is a sequence of extended real-valued measurable functions defined on X which satisfy the following property: There exists an integrable function g such that

$$g \leq f_n \quad \text{all } n\,. \tag{2.11}$$

Then the integrals of $\liminf f_n$ and f_n over X with respect to μ exist, $n = 1, 2 \ldots$, and we have

$$\int_X \liminf f_n\,d\mu \leq \liminf \int_X f_n\,d\mu\,. \tag{2.12}$$

Proof. Since by (2.11) $f_n - g \geq 0$, the integral of the functions $f_n - g$ over X with respect to μ is well-defined for $n = 1, 2, \ldots$; similarly for $\liminf f_n - g$.

Now, by Fatou's Lemma for nonnegative functions we have

$$\int_X \liminf (f_n - g)\, d\mu = \int_X (\liminf f_n - g)\, d\mu$$
$$\leq \liminf \int_X (f_n - g)\, d\mu\,. \qquad (2.13)$$

First we consider the left-hand side of (2.13). By Proposition 2.5 with $f = \liminf f_n - g$ and $g = g$ there, we get that $f + g = \liminf f_n$ has a well-defined integral over X with respect to μ which satisfies

$$\int_X \liminf f_n\, d\mu = \int_X (\liminf f_n - g)\, d\mu + \int_X g\, d\mu\,. \qquad (2.14)$$

Since g is integrable, by (2.14) it readily follows that the left-hand side of (2.13) equals

$$\int_X \liminf f_n\, d\mu - \int_X g\, d\mu\,. \qquad (2.15)$$

A similar argument gives that the integral of f_n over X with respect to μ exists, $n = 1, 2, \ldots$, and that the integral that appears on the right-hand side of (2.13) is equal to

$$\int_X f_n\, d\mu - \int_X g\, d\mu\,, \quad n = 1, 2, \ldots \qquad (2.16)$$

Thus combining (2.15) and (2.16) we may rewrite (2.13) as

$$\int_X \liminf f_n\, d\mu - \int_X g\, d\mu \leq \liminf \int_X f_n\, d\mu - \int_X g\, d\mu\,.$$

Since g is integrable we may now cancel $\int_X g\, d\mu$ in the above inequality. Whence (2.12) holds, and we have finished. ∎

There is a version of Theorem 2.6 with the inequality (2.12) reversed, but with the lim inf replaced by the lim sup.

Corollary 2.7. Let $(X, \mathcal{M}, \mu)$ be a measure space and suppose $\{f_n\}$ is a sequence of extended real-valued measurable functions defined on X

which have the following property: There exists an integrable function g such that

$$f_n \leq g\,, \quad \text{all } n\,. \tag{2.17}$$

Then the integrals of $\limsup f_n$ and f_n over X with respect to μ exist, $n = 1, 2, \ldots$, and we have

$$\limsup \int_X f_n \, d\mu \leq \int_X \limsup f_n \, d\mu\,. \tag{2.18}$$

Proof. Observe that (2.17) is equivalent to

$$-g \leq -f_n\,, \quad \text{all } n\,,$$

and that g is integrable iff $-g$ is integrable, in other words, the hypotheses of Fatou's Lemma are satisfied by $\{-f_n\}$ and $-g$. Since $\liminf(-f_n) = -\limsup f_n$, by Theorem 2.6 we have

$$\begin{aligned} -\int_X \limsup f_n \, d\mu &= \int_X \liminf\,(-f_n)\, d\mu \\ &\leq \liminf \int_X (-f_n)\, d\mu = \liminf \left(-\int_X f_n \, d\mu \right) \\ &= -\limsup \int_X f_n \, d\mu\,. \end{aligned} \tag{2.19}$$

Now, (2.19), whether involving finite quantities or not, is equivalent to (2.18), and we have finished. ■

Some remarks concerning these results: Clearly we may assume that (2.11) and (2.17) hold μ-a.e. and obtain the same conclusion; also strict inequality may occur in estimates (2.12) and (2.18). For instance, for the Lebesgue measure on $I = [0,1]$ and the sequence $f_n = \chi_{[0,3/4)}$, n odd, and $f_n = \chi_{[3/4,1]}$, n even, we have

$$\int_I \liminf f_n \, d\mu = 0 < \liminf \int_I f_n \, d\mu = 1/4\,,$$

and

$$\limsup \int_I f_n \, d\mu = 3/4 < \int_I \limsup f_n \, d\mu = 1\,.$$

We close this section with the Lebesgue dominated convergence theorem, or LDCT, which describes under what conditions we may pass to the limit under the integral sign.

Theorem 2.8 (LDCT). Let $(X, \mathcal{M}, \mu)$ be a measure space and suppose $\{f_n\}$ is a sequence of extended real-valued measurable functions defined on X such that

(i) $\lim_{n\to\infty} f_n = f$ exists μ-a.e.

(ii) There is an integrable function g so that for each $n, |f_n| \le g$ μ-a.e.

Then f is integrable and

$$\int_X f\, d\mu = \lim_{n\to\infty} \int_X f_n\, d\mu\,. \tag{2.20}$$

Proof. By (i) it readily follows that f is measurable and by (ii) that $|f| \le g$ μ-a.e., thus f is also integrable. As for the f_n's, by (ii) they are integrable and for any n we have

$$-g \le f_n \le g \quad \mu\text{-a.e.}$$

So, Fatou's Lemma and its corollary apply, and since $f = \liminf f_n = \limsup f_n$ μ-a.e., it readily follows that

$$\begin{aligned}\int_X f\, d\mu &\le \liminf \int_X f_n\, d\mu \\ &\le \limsup \int_X f_n\, d\mu \le \int_X f\, d\mu\,.\end{aligned}$$

Whence, all four quantities in the above inequality are equal (and finite), and (2.20) holds. ■

By the way, the example following Theorem 1.8 shows that, in the absence of an integrable majorant, the conclusion of LDCT may fail.

3. RIEMANN AND LEBESGUE INTEGRALS

Suppose g is a Riemann integrable function over an interval I. Does the integral of g over I with respect to the Lebesgue measure exist? If so, do both integrals coincide? In other words, we would like to know whether the notion of Lebesgue integral extends that of Riemann integral.

First some notations. Lebesgue measurable functions will be called measurable functions, Lebesgue integrable functions will be called integrable, Lebesgue a.e. will be denoted plainly by a.e. and the integral of g over I with respect to the Lebesgue measure is denoted by $\int_I g\, dx$.

We then have

Theorem 3.1. Let g be a bounded real-valued function defined on $I = [a,b]$ and suppose that $g \in \mathcal{R}(I)$. Then $g \in L(I)$ and

$$\int_a^b g(x)\,dx = \int_I g\,dx\,. \tag{3.1}$$

Proof. Let $\mathcal{P}_n = \{a = x_{1,n} < \ldots < x_{k_n,n} = b\}, n = 1,2,\ldots$, be a sequence of finite partitions of I such that $\mathcal{P}_{n+1}$ is a refinement of $\mathcal{P}_n$, $n \geq 1$, and so that the norm of $\mathcal{P}_n \to 0$ as $n \to \infty$. If $I_{k,n} = [x_{k,n}, x_{k+1,n}]$, $1 \leq k \leq k_n - 1$, are the intervals induced by $\mathcal{P}_n$, and

$$m_{k,n} = \inf_{I_{k,n}} g\,, \quad M_{k,n} = \sup_{I_{k,n}} g$$

note that the functions

$$L_n(x) = \sum_k m_{k,n}\chi_{I_{k,n}}(x), \quad \text{and} \quad U_n(x) = \sum_k M_{k,n}\chi_{I_{k,n}}(x)$$

are bounded and measurable, and hence integrable over I. Now, the sequence $\{L_n\}$ is nondecreasing and the sequence $\{U_n\}$ is nonincreasing, and consequently, the limits

$$L(x) = \lim_{n\to\infty} L_n(x)\,, \quad \text{and} \quad U(x) = \lim_{n\to\infty} U_n(x)\,, \tag{3.2}$$

exist and are finite everywhere on I. Furthermore, L and U are measurable, and

$$L(x) \leq g(x) \leq U(x)\,, \quad x \in I\,. \tag{3.3}$$

Next observe that for $n = 1,2,\ldots$ we have

$$\int_I L_n\,dx = \sum_k m_{k,n}|I_{k,n}| = s(g,\mathcal{P}_n)\,, \tag{3.4}$$

and

$$\int_I U_n\,dx = \sum_k M_{k,n}|I_{k,n}| = S(g,\mathcal{P}_n)\,. \tag{3.5}$$

Since g is bounded the L_n's and the U_n's are uniformly bounded on I, and consequently, by LDCT we get that

$$\int_I L\,dx = \lim_{n\to\infty}\int_I L_n\,dx\,, \quad \int_I U\,dx = \lim_{n\to\infty}\int_I U_n\,dx\,. \tag{3.6}$$

Moreover, since $g \in \mathcal{R}(I)$, by (3.4), (3.5) and (3.6) we have

$$\int_I L\,dx = \int_I U\,dx = \int_a^b g(x)\,dx\,. \tag{3.7}$$

From (3.7) it readily follows that the integral of the nonnegative function $U-L$ over I vanishes, and by Corollary 2.3 we have that $U = L$ a.e. Hence, by (3.3) we obtain that $g = U$ a.e., and since the Lebesgue measure is complete, by Theorem 1.4 in Chapter VI we get that g is also measurable. By (3.3) it now follows that (3.1) holds, and we have finished. ■

The converse to Theorem 3.1 is false, to wit, there are bounded integrable functions g defined on I which are not Riemann-integrable; the characteristic function of the rationals in I will do.

The notion of Riemann-integrability incorporates unbounded functions by means of the so-called "improper" convergence methods. For instance, suppose g is unbounded on I, but it satisfies the following properties:

(i) For each $0 < \varepsilon < b-a$, $g \in \mathcal{R}([a+\varepsilon, b])$.

(ii) $\lim_{\varepsilon\to 0^+} \int_{a+\varepsilon}^b g(x)\,dx = \int_{a+}^b g(x)\,dx$ exists.

Of course there are similar definitions for $\int_a^{b-} g(x)\,dx$ and $\int_{a+}^{b-} g(x)\,dx$. For instance, the function $g(x) = x^{-1/2}$ is unbounded on (0,1], but

$$\int_\varepsilon^1 g(x)\,dx = 2(1-\varepsilon^{1/2}) \to 2 \quad \text{as } \varepsilon \to 0^+.$$

Functions for which the improper Riemann integral exists are also integrable, as our next result shows.

Theorem 3.2. Suppose that the nonnegative function g is defined and finite on $I = (a,b]$ and that $\int_{a+}^b g(x)\,dx$ exists. Then $g \in L([a,b])$ and

$$\int_I g\,dx = \int_{a+}^b g(x)\,dx\,.$$

Proof. Let $0 < \varepsilon_n < b-a$ be a sequence which tends to 0 as $n \to \infty$. By Theorem 3.1 the functions $g_n = g\chi_{[a+\varepsilon_n,b]}$, $n = 1,2,\ldots$, are integrable on I and

$$\int_I g_n\,dx = \int_{a+\varepsilon_n}^b g(x)\,dx.$$

Moreover, since the g_n's increase to g on I (the value of g at a is irrelevant) by MCT it readily follows that

$$\int_{a+}^{b} g(x)\,dx = \lim_{n\to\infty} \int_I g_n\,dx = \int_I g\,dx\,. \quad \blacksquare$$

A similar result holds for the other improper integrals. On the other hand, it is also possible to consider the Riemann integral of a function g defined on R by simply letting

$$\int_{-\infty}^{\infty} g(x)\,dx = \lim_{N\to\infty} \int_{-N}^{N} g(x)\,dx\,,$$

whenever the limit exists. In this case the function $g(x) = \sin x/x$ is Riemann integrable over R, but not integrable there since, as is readily seen, $\int_R |g(x)|\,dx = \infty$.

With the aid of the Lebesgue measure we are also able to identify those functions which are Riemann integrable on finite intervals.

Theorem 3.3. Suppose g is a real-valued bounded function defined on $I = [a,b]$. Then $g \in \mathcal{R}(I)$ iff g is continuous a.e. on I.

Proof. First assume $g \in \mathcal{R}(I)$, and fix a sequence of partitions $\{\mathcal{P}_n\}$ of I as in Theorem 3.1. If L and U are defined by (3.2), let

$$N = \{\, x \in I : L(x) < g(x) \text{ or } g(x) < U(x)\,\}\,.$$

In Theorem 3.1 we actually proved that $|N| = 0$. Now let N' be the set consisting of all those points in I which belong to some $\mathcal{P}_n$; N' is countable and hence null. We claim that g is continuous off the null set $N \cup N'$; we prove this by contradiction. Given an interval $J \subseteq I$, consider the oscillation $\operatorname{osc}(g, J)$ of g over J given by

$$\operatorname{osc}(g, J) = \sup_J g - \inf_J g\,.$$

Now, if g is not continuous at $x \notin N \cup N'$, there exists $\varepsilon > 0$, such that the oscillation of g on any interval containing x is at least ε. Since $x \notin N'$, x is an interior point to one of the intervals of each $\mathcal{P}_n$, and consequently,

$$U_n(x) - L_n(x) \geq \varepsilon \quad \text{for all } n\,.$$

Fromfoll (3.2) it follows that if this is the case, then $U(x) - L(x) \geq \varepsilon$, and $x \in N$, which is the desired contradiction.

Conversely, suppose $\mathcal{P}_n$ is an arbitrary sequence of partitions of I such that $\lim_{n\to\infty}(\text{norm } \mathcal{P}_n) = 0$ and observe that by assumption,

$$\lim_{n\to\infty} L_n(x) = \lim_{n\to\infty} U_n(x) = g(x) \text{ a.e.}$$

By (3.4), (3.5) and LDCT we get that

$$\lim_{n\to\infty} s(g, \mathcal{P}_n) = \lim_{n\to\infty} S(g, \mathcal{P}_n) = \int_I g\, dx\,.$$

Whence, by Proposition 2.1 in Chapter III, $g \in \mathcal{R}(I)$. ■

4. PROBLEMS AND QUESTIONS

The first four problems describe how the concept of integral was viewed by different mathematicians. For simplicity we assume that μ is the Lebesgue measure defined on $(X, \mathcal{L})$, where X is a compact interval of R^n, and that f is a nonnegative measurable function defined on X.

4.1 Lebesgue defined the integral of a bounded measurable function f as follows: If $0 \le f \le M$ a.e. on X, put

$$\int_X f\, dx = \lim_{m\to\infty} \sum_{k=1}^{mM} \frac{k}{m} |\{k/m \le f < (k+1)/m\}|\,.$$

Show that this definition coincides with the one given in the text.

4.2 de la Vallée-Poussin (1866–1962) extended the definition of Lebesgue to include unbounded functions f as follows: Let $f_m = \wedge(f, m)$ be the truncation of f at level m. $\{f_m\}$ is a sequence of bounded measurable functions, cf. 4.5 in Chapter VI, and we put

$$\int_X f\, dx = \lim_{m\to\infty} \int_X f_m\, dx\,.$$

Show that this definition is equivalent to the one given in the text.

4.3 Saks (1897–1942) defines the integral in a manner reminiscent of the Riemann-Stieltjes integral. More precisely, let $\mathcal{P} = \{E_1, \ldots, E_k\}$ be a measurable partition of X, let $m_k = \inf_{E_k} f$, and put

$$\int_X f\, dx = \sup_{\mathcal{P}} \sum_k m_k |E_k|\,.$$

Show that this definition is also equivalent to the one given in the text.

4.4 Finally, there is the notion of integral as the "area under the graph." Carathéodory defines the integral of a bounded measurable function f as follows: Let $A(f) = \{(x, y) \in X \times R : 0 \leq y \leq f(x)\}$; show that $A(f)$ is a Lebesgue measurable subset of R^{n+1} and put

$$\int_X f\,dx = |A(f)|\,. \tag{4.1}$$

Show that (4.1) is equivalent to the definition given in the text. It is also interesting to interpret results such as MCT in terms of (4.1), as their meaning is quite apparent.

4.5 True or false: If f is a nonnegative function defined on R and $\int_R f\,dx < \infty$, then $\lim_{|x|\to\infty} f(x) = 0$.

4.6 Suppose f is integrable on R^n and for a fixed $h \in R^n$ let $g(x) = f(x+h)$ be a translate of f. Show that g is also integrable and that

$$\int_{R^n} g\,dx = \int_{R^n} f\,dx\,. \tag{4.2}$$

(4.2) is a restatement of the translation-invariance of the Lebesgue measure.

4.7 Let $(X, \mathcal{M}, \mu)$ be a measure space and $\{f_n\}$ a sequence of measurable functions such that $\sum_n \int_X |f_n|\,d\mu < \infty$. Show that $\sum_n f_n$ converges absolutely μ-a.e. and $\int_X (\sum_n f_n)\,d\mu = \sum_n \int_X f_n\,d\mu$. In particular, also $\lim_{n\to\infty} f_n = 0$ μ-a.e.

4.8 Let $r_1, r_2, \ldots, r_n, \ldots$ be an enumeration of the rational numbers in $I = [0,1]$, and let

$$f(x) = \sum_{\{n : x > r_n\}} 2^{-n}\,.$$

Compute $\int_I f(x)\,dx$.

4.9 Prove that the sum $\sum_{n=0}^{\infty} \int_{[0,\pi]} (1 - \sqrt{\sin x})^n \cos x\,dx$ converges to a finite limit, and find its value.

4.10 Let f be a nonnegative measurable function defined on R. Prove that if $\sum_{n=-\infty}^{\infty} f(x+n)$ is integrable, then $f = 0$ a.e. On the other hand, if f is integrable, then $\phi(x) = \sum_{n=-\infty}^{\infty} f(2^n x + 1/n)$ is finite a.e. and integrable, and $\int_R \phi(x)\,dx = \int_R f(x)\,dx$.

4.11 Let $(X, \mathcal{M}, \mu)$ be a measure space and $f \in L(\mu)$. Show that the set $\{f \neq 0\}$ is σ-finite, i.e., the at most countable union of sets of finite measure.

4.12 Referring to Proposition 1.7, decide when the measure ν introduced there is: (a) finite, and (b) σ-finite.

4.13 Referring to Proposition 1.7 again, suppose that g is a nonnegative measurable function defined on X. Show that $\int_X g\, d\nu = \int_X gf\, d\mu$.

4.14 Suppose that the assumptions of 4.27 in Chapter IV hold and let f be a nonnegative real-valued measurable function defined on Y. Prove that the following "change of variable" formula holds:

$$\int_Y f\, d\nu = \int_X f \circ \tau\, d\mu\,.$$

4.15 Show that Fatou's Lemma is also true for functions that depend on a continuous parameter. More precisely, under the relevant assumptions, the following is true: $\int_X \liminf_{i \in I} f_i\, d\mu \leq \liminf_{i \in I} \int_X f_i\, d\mu$.

4.16 Prove the following variant of Fatou's Lemma: If $\{f_n\}$ is a sequence of nonnegative measurable functions which converges to f μ-a.e. and $\int_X f_n\, d\mu \leq M < \infty$ for all n, then f is integrable and $\int_X f\, d\mu \leq M$.

4.17 Decide whether the following Fatou-like statements are true: (a) If $\{f_n\}$ is a sequence of nonnegative measurable functions and f_n converges to f in probability, then $\int_X f\, d\mu \leq \liminf \int_X f_n\, d\mu$, and, (b) Same result with convergence in probability replaced by convergence in measure.

4.18 Show that the following extension of Fatou's Lemma is true: Rather than assuming $g \in L(\mu)$, we may assume that $\int_X g^-\, d\mu < \infty$ in Theorem 2.6, and that $\int_X g^+\, d\mu < \infty$ in Corollary 2.7.

4.19 Describe the relation of 4.26 in Chapter IV to Fatou's Lemma.

4.20 Let $(X, \mathcal{M}, \mu)$ be a measure space and $\{\phi_n\}$, $\{\psi_n\}$, and $\{f_n\}$ be sequences of measurable functions defined on X that converge to functions ϕ, ψ, and f, respectively. Further, if $\phi_n(x) \leq f(x) \leq \psi_n(x)$ for all $x \in X$, and if $\lim_{n\to\infty} \int_X \phi_n\, d\mu = \int_X \phi\, d\mu < \infty$ and $\lim_{n\to\infty} \int_X \psi_n\, d\mu = \int_X \psi\, d\mu < \infty$, prove that the f_n's may be integrated to the limit, i.e., $\lim_{n\to\infty} \int_X f_n\, d\mu = \int_X f\, d\mu$.

4.21 Let $(X, \mathcal{M}, \mu)$ be a measure space and $\{f_n\}$ be a sequence of nonincreasing nonnegative measurable functions which converges to f.

Show that $\lim_{n\to\infty} \int_X f_n \, d\mu = \int_X f \, d\mu$ provided that $f_1 \in L(\mu)$, and that the conclusion may fail if $f_1 \notin L(\mu)$.

4.22 Let $(X, \mathcal{M}, \mu)$ be a finite measure space and f a μ-a.e. strictly positive measurable function. If $\{E_n\} \subseteq \mathcal{M}$ is such that $\lim_{n\to\infty} \int_{E_n} f \, d\mu = 0$, prove that $\lim_{n\to\infty} \mu(E_n) = 0$.

4.23 Let $(X, \mathcal{M}, \mu)$ be a finite measure space and $\{f_n\}$ a sequence of measurable functions that converges to a function f uniformly on X. Show that f is also integrable and that $\lim_{n\to\infty} \int_X f_n \, d\mu = \int_X f \, d\mu$. Is a similar result true if $\mu(X) = \infty$?

4.24 Let $(X, \mathcal{M}, \mu)$ be a measure space and $\{f_n\}$ a sequence of measurable functions that converges to f μ-a.e. If $f \in L(\mu)$, show that

$$\lim_{n\to\infty} \int_X |f_n| \, d\mu = \int_X |f| \, d\mu \quad \text{implies} \quad \lim_{n\to\infty} \int_X |f_n - f| \, d\mu = 0 \,,$$

and that the conclusion may fail if f is not integrable.

4.25 Let $(X, \mathcal{M}, \mu)$ be a measure space and assume $\{f_n\}$ is a sequence of nonnegative measurable functions that converges to f μ-a.e. If $\lim_{n\to\infty} \int_X f_n \, d\mu = \int_X f \, d\mu < \infty$, is it true that

$$\lim_{n\to\infty} \int_E f_n \, d\mu = \int_E f \, d\mu \,, \quad \text{for every } E \in \mathcal{M}?$$

4.26 Suppose that $I = [0,1]$, and let $(I, \mathcal{M}, \mu)$ be a measure space, and f a real-valued measurable function defined on I. If A is the subset of those x's in I where $f(x)$ assumes an integer value, prove that A is measurable and that $\lim_{n\to\infty} \int_I (\cos(\pi f(x)))^{2n} \, d\mu = \mu(A)$.

4.27 Let $([0,1], \mathcal{M}, \mu)$ be a measure space and suppose $f \in L(\mu)$. Show that $x^n f(x) \in L(\mu), n = 1, 2, \ldots$, and compute

$$\lim_{n\to\infty} \int_{[0,1]} x^n f(x) \, d\mu(x) \,.$$

4.28 Let $(X, \mathcal{M}, \mu)$ be a finite measure space, and f a nonnegative real-valued function defined on X. Prove that a necessary and sufficient condition that $\lim_{n\to\infty} \int_X f^n \, d\mu$ should exist as a finite number is that $\mu(\{f > 1\}) = 0$.

4.29 Let $I = [0,1]$ and suppose $\{f_n\}$ is a sequence of real-valued measurable functions defined on I such that $\lim_{n\to\infty} f_n = f$ a.e. Prove that $\lim_{n\to\infty} \int_I f_n(x) e^{-f_n(x)} \, dx = \int_I f(x) e^{-f(x)} \, dx$, and $\int_I \sum_n f_n(x)^2 \, dx$

$= \sum_n \int_I f_n(x)^2\, dx$. Furthermore, if the f_n's and f vanish only on a null set, show that

$$\lim_{n\to\infty} \int_I \frac{\sin(f_n(x))}{f_n(x)}\, dx = \int_I \frac{\sin(f(x))}{f(x)}\, dx\,.$$

4.30 Evaluate

$$\lim_{n\to\infty} \int_{[0,1]} \left(1 - e^{-x^2/n}\right) x^{-1/2}\, dx\,.$$

4.31 Evaluate

$$\lim_{n\to\infty} \int_{[0,n]} (1 - x/n)^n\, e^{x/2}\, dx\,.$$

4.32 Using the definition of derivative show that

$$f(t) = \int_{[0,1]} e^{tx} x^{-1/3}\, dt$$

is differentiable at every $t \in R$.

4.33 Let $I = [0,1]$ and suppose $f \in L(I)$. Show that the function $g(t) = \int_I \cos(tf(x))\, dx$ is a well-defined differentiable function of $t \in R$. What assumption on f will insure that g has two derivatives? three derivatives?

4.34 If

$$s_n(x) = \sum_{k=0}^{n} \frac{1 \cdot 3 \cdots (2k-1)}{k!\, 2^k}\, x^k\,, \quad -1 < x < 1\,,$$

is the sequence of the partial sums of the Maclaurin series of the function $f(x) = (1-x)^{-1/2}$, show that $\lim_{n\to\infty} \int_{(-1,1)} |s_n - f|\, dx = 0$.

4.35 Suppose $(X, \mathcal{M}, \mu)$ is a measure space and $\{f_n\}$ is a sequence of nonnegative integrable functions such that $\lim_{n\to\infty} \int_X f_n\, d\mu = 0$. If $g \in L(\mu)$ has the property that $gf_n \in L(\mu)$ for all n, does it follow that $\lim_{n\to\infty} \int_X gf_n\, d\mu = 0$?

4.36 Let $(X, \mathcal{M}, \mu)$ be a measure space, and f, g, f_n, g_n be integrable functions, $n = 1, 2, \ldots$ If $\lim_{n\to\infty} f_n = f$ μ-a.e., $|f_n| \le g_n$ for all n, and $\lim_{n\to\infty} \int_X g_n\, d\mu = \int_X g\, d\mu$, is it also true that $\lim_{n\to\infty} \int_X f_n\, d\mu = \int_X f\, d\mu$?

4.37 Assume $(X, \mathcal{M}, \mu)$ is a finite measure space, and for measurable functions f, g put

$$d'(f,g) = \int_X \frac{|f-g|}{1 + |f-g|}\, d\mu\,.$$

Show that f_n converges to f in probability iff $d'(f_n, f) \to 0$ as $n \to \infty$.
Is a similar result true if instead of convergence in probability we consider convergence in measure?

The concept of L^1 allows us to sharpen several of the results we discussed thus far; the next three results are an indication of what we have in mind.

4.38 Let $(X, \mathcal{M}, \mu)$ be a measure space and $f, f_n, n = 1, 2, \ldots$, measurable extended real-valued functions defined on X. Suppose that the f_n's converge to f μ-a.e. and that there exists $g \in L(\mu)$ so that $|f_n| \leq g$ μ-a.e. Prove that f_n converges to f in μ-measure.

4.39 Let $I = [0,1]$ and suppose $f \in L(I)$. Show that given $\varepsilon > 0$, there is a continuous function g defined on I with the property that

$$|\{x \in I : f(x) \neq g(x)\}| \leq \varepsilon \quad \text{and} \quad \int_I |f(x) - g(x)|\, dx \leq \varepsilon .$$

4.40 Show that the conclusion of Egorov's theorem is true under the following assumption: The fact that X is a compact interval may be replaced by $|f_n| \leq g$ a.e., $g \in L(R)$.

4.41 Show that the conclusion of LDCT still holds if we replace the assumption $\lim_{n\to\infty} f_n = f$ μ-a.e. there, by f_n converges to f in μ-measure.

4.42 For any space X describe $L(\delta_x), x \in X$, and explain what LDCT states in this case.

4.43 Let $(X, \mathcal{M}, \mu)$ be a finite measure space and f a real-valued measurable function defined on X such that

$$\int_X f^n \, d\mu = c \quad \text{for } n = 2, 3, 4 .$$

Show that $f = \chi_A$ μ-a.e. for some measurable set $A \subseteq X$.

4.44 Let μ be the counting measure on $(Z, \mathcal{P}(Z))$, $Z =$ integers. Characterize the real-valued measurable functions f defined on Z, decide when two such functions coincide μ-a.e., discuss under what conditions $\int_Z f\, d\mu$ exists and explain what MCT and LDCT state in this case.

4.45 Suppose $\{A_n\}$ is a sequence of Lebesgue measurable subsets of I with the property that $|A_n| \geq \eta > 0$ for all n and (a_n) is a sequence of real numbers such that $\sum_{n=1}^{\infty} |a_n| \chi_{A_n}(x) < \infty$ a.e. on I. Prove that $\sum_{n=1}^{\infty} |a_n| < \infty$.

4.46 Let I be a compact interval in R^n which contains the origin. Show that

$$\int_I |x|^{-\eta}\,dx < \infty \quad \text{iff} \quad \eta < n\,,$$

and that

$$\int_{R^n\setminus I} |x|^{-\eta}\,dx < \infty \quad \text{iff} \quad \eta > n\,.$$

4.47 Discuss for what values of η, ε,

$$\int_{0+}^{1-} x^{\eta}(\ln 1/x)^{\varepsilon}\,dx \quad \text{exists}\,.$$

4.48 Prove the Vitali-Carathéodory theorem: If $f \in L(R)$ and $\varepsilon > 0$, then there exist functions ϕ and ψ defined everywhere on R which satisfy:

(i) ϕ is upper continuous and bounded above, and ψ is lower semicontinuous and bounded below.

(ii) $\phi, \psi \in L(R)$.

(iii) At every x where f is defined we have

$$\phi(x) \le f(x) \le \psi(x)\,.$$

(iv) $\int_R(\phi(x) - \psi(x))\,dx \le \varepsilon$.

Do properties (i)–(iv) characterize integrable functions?

4.49 Let $(X, \mathcal{M}, \mu)$ be a measure space and f a complex-valued measurable function defined on X, cf. 4.43 in Chapter VI. Show that also the modulus $|f|$ is measurable and introduce $L(\mu) = \{f : f \text{ is measurable and } \int_X |f|\,d\mu < \infty\}$. This is another open ended question: Discuss the properties of $L(\mu)$.

CHAPTER **VIII**

More About L^1

In this chapter we discuss the metric properties of L^1, including completeness, and some local properties of integrable functions, such as the Lebesgue Differentiation Theorem.

1. METRIC STRUCTURE OF L^1

Let $(X, \mathcal{M}, \mu)$ be a measure space and $f, g \in L(\mu)$. We measure the distance from f to g by the expression

$$d(f,g) = \int_X |f - g| \, d\mu \,. \tag{1.1}$$

Is $d(f,g)$ a metric on $L(\mu)$? Clearly $d(f,g) \geq 0$ and by 4.33 in Chapter VII, $d(f,g) = 0$ iff $f = g$ μ-a.e. Since we identify those functions that coincide μ-a.e., it is true that $d(f,g) = 0$ iff $f = g$. Also $d(f,g) = d(g,f)$. Finally, since for $f, g, h \in L(\mu)$ we have

$$|f - g| \leq |f - h| + |h - g| \quad \mu\text{-a.e.},$$

it readily follows that $d(f,g) \leq d(f,h) + d(h,g)$, and d indeed is a distance function.

The interesting question to consider is whether endowed with this metric $L(\mu)$ is a complete metric space. The answer is affirmative.

Theorem 1.1. Let $(X, \mathcal{M}, \mu)$ be a measure space. The distance function introduced in (1.1) above turns $L(\mu)$ into a complete metric space.

Proof. Assuming $\{f_n\}$ is a Cauchy sequence of integrable functions, we must show that there is a function $f \in L(\mu)$ so that $\lim_{n\to\infty} d(f_n, f) = 0$. First observe that since $\{f_n\}$ is Cauchy we can find an increasing sequence $n_{k+1} > n_k$, such that

$$d(f_n, f_{n_k}) \le 1/2^k\,, \quad \text{all } n \ge n_k\,, \quad k = 1,2,\ldots$$

Whence, by Theorem 1.6 in Chapter VII we get

$$\int_X \sum_{k=1}^{\infty} |f_{n_{k+1}} - f_{n_k}|\, d\mu = \sum_{k=1}^{\infty} d(f_{n_{k+1}}, f_{n_k}) \le \sum_{k=1}^{\infty} 1/2^k = 1\,. \tag{1.2}$$

Now, by Corollary 2.3 in Chapter VII, the integrand on the left-hand side of (1.2) is finite μ-a.e., and consequently, the series with terms $f_{n_{k+1}} - f_{n_k}$, $k = 1,2,\ldots$, converges absolutely to a finite sum μ-a.e. In particular, we have that the limit

$$\lim_{m\to\infty} \sum_{k=1}^{m} (f_{n_{k+1}} - f_{n_k}) = g \tag{1.3}$$

exists, and is measurable and finite μ-a.e. Moreover, since the sum on the left-hand side of (1.3) telescopes to $f_{n_{m+1}} - f_{n_1}$, it readily follows that

$$\lim_{k\to\infty} f_{n_k} = g + f_{n_1} = f\,, \tag{1.4}$$

say, is measurable and finite μ-a.e.; we want to show that the convergence is also in the metric of $L(\mu)$.

Let $\phi = \sum_{k=1}^{\infty} |f_{n_{k+1}} - f_{n_k}|$; by (1.2), $\phi \in L(\mu)$. Also, since $f_{n_k} = f_{n_k} - f_{n_1} + f_{n_1}$, we get that

$$|f_{n_k}| \le \sum_{m=1}^{k-1} |f_{n_{m+1}} - f_{n_m}| + |f_{n_1}| \le \phi + |f_{n_1}| \in L(\mu)\,, \quad \text{all } k\,.$$

By (1.4) a similar estimate holds with $|f|$ on the left-hand side above, and consequently, by LDCT

$$\lim_{k\to\infty} \int_X |\, f_{n_k} - f\,|\, d\mu = \lim_{k\to\infty} d(f_{n_k}, f) = 0\,.$$

To complete the proof we invoke the well-known fact that if a Cauchy sequence in a metric space has a convergent subsequence, then the sequence itself converges to the same limit. ∎

Corollary 1.2. Let $(X, \mathcal{M}, \mu)$ be a measure space, $f, f_n \in L(\mu)$, $n = 1, 2, \ldots$, and suppose that $\lim_{n\to\infty} \int_X |f_n - f|\, d\mu = 0$. Then there is a subsequence $\{f_{n_k}\}$ such that $\lim_{k\to\infty} f_{n_k} = f$ μ-a.e.

Proof. Since the sequence $\{f_n\}$ converges it is Cauchy, and, as in the proof of Theorem 1.1, we can find a subsequence $\{f_{n_k}\}$ which converges pointwise μ-a.e., and in the $L(\mu)$ metric, to an integrable function g, say. But then it is clear that $d(f, g) = 0$ and $f = g$. Thus

$$\lim_{k\to\infty} f_{n_k} = f \ \mu\text{-a.e.} \quad \blacksquare$$

The converse to Corollary 1.2 is false, namely, there is a sequence $\{f_n\}$ of integrable functions and an $f \in L(\mu)$ such that $\lim_{n\to\infty} f_n = f$ μ-a.e., and yet $\lim_{n\to\infty} d(f_n, f) \neq 0$, cf. 4.24 in Chapter VII.

Suppose next that X is a topological space. It is natural to consider whether integrable functions can be approximated by continuous functions in the metric of $L(\mu)$. We only consider the case $X = R^n, \mathcal{M} = \mathcal{L}$ here, but the proof can be readily extended to more general settings.

Theorem 1.3. $C_0(R^n)$, the space of continuous functions which vanish off a compact set, is dense in $L(R^n)$. More precisely, for any $f \in L(R^n)$, given $\varepsilon > 0$, there is a continuous function g which vanishes off a bounded set, and such that

$$d(f, g) = \int_{R^n} |f - g|\, dx < \varepsilon\,.$$

Proof. Since $f = f^+ - f^-$ and $f^+, f^- \in L(R^n)$, we may think positive and assume that f is nonnegative. In this case, by Theorem 1.2 in Chapter VI, there is a sequence of simple functions $0 \le \phi_k \le f$, such that $\lim_{k\to\infty} \phi_k = f$ a.e. Whence, by MCT, $\lim_{k\to\infty} \int_{R^n} \phi_k\, dx = \int_{R^n} f\, dx$, and since all the quantities involved are finite we also have

$$0 \le \lim_{k\to\infty} \int_{R^n} (f - \phi_k)\, dx = 0\,.$$

Let now $\varepsilon > 0$ be given, and choose one of the ϕ_k's, call it ϕ, so that $0 \le \phi \le f$, and

$$\int_{R^n} (f - \phi)\, dx < \varepsilon/2\,.$$

We have thus reduced the problem at hand to one of approximating simple integrable functions by continuous functions in the metric of $L(R^n)$.

Suppose $\phi = \sum_{k=1}^{m} c_k \chi_{E_k}$, where for each k, $c_k > 0$ and E_k is a measurable set of finite measure, $1 \le k \le m$. It suffices now to approximate each summand that appears in the definition of ϕ, or equivalently, the characteristic function of a measurable set E, say, of finite measure. By the regularity of the Lebesgue measure, for any $\eta > 0$, there is an open set $\mathcal{O}$ of finite measure such that

$$|\mathcal{O} \setminus E| < \eta, \quad \mathcal{O} \supseteq E.$$

This estimate, in particular, implies

$$0 \le \int_{R^n} (\chi_{\mathcal{O}} - \chi_E)\, dx < \eta,$$

and since η above is arbitrary, we may assume that the set E in question is actually open. It is at this point that the geometry of the situation plays a role. Let $\mathcal{O} = \bigcup_{k=1}^{\infty} I_k$ be an open set of finite measure; here the I_k's are nonoverlapping closed intervals. Since $|\mathcal{O}| = \sum_{k=1}^{\infty} |I_k| < \infty$, it readily follows that

$$\int_{R^n} (\chi_{\mathcal{O}} - \chi_{\bigcup_{k=1}^{m} I_k})\, dx = |\mathcal{O} \setminus \textstyle\bigcup_{k=1}^{m} I_k|$$
$$= \sum_{k=m+1}^{\infty} |I_k| \to 0, \quad \text{as } m \to \infty,$$

and consequently it suffices to approximate the characteristic function of $\bigcup_{k=1}^{m} I_k$, all finite m. But then it is enough to consider χ_I, where I is a closed interval of R^n.

Suppose first that $n = 1$ and $I = [0,1]$, and given $\eta > 0$, let ψ_η be the continuous function

$$\psi_\eta(x) = \begin{cases} 0 & \text{if } x < -\eta \\ 1 + x/\eta & \text{if } -\eta \le x < 0 \\ 1 & \text{if } 0 \le x \le 1 \\ (\eta + 1 - x)/\eta & \text{if } 1 < x \le 1 + \eta \\ 0 & \text{if } x > 1 + \eta. \end{cases}$$

Then clearly $\psi_\eta \ge \chi_I$, and

$$\int_R (\psi_\eta - \chi_I)\, dx = \eta \to 0 \quad \text{with } \eta.$$

Now, if $I = [a,b]$, the function $\psi_\eta((x-a)/(b-a))$ does the job.

As for the general case, observe that if $I = [a_1, b_1] \times \cdots \times [a_n, b_n]$, then χ_I can be approximated in the metric of $L(R^n)$ as close as we want by the continuous function

$$\psi_\eta\left(\frac{x_1 - a_1}{b_1 - a_1}\right) \times \cdots \times \psi_\eta\left(\frac{x_n - a_n}{b_n - a_n}\right), \quad \eta \text{ small}. \quad \blacksquare$$

Theorem 1.3 indicates that simple functions are dense in $L(\mu)$. As for the Lebesgue integral, one of its important applications concerns the continuity of the translates of integrable functions.

Proposition 1.4. Integrable functions are continuous in the metric of $L(R^n)$. More precisely, for any $f \in L(R^n)$ we have

$$\lim_{|h| \to 0} \int_{R^n} |f(x+h) - f(x)|\, dx = 0.$$

Proof. We show that given $\varepsilon > 0$, there exists $\delta > 0$, such that

$$\int_{R^n} |f(x+h) - f(x)|\, dx \le \varepsilon, \quad \text{whenever } |h| \le \delta. \tag{1.5}$$

This is not hard. First let $g \in C_0(R^n)$ be such that $d(f, g) \le \varepsilon/3$, and observe that $|f(x+h) - f(x)|$ may be estimated by

$$|f(x+h) - g(x+h)| + |g(x+h) - g(x)| + |g(x) - f(x)|. \tag{1.6}$$

Whence, integrating (1.6) over R^n, we get that the integral on the left-hand side of (1.5) does not exceed

$$\int_{R^n} |f(x+h) - g(x+h)|\, dx + \int_{R^n} |g(x+h) - g(x)|\, dx + \int_{R^n} |f(x) - g(x)|\, dx$$
$$= A + B + C,$$

say. By 4.6 in Chapter VII we have $A = C$ for all h, and, by our choice of g, $A, C \le \varepsilon/3$.

As for B, note that g is actually a uniformly continuous function that vanishes off a bounded interval of R^n. Thus, given $\eta > 0$, there exists $\delta > 0$ such that

$$|g(x+h) - g(x)| \le \eta \quad \text{for all } |h| \le \delta, x \in R^n.$$

Moreover, since for any fixed h, $|h| \le 1$, also $g(x+h) - g(x)$ vanishes off a bounded interval I of R^n, it is clear that

$$B \le |I|\, \eta, \quad \text{whenever } |h| \le \min(1, \delta).$$

Thus, by choosing η small enough, we also have $B \le \varepsilon/3$ whenever $|h|$ is sufficiently small, and $A + B + C < \varepsilon$. $\blacksquare$

2. THE LEBESGUE DIFFERENTIATION THEOREM

Given $x = (x_1, \ldots, x_n) \in R^n$ and $r > 0$, let $I(x,r) = \{y : |x_i - y_i| < r, i = 1, 2, \ldots, n\}$ denote the open interval of sidelength $2r$ centered at x. The question we address in this section is: If f is an integrable function, for what x's does

$$\lim_{r\to 0} \frac{1}{|I(x,r)|} \int_{I(x,r)} f\,dy = f(x)\ ? \tag{2.1}$$

At those points x where (2.1) holds we say that the (indefinite) integral of f differentiates to $f(x)$. In case $n = 1$, the question is whether

$$\lim_{r\to 0} \frac{1}{2r} \int_{(x-r,x+r)} f\,dy = f(x)\,,$$

which is equivalent to

$$\lim_{h\to 0, h\neq 0} \frac{1}{h} \int_{[x,x+h)} f\,dy = f(x)\,. \tag{2.2}$$

If we set $F(x) = \int_{[0,x)} f\,dy$, (2.2) reads precisely $F'(x) = f(x)$, and this justifies the terminology.

In fact, there are two questions implicit in (2.1): When does the limit exist, and, if it exists, when does it equal $f(x)$. For instance, when $I = [0,1]$ and $f = \chi_I$, we have

$$\frac{1}{2r} \int_{(-r,r)} f\,dy = 1/2\,, \quad \text{all } r > 0\,. \tag{2.3}$$

Thus the limit of the left-hand side of (2.3) exists and it equals $1/2 \neq f(0) = 1$.

Some observations are in order. First, the question we posed is "local" in nature, i.e., since we take limits as $r \to 0$ only the values of f near x are relevant. Thus, we may assume that $x \in I(0,1)$ and that f vanishes off $I(0,2)$.

Next, since the example given in (2.3) is not very reassuring, we consider an instance where (2.1) is true. Suppose, then, that f is continuous at x and note that

$$\frac{1}{|I(x,r)|} \int_{I(x,r)} f\,dy - f(x) = \frac{1}{|I(x,r)|} \int_{I(x,r)} (f - f(x))\,dy\,. \tag{2.4}$$

Given $\varepsilon > 0$, let r be so small that $|f(x) - f(y)| < \varepsilon$ for $y \in I(x,r)$; clearly we may assume $r < 1$. By (2.4) it follows that

$$\left| \frac{1}{|I(x,r)|} \int_{I(x,r)} f\,dy - f(x) \right| \leq \frac{1}{|I(x,r)|} \int_{I(x,r)} |f - f(x)|\,dy \leq \varepsilon\,,$$

and (2.1) is true in this case.

Since continuous functions are dense in the metric of $L(R^n)$, we expect that the good behaviour of continuous functions will somehow translate into an a.e. good behaviour of integrable functions.

The idea of Hardy (1877–1947) and Littlewood (1885–1977) is to seek the control of all the averages of f. They devised this procedure to study the convergence of Fourier series and were inspired by the averages in the game of cricket.

To control the averages of f we introduce the so-called Hardy-Littlewood maximal function. Specifically, suppose f is an integrable function which vanishes off $I(0,2)$, and for $x \in R^n$ put

$$M(f) = \sup_{r>0} \frac{1}{|I(x,r)|} \int_{I(x,r)} |f|\,dy\,. \tag{2.5}$$

What can we say about Mf? We claim that Mf is a nonnegative lower-semicontinuous, and hence measurable, function, which tends to 0 as $|x| \to \infty$ at the rate of $|x|^{-n}$.

To show that Mf is lower-semicontinuous we must verify that for each $\lambda > 0$, the set $\{Mf > \lambda\}$ is open; this is not hard. Working with complements we show that for each $\lambda > 0$, $\{Mf \leq \lambda\}$ is closed. Fix $\lambda > 0$, then, and suppose $\{x_k\}$ is a sequence of points in $\{Mf \leq \lambda\}$ such that $x_k \to x$; we show that $x \in \{Mf \leq \lambda\}$ as well. In other words, we check that all the averages of $|f|$ about x are less than or equal to λ. First observe that since $x_k \to x$,

$$\lim_{k\to\infty} I(x_k, r) \,\Delta\, I(x,r) = \emptyset, \quad \text{all } r.$$

Therefore, if χ_k denotes the characteristic function of $I(x_k,r) \,\Delta\, I(x,r)$ and $f_k = f\chi_k$, it follows that

$$|f_k(y)| \leq |f(y)| \quad \text{and} \quad \lim_{k\to\infty} f_k = 0 \text{ a.e.}$$

Thus, by LDCT

$$\lim_{k\to\infty} \int_{R^n} |f_k|\,dy = 0\,. \tag{2.6}$$

Next consider the average $\frac{1}{|I(x,r)|}\int_{I(x,r)}|f|\,dy$. Since

$$I(x,r)\subseteq (I(x,r)\,\Delta\, I(x_k,r))\bigcup I(x_k,r) \quad \text{and} \quad |I(x,r)|=|I(x_k,r)|\,,$$

the average in question does not exceed

$$\frac{1}{|I(x,r)|}\int_{I(x,r)\Delta I(x_k,r)}|f|\,dy+\frac{1}{|I(x_k,r)|}\int_{I(x_k,r)}|f|\,dy=A+B\,,$$

say. By (2.6), $A\to 0$ as $k\to\infty$. As for B, since $Mf(x_k)\le\lambda$ for all k, we also have $B\le\lambda$. Thus all averages of $|f|$ about x are less than or equal to λ, and $Mf(x)\le\lambda$.

Next we show that $Mf(x)\sim|x|^{-n}$ for $|x|$ large. To see this take $x\in R^n$ with $|x|$ large, $x\in R^n\setminus I(0,10)$ will do, and observe that unless $r>c|x|$, where c is a dimensional constant independent of x, the average of $|f|$ about x vanishes. Thus, with a dimensional constant c which may differ at different occurrences even in the same chain of inequalities, we have

$$\frac{1}{|I(x,r)|}\int_{I(x,r)}|f|\,dy\le\frac{c}{r^n}\int_{I(0,1)}|f|\,dy\le\frac{c}{|x|^n}\int_{I(0,1)}|f|\,dy\,,$$

and consequently,

$$Mf(x)\le\frac{c}{|x|^n}\int_{I(0,1)}|f|\,dy\,. \tag{2.7}$$

Since there is a dimensional constant c such that for $|x|$ large we have $I(x,c|x|)\supseteq I(0,1)$, it readily follows that

$$Mf(x)\ge\frac{1}{|I(x,c|x|)|}\int_{I(x,c|x|)}|f|\,dy\ge\frac{c}{|x|^n}\int_{I(0,1)}|f|\,dy\,,$$

the inequality opposite to (2.7) holds, and, as asserted, $Mf(x)\sim|x|^{-n}$ for $|x|$ large.

It is then apparent that Mf is not integrable, cf. 4.46 in Chapter VII, but just barely. As for the function $|x|^{-n}$, it satisfies a weak integrability condition reminiscent of Chebychev's inequality. More precisely, there is a constant c such that

$$\lambda|\{|x|^{-n}>\lambda\}|\le c\,,\quad \text{all } \lambda>0\,. \tag{2.8}$$

Indeed, if $|x|^{-n}>\lambda$, then there is a dimensional constant c so that $x\in I\left(0,c\lambda^{-1/n}\right)$, and

$$|\{|x|^{-n}>\lambda\}|\le|I(0,c\lambda^{-1/n})|=c\lambda^{-1}\,.$$

The class of those measurable functions f which satisfy the estimate

$$\lambda|\{|f| > \lambda\}| \leq c, \quad \text{all } \lambda > 0, \tag{2.9}$$

was studied by Marcinkiewicz (1910–1940). It is called the weak-L^1 class of Marcinkiewicz and it is denoted by wk-$L(R^n)$.

By Chebychev's inequality, $L(R^n) \subseteq$ wk-$L(R^n)$, and for integrable functions f, (2.9) is true with $c = \int_{R^n} |f(y)|\, dy$. On the other hand, $|x|^{-n} \in$ wk-$L(R^n) \setminus L(R^n)$.

The remarkable fact that Hardy and Littlewood proved is that although for f integrable Mf is not necessarily integrable, it belongs to wk-$L(R^n)$; in a sense this gives the next best result.

Theorem 2.1 (Hardy-Littlewood). Suppose f is an integrable function which vanishes off $I(0,2)$. Then $Mf \in$ wk-$L(R^n)$, and for any $\lambda > 0$ we have

$$\lambda|\{Mf > \lambda\}| \leq 3^n \int_{R^n} |f|\, dy. \tag{2.10}$$

Proof. Given $\lambda > 0$, let $\mathcal{O}_\lambda = \{Mf > \lambda\}$; we want to show that the open set $\mathcal{O}_\lambda$ has finite measure and that (2.10) holds. Since by (2.7) $Mf(x) \to 0$ as $|x| \to \infty$, $\mathcal{O}_\lambda$ is a bounded set of finite measure; to show that (2.10) holds requires some work. The following line of reasoning is a prototype of the so-called "covering arguments" and it is due to Wiener (1894-1964).

If $\mathcal{O}_\lambda = \emptyset$ there is nothing to prove. Otherwise, let $x \in \mathcal{O}_\lambda$, and observe that by the definition of $Mf(x)$ there exists $r = r_x$ such that

$$\frac{1}{|I(x, r_x)|} \int_{I(x,r_x)} |f|\, dy > \lambda. \tag{2.11}$$

Clearly

$$\mathcal{O}_\lambda \subseteq \bigcup_{x \in \mathcal{O}_\lambda} I(x, r_x). \tag{2.12}$$

Although the set on the right-hand side of (2.12) appears to be quite cumbersome, with the overlaps and all, things are not as complicated as a first impression might indicate. Since by the regularity of the Lebesgue measure, cf. 3.37 in Chapter V,

$$|\mathcal{O}_\lambda| = \sup\{|K| : K \subset \mathcal{O}_\lambda, K \text{compact}\}, \tag{2.13}$$

it suffices to estimate $|K|$ for each compact subset K of $\mathcal{O}_\lambda$. Now, for each such a compact set of K we also have

$$K \subseteq \bigcup_{x \in \mathcal{O}_\lambda} I(x, r_x)$$

and, since the $I(x, r_x)$'s are open, by the Heine-Borel Theorem there exist finitely many intervals $I(x_1, r_1), \ldots, I(x_m, r_m)$, say, so that

$$K \subseteq \bigcup_{i=1}^{m} I(x_i, r_i) . \tag{2.14}$$

Clearly we may assume that no interval that appears on the right-hand side of (2.14) is contained in the union of all the others, that is to say, each interval there contributes something to the union.

Let $r = \max\{r_1, \ldots, r_m\}$ and, by renaming the intervals if necessary, suppose that $r_1 = r$; if more than one r_i equals r just choose any. At this juncture of the argument the geometry of the situation takes over: Observe that if

$$I(x_1, r_1) \cap I(x_j, r_j) \neq \emptyset , \quad \text{then} \quad I(x_j, r_j) \subseteq I(x_1, 3r_1) .$$

Because of this property we discard all the intervals $I(x_j, r_j)$, $j \neq 1$, which intersect $I(x_1, r_1)$, and repeat the same procedure with the remaining intervals, i.e., the family of those intervals $I(x_j, r_j)$ which are disjoint with $I(x_1, r_1)$. In other words, we separate an interval with largest sidelength, and then discard all the intervals which intersect it. Since the original family of intervals is finite, after a finite number of steps we are left with a pairwise disjoint family of open intervals $I(x_1, r_1), \ldots, I(x_k, r_k)$, say, which by (2.14) has the property that

$$K \subseteq \bigcup_{i=1}^{k} I(x_i, 3r_i) .$$

Whence

$$|K| \leq \sum_{i=1}^{k} |I(x_i, 3r_i)| = 3^n \sum_{i=1}^{k} |I(x_i, r_i)| . \tag{2.15}$$

But the intervals that appear on the right-hand side of (2.15) are special: They are pairwise disjoint and they all satisfy (2.11). Therefore the sum on the right-hand side of (2.15) is less than or equal to

$$\sum_{i=1}^{k} \frac{1}{\lambda} \int_{I(x_i, r_i)} |f| \, dy = \frac{1}{\lambda} \int_{\bigcup_{i=1}^{k} I(x_i, r_i)} |f| \, dy \leq \frac{1}{\lambda} \int_{R^n} |f| \, dy ,$$

and consequently,

$$|K| \le \frac{3^n}{\lambda} \int_{R^n} |f|\, dy\,. \tag{2.16}$$

Taking the sup of the left-hand side of (2.16) over those $K \subset \mathcal{O}_\lambda$, by (2.13) it follows that (2.10) holds. ■

We are now ready to address the questions raised in (2.1), to wit, the existence of the limit there, and its precise value.

Theorem 2.2 (Lebesgue Differentiation Theorem). Suppose f is an integrable function which vanishes off $I(0,2)$. Then

$$\lim_{r\to 0} \frac{1}{|I(x,r)|} \int_{I(x,r)} f(y)\, dy = f(x) \quad \text{a.e. on } I(0,1)\,. \tag{2.17}$$

Proof. First we show that the limit on the left-hand side of (2.17) exists a.e. For this purpose consider the function

$$\Phi(f,x) = \limsup_{r\to 0} \frac{1}{|I(x,r)|} \int_{I(x,r)} f\, dy - \liminf_{r\to 0} \frac{1}{|I(x,r)|} \int_{I(x,r)} f\, dy\,.$$

Although Φ is measurable, we need not make use of this fact to proceed with the proof. However we point out that Φ is a well-defined nonnegative function, and show that $\Phi(f,x) = 0$ a.e. on $I(0,1)$.

The idea is to control Φ by the Hardy-Littlewood maximal function. Now, if g is continuous it is clear that $\Phi(g,x) = 0$ everywhere and consequently, $\Phi(f,x) = \Phi(f-g,x)$. Whence it readily follows that for any continuous function g we have

$$\Phi(f,x) = \Phi(f-g,x) \le 2M(f-g)(x)\,, \quad \text{all } x\,.$$

Let now $\lambda > 0$, and note that the above inequality gives

$$\{x \in R^n : \Phi(f,x) > \lambda\} \subseteq \{x \in R^n : M(f-g)(x) > \lambda/2\}\,.$$

Thus, by the monotonicity of the Lebesgue outer measure and Theorem 2.1, we obtain

$$|\{\Phi(f,\cdot) > \lambda\}|_e \le |\{M(f-g) > \lambda/2\}| \le \frac{2\cdot 3^n}{\lambda} \int_R |f-g|\, dy\,. \tag{2.18}$$

Now, since continuous functions are dense in the metric of $L(R^n)$, the integral on the right-hand side of (2.18) may be made arbitrarily small, and consequently,

$$|\{x \in R^n : \Phi(f,x) > \lambda\}|_e = 0\,, \quad \text{all } \lambda > 0\,.$$

This can only be true if $\Phi(f,x) = 0$ a.e., in other words the lim sup is equal to the lim inf and the limit on the left-hand side of (2.17) exists a.e.

Next we show that the limit equals f a.e. Let

$$\Psi(f,x) = \left|\lim_{r\to 0}\frac{1}{|I(x,r)|}\int_{I(x,r)} f(y)\,dy - f(x)\right|.$$

Ψ is an a.e. well-defined nonnegative function, and it is apparent that for continuous g

$$\Psi(f,x) = \Psi(f-g,x)\,, \quad \text{all } x\,.$$

Whence it readily follows that $\Psi(f,x) \le M(f-g,x) + |f(x)-g(x)|$, and consequently, for $\lambda > 0$ we have

$$\{x \in R^n : \Psi(f,x) > \lambda\} \subseteq \{x \in R^n : M(f-g)(x) > \lambda/2\} \\ \bigcup \{x \in R^n : |f(x)-g(x)| > \lambda/2\}\,.$$

Thus by the monotonicity of the Lebesgue outer measure, the Hardy-Littlewood maximal theorem and Chebychev's inequality, it follows that

$$|\{\Psi(f,\cdot) > \lambda\}|_e \le |\{M(f-g) > \lambda/2\}| + |\{|f-g| > \lambda/2\}| \\ \le \frac{2\cdot 3^n}{\lambda}\int_{R^n}|f-g|\,dy + \frac{2}{\lambda}\int_{R^n}|f-g|\,dy\,.$$

Since g is an arbitrary continuous function, we get that

$$|\{x \in R^n : \Psi(f,x) > \lambda\}|_e = 0\,, \quad \text{all } \lambda > 0\,.$$

Whence $\Psi(f,x) = 0$ a.e., (2.17) holds. ∎

Although we have only discussed a "local" version of (2.1), it is not hard to obtain the "global" version as well. One way to go about this is to use the general Hardy-Littlewood maximal theorem, cf. 3.23 below, but a simpler way to proceed is this: First observe that $R^n = \bigcup_{k=1}^\infty I(0,2k)$. Now, an argument entirely analogous to Theorems 2.1 and 2.2 gives that if $f \in L(I(0,2k))$, then the integral of f differentiates to $f(x)$ a.e. on $I(0,k)$. Given an arbitrary integrable function f note that $f_k = f\chi_{I(0,2k)}$ is integrable and it vanishes off $I(0,2k)$, and consequently there exists a null set N_k such that the integral of f_k differentiates to $f_k(x) = f(x)$ a.e. on $I(0,k)$, $k = 1,2,\ldots$ Clearly the integral of f differentiates to $f(x)$ off the null set $N = \bigcup_{k=1}^\infty N_k$, and the global version of the differentiation theorem also holds.

3. PROBLEMS AND QUESTIONS

In what follows $(X, \mathcal{M}, \mu)$ is a measure space.

3.1 By means of an example show that if $f, g \in L(\mu)$ it is not necessarily true that fg is integrable. However, if $f^2, g^2 \in L(\mu)$, prove that $fg \in L(\mu)$ and

$$\int_X |fg|\, d\mu \le \left(\int_X f^2\, d\mu\right)^{1/2} \left(\int_X g^2\, d\mu\right)^{1/2} . \tag{3.1}$$

(3.1) is known as the Cauchy-Schwarz inequality and it has many interesting applications. One of them is the following: Show that if $\{f_n\}, \{g_n\}$ are sequences with the property that $f_n^2, g_n^2 \in L(\mu)$, $n = 1, 2, \ldots$, and if

$$\lim_{n\to\infty} \int_X (f_n - f)^2\, d\mu = \lim_{n\to\infty} \int_X (g_n - g)^2\, d\mu = 0 ,$$

then

$$\lim_{n\to\infty} \int_X f_n g_n\, d\mu = \int_X fg\, d\mu .$$

3.2 In the spirit of 3.1, show that if $f \in L(\mu)$ and $\{f_n\}$ is a sequence of integrable functions so that $\lim_{n\to\infty} d(f_n, f) = 0$, and if $\{g_n\}$ is a sequence of measurable functions such that

$$|g_n| \le M \ \mu\text{-a.e.}, \quad \lim g_n = g \ \mu\text{-a.e.},$$

then

$$\lim_{n\to\infty} \int_X f_n g_n\, d\mu = \int_X fg\, d\mu .$$

3.3 Suppose that $f, f_n \in L(\mu), n = 1, 2, \ldots$, and $\lim_{n\to\infty} d(f_n, f) = 0$. Show that there exist an integrable function F and a sequence $\{n_k\}$ such that

$$|f_{n_k}| \le F \ \mu\text{-a.e.} \quad \text{and} \quad \lim_{k\to\infty} f_{n_k} = f \ \mu\text{-a.e.}$$

3.4 Construct a Lebesgue integrable function $f \in L(R)$ with the following property: For any interval $I \subseteq R$ and $M > 0$,

$$|\{x \in I : |f(x)| > M\}| > 0 .$$

3.5 Suppose f is a Lebesgue integrable function on R^+ and put

$$g(x) = \int_{[0,\infty)} \frac{f(t)}{x+t}\, dt\,, \quad x > 0\,.$$

Is g continuous? Does g have a limit as $x \to \infty$? Is g differentiable?

3.6 Suppose $f \in L([a,b])$ and put

$$F(x) = \int_{[a,x]} f\, dy\,, \quad a \le x \le b\,.$$

Show that F is a continuous BV function on $[a,b]$ that satisfies the following property: Given $\varepsilon > 0$, there exists $\delta > 0$, such that for any finite collection $\{[a_i,b_i]\}$ of nonoverlapping subintervals of $[a,b]$, we have

$$\sum_i |F(b_i) - F(a_i)| < \varepsilon\,, \quad \text{whenever } \sum_i (b_i - a_i) < \delta\,. \tag{3.2}$$

Functions which satisfy (3.2) are said to be "absolutely continuous" in the sense of Vitali; we will have more to say about absolutely continuous functions in Chapter X.

3.7 Show that the integral of an arbitrary $f \in L(\mu)$ is "absolutely continuous" in the following sense: Given $\varepsilon > 0$, there exists $\delta > 0$, such that

$$\int_E |f|\, d\mu < \varepsilon\,, \quad \text{whenever } \mu(E) < \delta\,.$$

We will have more to say about this notion of absolute continuity in Chapter XI.

3.8 Let $(X, \mathcal{M}, \mu)$ be a finite measure space, and f a measurable extended real-valued function defined on X. Show that $f \in L(\mu)$ iff $\sum_{k=1}^{\infty} \mu(\{|f| \ge k\}) < \infty$.

3.9 Show that if $f \in L(\mu)$, then

$$\lim_{\lambda\to\infty} \int_{\{|f|>\lambda\}} |f|\, d\mu = 0\,. \tag{3.3}$$

3.10 A sequence $\{f_n\}$ of integrable functions is said to be "uniformly integrable" if (3.3) holds uniformly in n. More precisely, $\{f_n\}$ is uniformly integrable iff

$$\lim_{\lambda\to\infty} \sup_n \int_{\{|f_n|>\lambda\}} |f_n|\, d\mu = 0\,.$$

Show that if $\mu(X) < \infty$ and $\{f_n\}$ is uniformly integrable and $\lim_{n\to\infty} f_n = f$ μ-a.e., then $f \in L(\mu)$ and

$$\lim_{n\to\infty} \int_X f_n \, d\mu = \int_X f \, d\mu .$$

3.11 If

$$\sup \int_X |f_n|^{1+\eta} \, d\mu < \infty , \quad \eta > 0 ,$$

show that $\{f_n\}$ is uniformly integrable.
The same conclusion is true if there exists $g \in L(\mu)$ so that $|f_n| \le g$ μ-a.e.; on the other hand the sequence

$$f_n = (n/\ln n)\chi_{(0,1/n)} , \quad n = 2, 3, \ldots$$

is uniformly integrable with respect to the Lebesgue measure of $I = [0,1]$. In fact, $\int_I f_n \, dx \to 0$, and yet the f_n's are not dominated by any integrable function.

3.12 Suppose $\mu(X) < \infty$ and show that uniformly integrable sequences are precisely those sequences with uniformly bounded integrals that are uniformly absolutely continuous. More precisely, $\{f_n\}$ is uniformly integrable iff (i) There is a constant M such that $\int_X |f_n| \, d\mu \le M$ for all n, and, (ii) Given $\varepsilon > 0$, there exists $\delta > 0$ such that $\mu(E) < \delta$ implies $\int_E |f_n| d\mu < \varepsilon$ for all n.

3.13 Suppose $\mu(X) < \infty$ and $f_n \ge 0$ μ-a.e. for all n. Show that if the sequence $\{f_n\}$ is uniformly integrable, then

$$\limsup \int_X f_n \, d\mu \le \int_X \limsup f_n \, d\mu .$$

3.14 Suppose $f, f_n \in L(R^n), n = 1, 2, \ldots$, and $\lim_{n\to\infty} d(f_n, f) = 0$. Show that if $|h_n| \to 0$, then $\lim_{n\to\infty} \int_{R^n} |f_n(y + h_n) - f(y)| \, dy = 0$.

3.15 Suppose that $f \in L(R^n)$ and compute

$$\lim_{|h|\to\infty} \int_{R^n} |f(y + h) + f(y)| \, dy .$$

3.16 Prove that if A is a Lebesgue measurable subset of R^n of positive measure, then the difference set $A - A = \{x \in R^n : x = y_1 - y_2, y_1, y_2 \in A\}$ contains a neighbourhood of the origin.

State and prove a similar statement for $A+A$, and for $A \pm B$, where B is another Lebesgue measurable set of positive measure.

3.17 Suppose f is a measurable function defined on R which assumes finite values on a set of positive measure and such that $f(x+y) = f(x) + f(y)$ for all real x, y. Show that f is of the form $f(x) = cx$.

3.18 Prove that $C_0^1(R^n) = \{g : g$ is real-valued, it vanishes off a compact set, and its first order partial derivatives are continuous$\}$ is dense in $L(R^n)$.
Also, $C_0^k(R^n) = \{g : g$ and all its partial derivatives of order $\leq (k-1)$ belong to $C_0^1(R^n)\}$, $k = 2, 3, \ldots$, and $C_0^\infty(R^n) = \bigcap_{k=1}^\infty C_0^k(R^n)$, are dense in $L(R^n)$.

3.19 If $f \in L(R^n)$, prove that there exists a sequence $\{f_k\}$ of continuous functions such that $\lim_{k\to\infty} f_k = f$ a.e.
Further show that we may also require that each f_k vanish off a compact set, and that it belong to $C_0^1(R^n)$, or even to $C_0^\infty(R^n)$.

3.20 Given $f \in L(R^n)$, put

$$F(x,r) = \frac{1}{|I(x,r)|} \int_{I(x,r)} f(y)\,dy\,, \quad x \in R^n\,, r > 0\,.$$

Is F continuous as a function of x, for each r fixed? Is F continuous as a function of r, for each x fixed?

3.21 In the notation of 3.17, show that

$$\limsup_{r\to 0} F(x,r) \quad \text{and} \quad \liminf_{r\to 0} F(x,r)$$

are Lebesgue measurable.

3.22 Prove that Theorem 2.1 is true if we replace intervals by balls, i.e., (2.10) holds with Mf there replaced by

$$M_1 f(x) = \sup_{r>0} \frac{1}{|B(x,r)|} \int_{B(x,r)} |f|\,dy\,,$$

where $B(x,r) = \{y \in R^n : |x-y| < r\}$ denotes the ball of radius r centered at x.

3.23 Prove the general version of the Hardy-Littlewood maximal theorem, i.e., remove the assumption that the integrable function f vanishes off a bounded set.

3.24 A point x at which

$$\lim_{r\to 0}\frac{1}{|I(x,r)|}\int_{I(x,r)}|f(y)-f(x)|\,dy=0$$

is called a Lebesgue point of f, and the collection of all such points is called the Lebesgue set of f.
Prove that if f is integrable, then almost every point is in the Lebesgue set of f. This notion is extremely important in the convergence of Fourier series, cf. Theorem 3.1 in Chapter XVII.

3.25 A family $\mathcal{R}=\{R\}$ of subsets of R^n is said to be regular at x provided that: (i) The diameters of the sets R tend to 0, and, (ii) There is a constant c such that if $I(x,r)$ denotes the smallest interval centered at x containing R, then $|I(x,r)|\le c|R|$; the sets R need not contain x.
Show that if f is integrable, $\mathcal{R}$ is regular at x, and x is in the Lebesgue set of f, then

$$\lim_{\operatorname{diam}(R)\to 0}\frac{1}{|R|}\int_R|f(y)-f(x)|\,dy=0\,.$$

3.26 Let E be a measurable subset of R^n; a point $x\in R^n$ for which

$$\lim_{r\to 0}\frac{|E\cap I(x,r)|}{|I(x,r)|}=1$$

is called a point of density of E. If the above limit equals 0, x is called a point of dispersion of E.
Prove that almost every point of E is a point of density of E and a point of dispersion of $R^n\setminus E$.

3.27 Suppose that

$$\int_I f(y)\,dy=0 \tag{3.4}$$

for every subinterval I of R, and show that $f=0$ a.e. In fact, the same conclusion is true if (3.4) holds for every I with $|I|=c>0$, c a fixed constant.

3.28 Suppose that $f\in L(R)$ vanishes off a bounded interval $I=[a,b]$, and let

$$F(x)=\int_{[a,x]}f\,dy\,,\quad a\le x\le b\,.$$

Is is true that

$$\int_{[a,b]} \left| \lim_{h\to 0} \frac{F(x+h)-F(x)}{h} - f(x) \right| dx = 0?$$

Calderón and Zygmund, while considering problems related to the "norm" convergence of Fourier series of functions of several variables, introduced a decomposition of an integrable function into an essentially bounded "good" part and a "bad" part. The decomposition is described in the next four problems.

3.29 Let I be a finite interval in R and suppose f is a nonnegative integrable function which vanishes off I. Show that for any λ satisfying

$$\frac{1}{|I|}\int_I f\,dy \le \lambda,$$

there is a sequence $\{I_k\}$ of nonoverlapping intervals contained in I such that

(i) $\lambda < \frac{1}{|I_k|}\int_{I_k} f\,dy \le 2\lambda, \quad k = 1, 2\ldots$

(ii) $f \le \lambda$ a.e. on $I \setminus \bigcup_k I_k$.

(iii) $|\bigcup_k I_k| \le \frac{1}{\lambda}\int_{\bigcup I_k} f\,dy \le \frac{1}{\lambda}\int_I f\,dy$.

3.30 Referring to 3.26, consider the "good" function

$$g = f\chi_{I\setminus\bigcup I_k} + \sum_k \left(\frac{1}{|I_k|}\int_{I_k} f\,dy\right)\chi_{I_k},$$

and the "bad" function

$$b = f - g.$$

Show that these functions enjoy the following properties

(i) $0 \le g \le \lambda$ a.e. on $I \setminus \bigcup I_k$.

(ii) $0 \le g \le 2\lambda$ on $\bigcup I_k$.

(iii) $b = 0$ in $I \setminus \bigcup I_k$; $\int_{I_k} b = 0$ for all k.

(iv) $|b| \le f + 2\lambda$.

3.31 Show that 3.29 and 3.30 are valid for $I = R$, $f \in L(R)$, and any $\lambda > 0$

3.32 The reader is invited to prove the n-dimensional version of 3.29, 3.30 and 3.31.

CHAPTER IX

Borel Measures

In this chapter we study Borel measures on Euclidean space, their regularity properties, and the distribution functions associated with them.

1. REGULAR BOREL MEASURES

A measure μ on $(R^n, \mathcal{B}_n)$ is called a Borel measure; the restriction of the Lebesgue measure to $\mathcal{B}_n$ is a familiar example of a Borel measure. In working with measures it is apparent that "regularity" plays an essential role. Now, in the case of the Lebesgue measure, regularity is built into its definition. We begin by showing that the same is true for those Borel measures which are finite on bounded sets; first the precise definition of regularity.

A Borel measure μ is said to be regular if for any $E \in \mathcal{B}_n$, $\mu(E)$ may be computed by either of the expressions

$$\mu(E) = \sup\{\mu(K) : K \text{ is compact, and } K \subseteq E\}, \tag{1.1}$$

$$\mu(E) = \inf\{\mu(\mathcal{O}) : \mathcal{O} \text{ is open, and } \mathcal{O} \supseteq E\}. \tag{1.2}$$

These conditions roughly state that μ is determined by the compact, or open, sets in R^n.

A convenient way to verify these conditions is to consider the following equivalent formulations. For (1.1) we have: If $\mu(E) < \infty$, given $\varepsilon > 0$, there exists a compact set $K \subseteq E$ such that

$$\mu(E \setminus K) = \mu(E) - \mu(K) \leq \varepsilon, \tag{1.3}$$

and if $\mu(E) = \infty$, given $M > 0$, we can find a compact set $K \subseteq E$ such that

$$\mu(K) \geq M. \tag{1.4}$$

As for (1.2), if $\mu(E) < \infty$, given $\varepsilon > 0$, there exists an open set $\mathcal{O} \supseteq E$ such that

$$\mu(\mathcal{O} \setminus E) = \mu(\mathcal{O}) - \mu(E) \le \varepsilon\,, \tag{1.5}$$

and if $\mu(E) = \infty$, by monotonicity any open set containing E also has infinite measure.

We consider the regularity of finite Borel measures first; although the idea for proving this assertion is clear, it takes some time to carry out the details.

Theorem 1.1. Suppose μ is a finite Borel measure, then μ is regular.

Proof. Let

$$\mathcal{A} = \{E \in \mathcal{B}_n : \text{(1.1) and (1.2) are true for } E\}\,.$$

The idea of the proof is to show that $\mathcal{A}$ is a σ-algebra that contains the closed intervals, and which therefore coincides with $\mathcal{B}_n$.

It is not hard to check that $\mathcal{A}$ contains the closed intervals of R^n: Let I be a closed interval, then I is also compact and (1.1) holds. As for (1.2), let $\{I_k\}$ be a decreasing sequence of open intervals that converges to I. By (iv) in Proposition 3.1 in Chapter IV it follows that

$$\mu(I) = \lim_{k\to\infty} \mu(I_k)\,,$$

and (1.2) holds as well.

Next, if $\{E_m\} \subseteq \mathcal{A}$ and $E = \bigcup_m E_m$, then we have $E \in \mathcal{A}$. Indeed, suppose that $\varepsilon > 0$ has been given, and invoke (1.3) to find a sequence $\{K_m\}$ of compact sets such that

$$K_m \subseteq E_m\,, \quad \mu(E_m \setminus K_m) \le \varepsilon/2^{m+1}\,, \quad m = 1, 2, \ldots$$

Furthermore, since $E \setminus \bigcup_m K_m \subseteq \bigcup_m (E_m \setminus K_m)$, we have

$$\mu\left(E \setminus \textstyle\bigcup_m K_m\right) \le \sum_m \mu(E_m \setminus K_m) \le \varepsilon/2\,. \tag{1.6}$$

Now, since $\mu\left(\bigcup_m K_m\right) < \infty$, by (iii) in Proposition 3.1 in Chapter IV we get at once that $\lim_{M\to\infty} \mu\left(\bigcup_{m=1}^M K_m\right) = \mu\left(\bigcup_m K_m\right)$. Whence, for M sufficiently large, the compact set $K = \bigcup_{m=1}^M K_m \subseteq E$ satisfies

$$\mu\left(\bigcup_m K_m\right) \le \mu(K) + \varepsilon/2\,. \tag{1.7}$$

Moreover, since

$$E \setminus K = (E \setminus \bigcup_m K_m) \cup (\bigcup_m K_m \setminus K)\,,$$

by (1.6) and (1.7) it follows that $\mu(E) \le \mu(K) + \varepsilon$. As noted in (1.3) above, this estimate gives that (1.1) is true for E.

Along similar lines, let $\{\mathcal{O}_m\}$ be a sequence of open sets such that

$$E_m \subseteq \mathcal{O}_m\,, \quad \mu(\mathcal{O}_m \setminus E_m) \le \varepsilon/2^m\,, \quad m = 1, 2, \ldots$$

Since $\mathcal{O} = \bigcup_m \mathcal{O}_m$ is an open set containing E and $\mathcal{O}\setminus E \subseteq \bigcup_m (\mathcal{O}_m \setminus E_m)$, we get that

$$\mu(\mathcal{O} \setminus E) \le \sum_m \mu(\mathcal{O}_m \setminus E_m) \le \varepsilon\,,$$

and consequently, (1.2) is also true for E. Thus $E \in \mathcal{A}$ and $\mathcal{A}$ is closed under countable unions.

Finally we check that $\mathcal{A}$ is closed under complementation. Suppose $E \in \mathcal{A}$ and $\varepsilon > 0$ is given. Since μ is finite, there exist an open set $\mathcal{O} \supseteq E$ and a compact set $K \subseteq E$ such that

$$\mu(\mathcal{O} \setminus E) \le \varepsilon/2\,, \quad \text{and} \quad \mu(E \setminus K) \le \varepsilon\,.$$

Moreover, since $E \setminus K = (R^n \setminus K) \setminus (R^n \setminus E)$ and $R^n \setminus K = \mathcal{O}'$ is open, we also have

$$\mu(\mathcal{O}' \setminus (R^n \setminus E)) \le \varepsilon\,, \quad \mathcal{O}' \supseteq (R^n \setminus E)\,,$$

and (1.2) holds for $R^n \setminus E$. A similar argument gives that

$$\mu((R^n \setminus E) \setminus (R^n \setminus \mathcal{O})) \le \varepsilon/2\,, \quad R^n \setminus \mathcal{O} \subseteq R^n \setminus E\,,$$

but we can only assert that $R^n \setminus \mathcal{O}$ is closed. This is not a major difficulty: Since

$$R^n = \bigcup_{m=1}^{\infty} \{x \in R^n : |x| \le m\} = \bigcup_{m=1}^{\infty} B_m\,,$$

say, the sequence of compact sets $\{(R^n \setminus \mathcal{O}) \cap B_m\}$ converges to $R^n \setminus \mathcal{O}$. Whence, by Proposition 3.1 (iii) in Chapter IV, we get that

$$\lim_{m \to \infty} \mu((R^n \setminus \mathcal{O}) \cap B_m) = \mu(R^n \setminus \mathcal{O})\,,$$

and consequently, for m sufficiently large it follows that

$$\mu(R^n \setminus \mathcal{O}) \le \mu((R^n \setminus \mathcal{O}) \cap B_m) + \varepsilon/2\,.$$

Let $K' = (R^n \setminus \mathcal{O}) \cap B_m$; K' is a compact subset of $R^n \setminus E$ and since

$$(R^n \setminus E) \setminus K' = ((R^n \setminus E) \setminus (R^n \setminus \mathcal{O})) \cup ((R^n \setminus \mathcal{O}) \setminus K'),$$

by the above estimates it follows that $\mu(R^n \setminus E) - \mu(K') \le \varepsilon$. Thus, (1.1) holds for $R^n \setminus E$, and $\mathcal{A}$ is closed under the taking of complements. Whence $\mathcal{A}$ is a σ-algebra that coincides with $\mathcal{B}_n$, and we have finished. ■

Since R^n is σ-compact it is possible to extend Theorem 1.1 to more general Borel measures. More precisely, we have

Theorem 1.2. Suppose μ is a Borel measure which is finite on bounded subsets of R^n. Then μ is regular.

Proof. Since $R^n = \bigcup_m \{x : |x| \le m\} = \bigcup_m B_m$, say, for each E in $\mathcal{B}_n$ we have

$$E = \bigcup_m (E \cap B_m), \quad \text{with} \quad E \cap B_m \in \mathcal{B}_n, \quad m = 1, 2 \ldots$$

Thus, from Proposition 3.1 (iii) in Chapter IV we get

$$\mu(E) = \lim_{m \to \infty} \mu(E \cap B_m). \tag{1.8}$$

The idea is to approximate the measure of the sets that appear on the right-hand side of (1.8), and this will be achieved by "restricting" μ to B_m. More precisely, consider the sequence of Borel measures given by

$$\mu_m(E) = \mu(E \cap B_m), \quad m = 1, 2, \ldots \tag{1.9}$$

Since μ is finite on bounded sets, the μ_m's are finite Borel measures, and, by Theorem 1.1, they are regular. Fix now $E \in \mathcal{B}_n$, let $\varepsilon > 0$ be given, and put $E_m = E \cap B_m, m = 1, 2, \ldots$ By regularity, there exist compact sets F_m and open sets G_m such that

$$F_m \subseteq E_m, \quad \mu_m(E_m) \le \mu_m(F_m) + \varepsilon/2^m, \quad m = 1, 2, \ldots \tag{1.10}$$

and

$$G_m \supseteq E_m, \quad \mu_m(G_m) \le \mu_m(E_m) + \varepsilon/2^m, \quad m = 1, 2, \ldots \tag{1.11}$$

We rewrite (1.10) and (1.11) in terms of μ. Since $K_m = F_m$ is a compact subset of E_m, (1.10) reads

$$K_m \subseteq E_m, \quad \mu(E_m) \le \mu(K_m) + \varepsilon/2^m, \quad m = 1, 2, \ldots \tag{1.12}$$

As for (1.11), let $\{I_{m,k}\}$ be a decreasing sequence of bounded open balls which converges to B_m. Since $\mu(I_{m,1}) < \infty$, by (iv) in Proposition 3.1 in Chapter IV, we have

$$\mu(G_m \cap B_m) = \lim_{k\to\infty} \mu(G_m \cap I_{m,k}), \quad m = 1, 2, \ldots$$

and consequently we can find a sequence of k_m's such that

$$\mu(G_m \cap I_{m,k_m}) \leq \mu(G_m \cap B_m) + \varepsilon/2^m, \quad m = 1, 2, \ldots$$

Now, $\mathcal{O}_m = G_m \cap I_{m,k_m}$ is an open set that contains E_m, and (1.11) becomes

$$\mathcal{O}_m \supseteq E_m, \quad \mu(\mathcal{O}_m) \leq \mu(E_m) + \varepsilon/2^m, \quad m = 1, 2, \ldots \tag{1.13}$$

We are now ready to show that $\mu(E)$ may be computed by both (1.1) and (1.2); we do (1.1) first. Combining (1.8) and (1.12) it readily follows that we may find a sequence of compact sets $\{K_m\}$ with the property that

$$K_m \subseteq E_m \subseteq E, \quad \lim_{m\to\infty} \mu(K_m) = \mu(E);$$

this gives (1.1) whether $\mu(E)$ is finite or not.

As for (1.2), we must only do the case $\mu(E) < \infty$. If $\{\mathcal{O}_m\}$ is the sequence of open sets introduced in (1.13), put $\mathcal{O} = \bigcup_m \mathcal{O}_m$ and observe that $\mathcal{O}$ is an open set which contains E. Furthermore, since $\mathcal{O} \setminus E \subseteq \bigcup_m (\mathcal{O}_m \setminus E_m)$, by (1.13) it follows that

$$\mu(\mathcal{O}) - \mu(E) = \mu(\mathcal{O} \setminus E) \leq \sum_m \mu(\mathcal{O}_m \setminus E_m) \leq \varepsilon.$$

Thus (1.2) is also true, and μ is regular. ∎

2. DISTRIBUTION FUNCTIONS

Borel measures on the line which are finite on bounded sets are important in applications and there is a useful way to describe them. Let $\mathcal{BB}$ denote the collection of those Borel measures which are finite on bounded sets, assume that $\mu \in \mathcal{BB}$ is finite and, referring to 4.28 in Chapter IV, let F_y be a distribution function induced by μ. Since μ is finite a way to normalize the F_y's is to consider not the expression given there but rather the distribution function F corresponding to $y = -\infty$, namely

$$F(x) = \mu((-\infty, x]). \tag{2.1}$$

F is called the distribution function of μ, it is nondecreasing and right-continuous, and it satisfies

$$\lim_{x \to -\infty} F(x) = 0, \quad \lim_{x \to \infty} F(x) < \infty. \tag{2.2}$$

Also, as a consequence of (2.1), it follows that for $-\infty < x < y < \infty$ we have

$$\mu((x,y]) = F(y) - F(x), \qquad \mu([x,y]) = F(y) - F(x^-)$$
$$\mu([x,y)) = F(y^-) - F(x^-), \qquad \mu((x,y)) = F(y^-) - F(x).$$

Furthermore, if D is a dense subset of R, then the relation (2.1) is already determined by $x \in D$, or by any of the four above relations when $x, y \in D$.

The remarkable fact is that, conversely, any nondecreasing right-continuous function F that satisfies (2.2) determines a unique finite Borel measure μ such that (2.1) is true. Rather than proving this result we discuss a more general one that also includes $F(x) = x$, which intuitively corresponds to the Lebesgue measure on the line. The precise statement is

Theorem 2.1. Let $\mathcal{BB} = \{\mu : \mu$ is a Borel measure on the line which is finite on bounded sets$\}$, and $\mathcal{D} = \{F : F$ is nondecreasing and right-continuous$\}$; we identify those functions in $\mathcal{D}$ that differ by a constant.

Then, there is an injective mapping T from $\mathcal{BB}$ onto $\mathcal{D}$ which satisfies the following property: If $T\mu = F$ and c is an arbitrary constant, we have

$$F(x) = \begin{cases} c + \mu((0,x]) & \text{if } x > 0 \\ c & \text{if } x = 0 \\ c - \mu((x,0]) & \text{if } x < 0. \end{cases} \tag{2.3}$$

Clearly (2.3) is equivalent to

$$F(y) - F(x) = \mu((x,y]), \quad \text{all real } x < y. \tag{2.4}$$

Proof. It is not hard to check that if T is given by (2.4), then T is one-to-one. Indeed, if $T\mu$ is constant, from (2.4) it follows that μ vanishes on all half-open intervals $(x,y]$. Since any open subset of R is a countable union of such intervals, μ also vanishes on all open sets. Furthermore, since μ is regular, by (1.2) we get $\mu(E) = 0$ for all $E \in \mathcal{B}_1$. Thus μ is the 0 measure, and T is injective.

To show that T is onto requires some work. Suppose $F \in \mathcal{D}$ and observe that since F is nondecreasing the (bad) set

$$B = \{r : F^{-1}(\{r\}) \text{ consists of more than one value } x \in R\}$$

is at most countable; B consists of those points in the range of F which correspond to the intervals of constancy of F.

Now, let Φ be the interval-valued function defined as follows: Since $F \in \mathcal{D}, F(x^-)$ and $F(x^+) = F(x)$ exist for each $x \in R$; then put

$$\Phi(x) = \left[F(x^-), F(x)\right], \quad x \in R. \tag{2.5}$$

Thus, for each real $x, \Phi(x)$ is a closed interval, degenerate if F is continuous at x, and closed and bounded otherwise. For each subset E of R let

$$\Phi(E) = \bigcup_{x \in E} \Phi(x), \tag{2.6}$$

and put $J = \Phi(R)$; clearly J is also an interval. Now set

$$\mathcal{A} = \{E \in \mathcal{B}_1 : \Phi(E) \in \mathcal{L}\};$$

we claim that $\mathcal{A}$ is a σ-algebra which contains all intervals, and which consequently coincides with $\mathcal{B}_1$.

First observe that if E is an interval, then so is $\Phi(E)$; thus $\mathcal{A}$ contains all intervals. Next we verify that $\mathcal{A}$ is closed under complementation; this is clear if F is strictly increasing for then

$$\Phi(R \setminus E) = J \setminus \Phi(E) \in \mathcal{L}, \quad \text{whenever } E \in \mathcal{A}. \tag{2.7}$$

In the general case, i.e., when F is only monotone nondecreasing, a slight complication occurs when E includes an interval of constancy of F. For instance, if $F(x) = r$ for $x \in I = [a,b]$ and $E = [a,(a+b)/2]$, then $\Phi(R \setminus E) = J$, but $J \setminus \Phi(E) = J \setminus \{r\}$, and equality does not hold in (2.7). Nevertheless, equality will hold there if we add a subset of B, namely $\{r\}$, to $J \setminus \Phi(E)$. The same reasoning applies to the general case. More precisely, for any $E \in \mathcal{A}$ there exists an (at most countable) subset B_1 of B such that

$$\Phi(R \setminus E) = (J \setminus E) \cup B_1. \tag{2.8}$$

From (2.8) it is clear that $\Phi(R \setminus E) \in \mathcal{L}$, which in turn implies that $R \setminus E \in \mathcal{A}$, and consequently, $\mathcal{A}$ is closed under the taking of complements.

Finally we check that if $\{E_n\} \subseteq \mathcal{A}$ and $E = \bigcup_n E_n$, then E also belongs to $\mathcal{A}$. This is a simple consequence of the readily verified identity

$$\Phi(E) = \bigcup_n \Phi(E_n).$$

Thus, $\mathcal{A}$ is the σ-algebra $\mathcal{B}_1$.

Let now μ be the set function given by

$$\mu(E) = |\Phi(E)|, \quad E \in \mathcal{B}_1. \tag{2.9}$$

Since for each Borel set E, $\Phi(E) \in \mathcal{L}$, μ is well-defined. We claim that μ is a Borel measure which is finite on the bounded sets of R. That $\mu(\emptyset) = 0$ is obvious. Next let $\{E_n\}$ be a sequence of pairwise disjoint Borel subsets of the line and let E denote their union. In general the $\Phi(E_n)$'s are not pairwise disjoint, as there may be overlaps created by the intervals of constancy of F; this is not a serious inconvenience.

First recall that since B is at most countable, for any $A \in \mathcal{L}$ we have $|A \setminus B| = |A|$, cf. 3.2 in Chapter V. Next observe that the sets $\Phi(E_n) \setminus B$, $n = 1, 2, \ldots$, are pairwise disjoint and Lebesgue measurable. Thus

$$\begin{aligned} \mu(E) &= |\Phi(E)| = |\textstyle\bigcup_n \Phi(E_n)| \\ &= |(\textstyle\bigcup_n \Phi(E_n)) \setminus B| = |\textstyle\bigcup_n (\Phi(E_n) \setminus B)| \\ &= \textstyle\sum_n |\Phi(E_n) \setminus B| = \sum_n |\Phi(E_n)| = \sum_n \mu(E_n). \end{aligned}$$

Whence, μ is a Borel measure on the line. It only remains to show that μ is finite on bounded sets, and that $T\mu = F$, i.e., that (2.3) holds.

Let $E \subseteq [-n,n]$ be a bounded Borel set. Since $\Phi(E) \subseteq \Phi([-n, n])$ we obtain

$$\mu(E) = |\Phi(E)| \le |\Phi([-n, n])| < \infty,$$

and μ is finite on bounded sets. Finally, let $I = (x,y]$. Since

$$\lim_{z \to x^+} F(z^-) = F(x^+) = F(x)$$

by (2.5) it is apparent that either

$$\Phi(I) = (F(x), F(y)], \quad \text{or} \quad \Phi(I) = [F(x), F(y)].$$

In either case we have

$$\mu(I) = |\Phi(I)| = F(y) - F(x),$$

and (2.4) holds. It is now easy to see that (2.3) holds as well, with $c = F(0)$ there. ∎

To emphasize the fact that $T\mu = F$, and that, consequently, μ and F are related by (2.3), or (2.4), we denote μ by μ_F; μ_F is called the Lebesgue-Stieltjes measure induced by F, and

$$\int_R f \, d\mu_F, \quad f \text{ Borel measurable},$$

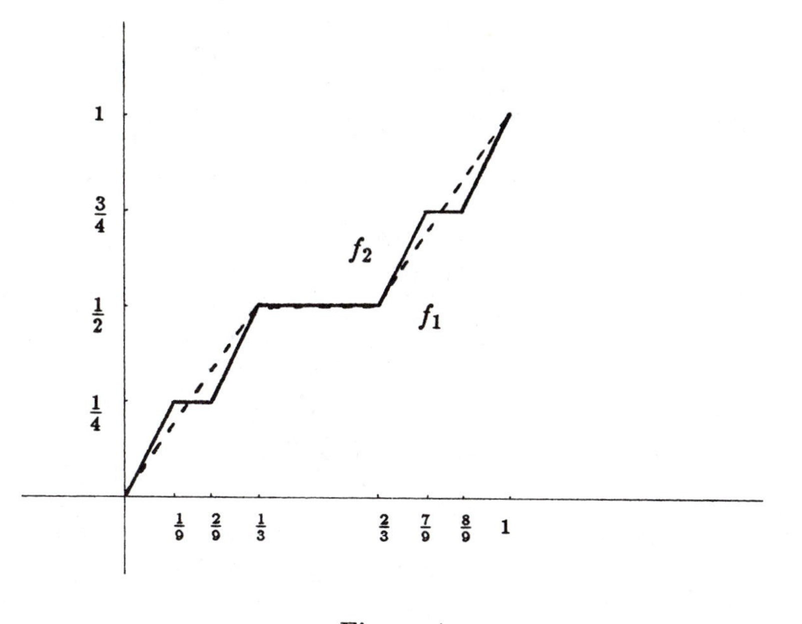

Figure 4

is called the Lebesgue-Stieltjes integral of f over R with respect to $d\mu_F$.

The following is a natural question to ponder: If $F \in \mathcal{D}$ has a measurable derivative F' a.e., under what circumstances is $d\mu_F(x) = F'(x)\,dx$?

That equality does not always hold follows by considering the Cantor Lebesgue function, which we construct next.

Referring to the construction of the Cantor set and to the construction in Section 1 of Chapter VI, let I_n^k be the $2^n - 1$ open intervals whose union is D_n, and for $n = 1, 2, \ldots$, let f_n be the continuous function defined on $I = [0,1]$ which satisfies $f_n(0) = 0$, $f_n(1) = 1, f_n(x) = k2^{-n}$ for $x \in I_n^k$, and which is linear on each interval of C_n.

It is clear that each f_n is monotone nondecreasing, and since the values only change on the C_n's, that $f_n = f_{n+1}$ on I_n^k, $k = 1, \ldots, 2^n - 1$, and

$$|f_n(x) - f_{n+1}(x)| \leq 2^{-n}, \quad \text{all } x \text{ in } I.$$

This estimate allows us to show that the f_n's converge uniformly on I. Indeed, for any $k < m$ we have

$$|f_m(x) - f_k(x)| = \left| \sum_{n=k}^{m-1} (f_n(x) - f_{n+1}(x)) \right|$$

$$\leq \sum_{n=k}^{\infty} |f_n(x) - f_{n+1}(x)| \leq \sum_{n=k}^{\infty} 2^{-n} = 2^{-k+1}.$$

Thus, given $\varepsilon > 0$, we may choose N so large that

$$|f_m(x) - f_k(x)| \leq \varepsilon, \quad m, k \geq N, \text{all } x \in I,$$

and the sequence $\{f_n\}$ is uniformly Cauchy on I. Whence $\lim_{n\to\infty} f_n(x) = f(x)$ exists on I, and since I is compact and the f_n's continuous and nondecreasing, f is also continuous and nondecreasing. Also note that $f(0) = 0,\ f(1) = 1$, and that f is constant on every interval removed in the construction of C; f is called the Cantor-Lebesgue function.

Let now

$$F(x) = \begin{cases} 0 & \text{if } x < 0 \\ f(x) & \text{if } 0 \leq x \leq 1 \\ 1 & \text{if } x > 1. \end{cases}$$

Clearly $F \in \mathcal{D}$, and since F is not constant, μ_F is not the 0 measure. On the other hand it is apparent that $F'(x) = 0$ for $x \notin C$, that is a.e. Whence it readily follows that

$$\begin{aligned} \int_R d\mu_F &= \mu_F([0,1]) = F(1) - F(0) \\ &= 1 \neq \int_R F'\, dx = 0. \end{aligned}$$

It is also clear that for the Cantor-Lebesgue function f the following is true:

$$\int_{[0,1]} f'\, dx = 0 \neq f(1) - f(0) = 1.$$

This puzzling fact will be discussed in full detail in Chapter X, where we characterize those functions which coincide with the integral of their derivatives.

We close this section with two interesting remarks; since the proofs follow along familiar lines we leave it to the reader to carry them out, cf. Section 3 in Chapter VII.

Theorem 2.2. Let g be a bounded real-valued function defined on $I = [a,b]$, f a nondecreasing right-continuous function defined on I, and put

$$F(t) = \begin{cases} f(a) & \text{if } t \leq a \\ f(t) & \text{if } a < t < b \\ f(b) & \text{if } t \geq b. \end{cases}$$

Then, $g \in \mathcal{R}(I)$ iff

$$\mu_F(\{x \in I : g \text{ is not continuous at } x\}) = 0\,.$$

Furthermore, if $\int_a^b g\,df$ exists, then $g \in L(\mu_F)$, and

$$\int_a^b g\,df = \int_R g\,d\mu_F\,.$$

3. PROBLEMS AND QUESTIONS

3.1 What is the cardinal number of $\mathcal{B}_n$?

3.2 Suppose μ is a Borel measure on the line such that $\mu([0,1]) = 1$, and for each real x,

$$\mu(E) = \mu(x + E)\,, \quad \text{all } E \in \mathcal{B}_1\,.$$

Show that μ coincides with the restriction of the Lebesgue measure to $\mathcal{B}_1$.

3.3 Suppose that μ is a probability Borel measure on $[0,1]$, and that for each Borel set $E \subset [0,1]$ with $|E| = 1/2$ we also have $\mu(E) = 1/2$. Does it follow that μ coincides with the restriction of the Lebesgue measure to $\mathcal{B}_1$?

3.4 Suppose that μ is a finite Borel measure on R^n, and that $A \subset R^n$ is closed. Show that $\phi(x) = \mu(x + A)$ is upper semicontinuous, and consequently, measurable.

3.5 Let μ be a finite Borel measure on a bounded interval of the line such that for each real x, $\mu(\{x\}) = 0$. Show that given $\varepsilon > 0$, there exists $\delta = \delta(\varepsilon) > 0$ with the property that if $E \in \mathcal{B}_1$ and $\operatorname{diam}(E) < \delta$, then $\mu(E) < \varepsilon$.

3.6 Suppose μ is a nonzero Borel measure on the line which is finite on bounded sets. Show that if

$$\int_R fg\,d\mu = \left(\int_R f\,d\mu\right)\left(\int_R g\,d\mu\right)\,, \quad \text{all } f, g \in C_0(R)\,,$$

then μ is a Dirac measure.

3.7 Let ϕ be a nonnegative additive set function defined on $\mathcal{B}_n$, and suppose that for E in $\mathcal{B}_n$ we have

$$\phi(E) = \sup\{\phi(K) : K \text{compact}, K \subseteq E\}.$$

Show that ϕ is σ-additive, and hence a Borel measure.

3.8 Suppose μ is a regular Borel measure on R^n, and let $E \in \mathcal{B}_n$. Show that there exist a G_δ set U and an F_σ set V such that

$$V \subseteq E \subseteq U\,, \quad \mu(U \setminus V) = 0\,.$$

3.9 Suppose μ is a regular Borel measure on R^n, and let f be a nonnegative integrable function. Show that the set function

$$\lambda(E) = \int_E f\,d\mu\,, \quad E \in \mathcal{B}_n\,,$$

is also a regular Borel measure on R^n.

3.10 Suppose μ, λ, are Borel measures on R^n and let

$$(\mu \vee \lambda)(E) = \sup\{\mu(A) + \lambda(E \setminus A) : A \subseteq E, A \in \mathcal{B}_n\}\,,$$

and

$$(\mu \wedge \lambda)(E) = \inf\{\mu(A) + \lambda(E \setminus A) : A \subseteq E, A \in \mathcal{B}_n\}\,.$$

Show that $\mu \vee \lambda$ and $\mu \wedge \lambda$ are Borel measures and that

$$(\mu \vee \lambda)(E) + (\mu \wedge \lambda)(E) = \mu(E) + \lambda(E)\,, \quad \text{all } E \in \mathcal{B}_n\,.$$

If μ and λ are regular, are $\mu \vee \lambda$ and $\mu \wedge \lambda$ also regular?

3.11 An atom of a Borel measure μ is a singleton $\{x\}$ such that $\mu(\{x\}) > 0$. Show that the number of atoms of a σ-finite Borel measure μ is at most countable.

3.12 Suppose μ_F is the Borel measure on the line induced by $F \in \mathcal{D}$. Show that

$$\mu_F(\{x\}) = 0 \quad \text{iff} \quad F \text{ is continuous at } x\,.$$

Moreover, if $\{x\}$ is an atom for μ_F, then we have

$$\mu_F(\{x\}) = F(x) - F(x^-)\,.$$

3.13 Show that a regular Borel measure μ on the line is a probability measure iff there exists $F \in \mathcal{D}$ such that $\mu = \mu_F$ and

$$\lim_{x \to -\infty} F(x) = 0\,, \quad \text{and} \quad \lim_{x \to \infty} F(x) = 1\,.$$

3.14 We say that a Borel measure μ_F is atomic, or discrete, if $\mu_F(E) = 0$ whenever $E \in \mathcal{B}_1$ does not contain any atom of μ_F; in this case the associated distribution function F is said to be discrete. On the other hand, if F is continuous, or equivalently when μ_F has no atoms, we say that the Borel measure is continuous. Show that if μ is a regular Borel measure on the line, then there exist a discrete measure μ_d and a continuous measure μ_c such that $\mu = \mu_d + \mu_c$. Is this decomposition unique?

3.15 Suppose μ is a regular Borel measure on the line with no atoms, and let $0 < \eta < \mu(R)$. Show that there exists $E \in \mathcal{B}_1$ such that $\mu(E) = \eta$.

3.16 Let μ be a regular Borel measure on R^n. Show that there exists a unique closed subset C of R^n with the following two properties: (a) $\mu(R^n \setminus C) = 0$, and (b) If $\mathcal{O}$ is an open set such that $C \cap \mathcal{O} \neq \emptyset$, then $\mu(C \cap \mathcal{O}) \neq 0$. This closed set C is called the support of μ, and we also say that μ is supported in C, and denote this relation by $\operatorname{supp} \mu = C$.
What is the support of the Dirac δ_x measure?
Given a compact subset K of R, construct a measure μ such that $\operatorname{supp} \mu = K$.

3.17 Show that

$$\operatorname{supp}(\mu \vee \lambda) = \operatorname{supp} \mu \cup \operatorname{supp} \lambda\,,$$

and that

$$\operatorname{supp}(\mu \wedge \lambda) \subseteq \operatorname{supp} \mu \cap \operatorname{supp} \lambda\,. \tag{3.1}$$

By means of an example show that the inclusion in (3.1) may be proper.

3.18 Suppose μ is a regular Borel measure on the plane such that for any horizontal or vertical line L we have $\mu(L) = 0$. Show that the function

$$\phi(x) = \mu(I(x,1))\,, \quad x \in R^2\,,$$

is continuous.
Can you think of n-dimensional extensions?

3.19 Let μ be a finite Borel measure on the plane, and suppose that for any line L we have $\mu(L) = 0$. Show that if $E \in \mathcal{B}_2$ and $0 < \eta < \mu(E)$, there is a Borel set $A \subset E$ such that $\mu(A) = \eta$.

3.20 Suppose μ_F is the Borel measure induced by the distribution function $F \in \mathcal{D}$ given by

$$F(x) = \begin{cases} 0 & \text{if } x \leq 0 \\ x^2 & \text{if } 0 < x < 1 \\ 2 & \text{if } x \geq 1. \end{cases}$$

Find the measures $\mu_{F,c}$, $\mu_{F,d}$ associated to μ_F as in 3.14 above, and compute

$$\int_R x \, d\mu_F(x) \, .$$

3.21 Suppose $F \in \mathcal{D}$ satisfies: $F(x) = 0$ for $x \leq 0$, and $F(x) = 1$ for $x \geq 1$. Compare

$$\int_{[0,1]} (1 - F(x)) \, dx \quad \text{and} \quad \int_R x \, d\mu_F(x) \, .$$

3.22 Given a distribution function F such that $\lim_{x \to -\infty} F(x) = 0$ and $\lim_{x \to \infty} F(x) = 1$, show that $\int_{-\infty}^{\infty} |x| \, d\mu_F(x) < \infty$ iff the integrals $\int_{(-\infty,0]} F(x) \, dx$ and $\int_{[0,\infty)} (1 - F(x)) \, dx$ are finite.

3.23 Show that if $F \in \mathcal{D}$ is odd, then for any $g \in C_0(R)$ we have

$$\int_R g(x) \, d\mu_F(x) = \int_R g(-x) \, d\mu_F(x) \, .$$

3.24 (Change of Variable). Let λ be a finite Borel measure on $\mathcal{B}_n$, τ a continuous real-valued function defined on R^n, and set

$$F(x) = \lambda(\tau^{-1}(-\infty, x]) \, , \quad x \in R \, .$$

Show that $F \in \mathcal{D}$, and that for any $g \in C_0(R)$ we have

$$\int_{R^n} g \circ \tau \, d\lambda = \int_R g \, d\mu_F \, .$$

In what follows we assume that the distribution functions $F \in \mathcal{D}$ we work with are normalized so that

$$\lim_{x \to -\infty} F(x) = 0 \, , \quad \text{and} \quad \lim_{x \to \infty} F(x) = 1 \, .$$

3.25 Given $F, F_n \in \mathcal{D}, n = 1, 2, \ldots$ we say that F_n converges weakly to F if at each point x at which F is continuous we have

$$\lim_{n\to\infty} F_n(x) = F(x).$$

Prove that F_n converges weakly to F iff

$$\lim_{n\to\infty} \mu_{F_n}((-\infty,x]) = \mu_F((-\infty,x]),$$

at every point x for which $\mu(\{x\}) = 0$.

3.26 Among other reasons, the notion of weak convergence is important in Probability because of the following approximation property: For every $F \in \mathcal{D}$ there exists a sequence $\{F_n\} \subseteq \mathcal{D}$ such that

(i) F_n converges weakly to F.

(ii) F_n is continuous everywhere on R.

(iii) F_n is constant on each interval of the form $((k-1)/n, k/n]$, $k = 0, \pm 1, \pm 2, \ldots$

3.27 The notion of weak convergence corresponds to a "metric" convergence. The Lévy distance $d(F,G)$ between $F, G \in \mathcal{D}$ is defined as the infimum of those $\varepsilon > 0$ for which

$$G(x-\varepsilon) - \varepsilon \le F(x) \le G(x+\varepsilon) + \varepsilon, \quad \text{all } x \in R.$$

Prove that a necessary and sufficient condition for F_n to converge weakly to F is that $d(F_n, F) \to 0$.

3.28 Show that if $F \in \mathcal{D}$ is everywhere continuous, then F is uniformly continuous.

3.29 Given a real-valued function G defined on R, we define its modulus of continuity $\omega(G, \varepsilon)$ by the expression

$$\omega(G,\varepsilon) = \sup\{G(x) - G(y) : |x-y| \le \varepsilon\}.$$

Show that if $F, G \in \mathcal{D}$, and $d(F,G) < \varepsilon$, then we have

$$\sup |F(x) - G(x)| \le \varepsilon + \omega(F,\varepsilon).$$

As a consequence of this prove that if F_n converges weakly to an everywhere continuous distribution F, then F_n converges uniformly to F.

Is this statement true if F is only continuous on a bounded closed interval?

3.30 Suppose $\mu, \mu_n \in \mathcal{BB}$, $\mu(R) = \mu_n(R) = 1$ for all n. We say that μ_n converges weakly to μ if

$$\lim_{n\to\infty} \mu_n((-\infty,x]) = \mu((-\infty,x])$$

at every point x at which $\mu(\{x\}) = 0$. For instance, the statements F_n converges weakly to F and μ_{F_n} converges weakly to μ_F are only different expressions of the same fact.
Prove that if μ_n is the Dirac measure at x_n, and μ the Dirac measure at x, then μ_n converges weakly to μ iff $x_n \to x$.

3.31 Let μ, μ_n be Borel measures on the line, $n = 1, 2, \ldots$ Then the following conditions are equivalent:
(i) μ_n converges weakly to μ.
(ii) For every $f \in C_0(R)$ we have

$$\lim_{n\to\infty} \int_R f\, d\mu_n = \int_R f\, d\mu\,.$$

(iii) For every Borel set E with the property that its boundary $\partial E = \overline{E} \cap \overline{(R \setminus E)}$ is μ-null, we have

$$\lim_{n\to\infty} \mu_n(E) = \mu(E)\,.$$

3.32 If f denotes the Cantor-Lebesgue function in $I = [0,1]$ and $x \in C$ is of the form $x = \sum_{n=1}^{\infty} 2a_n/3^n$, where each $a_n = 0$ or 1, show that $f(x) = \sum_{n=1}^{\infty} a_n/2^n$.
Also compute

$$\int_I x^n f(x)\, dx\,, \quad n = 0, 1, 2\,.$$

CHAPTER **X**

Absolute Continuity

In this chapter we discuss the class of absolutely continuous functions, namely, those functions which may be recovered by integrating their derivatives.

1. VITALI'S COVERING LEMMA

In dealing with the question of whether the indefinite integral of f differentiates to $f(x)$ it was essential to handle families of intervals. The same is true in general problems of differentiation, or any other area of Analysis where intervals are sorted out. For the problem at hand this process is carried out by means of Vitali's covering lemma; first a definition.

A family $\mathcal{V}$ of closed intervals of R is said to be a covering of E in the sense of Vitali if for any $x \in E$ and $\varepsilon > 0$ there is an interval I in $\mathcal{V}$ which contains x and so that $|I| \leq \varepsilon$. In other words, every point of E belongs to arbitrarily small intervals of $\mathcal{V}$.

Such coverings satisfy the following remarkable property.

Theorem 1.1 (Vitali's Covering Lemma). Suppose E is a subset of the line with $|E|_e < \infty$ and $\mathcal{V}$ is a covering of E in the sense of Vitali. Then there exists an at most countable family $\{I_k\}$ of pairwise disjoint intervals of $\mathcal{V}$ such that

$$|E \setminus \bigcup_k I_k| = 0\,. \tag{1.1}$$

Proof. Let $\mathcal{O}$ be an open set of finite measure which contains E, and discard from $\mathcal{V}$ those intervals that are not totally contained in $\mathcal{O}$. It is clear that this new family, which we call $\mathcal{V}$ again for simplicity, is also a Vitali covering of E.

Having done this, pick an interval I_1, say, of $\mathcal{V}$. If $|E \setminus I_1| = 0$ we are done, otherwise we choose recursively a family of intervals of $\mathcal{V}$ according to the following rule: Suppose that the pairwise disjoint intervals $I_1, \ldots, I_n$ of $\mathcal{V}$ have been chosen and that $|E \setminus \bigcup_{k=1}^{n} I_k|_e > 0$. Consider then the open set

$$G_n = \mathcal{O} \setminus \bigcup_{k=1}^{n} I_k \neq \emptyset \,,$$

and the class of those intervals of $\mathcal{V}$ totally contained in G_n; the idea is to select I_{n+1} as a largest interval in this class. Thus, if

$$k_n = \sup\{|I| : I \in \mathcal{V} \text{ and } I \subset G_n\} > 0 \,,$$

let I_{n+1} be any subinterval of $\mathcal{V}$ contained in G_n such that

$$|I_{n+1}| > k_n/2 \,;$$

by construction it is clear that $I_{n+1} \cap (I_1 \cup \ldots \cup I_n) = \emptyset$.

Either the selection process stops after a finite number of steps, and if this is the case we have finished, or else there exists a pairwise disjoint sequence $\{I_k\}$ of intervals of $\mathcal{V}$ such that

$$\bigcup_k I_k \subseteq \mathcal{O} \,, \quad \text{and} \quad \sum_k |I_k| \le |\mathcal{O}| < \infty \,.$$

In this case, given $\eta > 0$, we may find N such that $\sum_{k=N+1}^{\infty} |I_k| < \eta$. We consider $R_N = E \setminus \bigcup_{k=1}^{N} I_k$ and estimate its Lebesgue outer measure in terms of η. Since each x in R_N belongs to the open set G_N, by assumption there is an interval $I \in \mathcal{V}$ containing x such that $I \cap (I_1 \cup \ldots \cup I_N) = \emptyset$. We claim that there is an index $n > N$ so that $I \cap I_n \neq \emptyset$. Indeed, if $I \in \mathcal{V}$ and for all m we have $I \cap I_m = \emptyset$, it follows that

$$|I| \le k_m < 2|I_{m+1}| \to 0 \,, \quad \text{as } m \to \infty \,,$$

which is impossible. Let n be smallest index so that $I \cap I_n \neq \emptyset$; clearly $n > N$. Furthermore, since by the way the I_n's were selected we have $|I| \le k_n < 2|I_{n+1}|$, by simple geometric considerations we obtain

$$\begin{aligned} d(x, \text{midpoint of } I_{n+1}) &\le |I| + |I_{n+1}|/2 \\ &< 2|I_{n+1}| + |I_{n+1}|/2 = 5|I_{n+1}|/2 \,. \end{aligned}$$

Let J_{n+1} denote the interval concentric with I_{n+1} with sidelength 5 times that of I_{n+1}. By the above estimate it follows that $x \in J_{n+1}$ and consequently, $R_n \subseteq \bigcup_{k=N+1}^{\infty} J_k$. Thus

$$|R_N|_e \le \sum_{k=N+1}^{\infty} |J_k| = 5 \sum_{k=N+1}^{\infty} |I_k| = 5\eta \,. \tag{1.2}$$

Whence by (1.2) we have

$$|E \setminus \bigcup_{k=1}^{\infty} I_k|_e \le \Big| E \setminus \bigcup_{k=1}^{N} I_k \Big|_e < 5\eta \,,$$

and, since η is arbitrary, (1.1) holds. ∎

Corollary 1.2. Under the assumptions of Theorem 1.1, given $\varepsilon > 0$, there exists a finite family $I_1, \dots, I_N$ of pairwise disjoint intervals of $\mathcal{V}$ such that

$$\left| E \setminus \bigcup_{k=1}^{N} I_k \right|_e < \varepsilon . \tag{1.3}$$

Proof. Pick $\eta = \varepsilon/5$ in the proof of Theorem 1.1; then (1.2) gives the desired conclusion. ■

Note that whereas the validity of Corollary 1.2 requires that $|E|_e < \infty$, the conclusion of the Vitali covering lemma is true for an arbitrary subset E of R.

2. DIFFERENTIABILITY OF MONOTONE FUNCTIONS

Suppose f is a real-valued function defined on $I = (a,b)$, and for $x \in I$ and $h \neq 0$ with $x + h \in I$ put

$$Df(x,h) = \frac{f(x+h) - f(x)}{h} .$$

Whether f is differentiable at $x \in I$, or has a one-sided derivative at x, or not, the following four quantities, called the Dini numbers of f at x, are well-defined:

$$D^+ f(x) = \limsup_{h \to 0+} Df(x,h) , \quad D_+ f(x) = \liminf_{h \to 0+} Df(x,h) ,$$
$$D^- f(x) = \limsup_{h \to 0-} Df(x,h) , \quad D_- f(x) = \liminf_{h \to 0-} Df(x,h) .$$

Clearly $D_+ f(x) \leq D^+ f(x)$ and $D_- f(x) \leq D^- f(x)$ and $f'(x)$ exists iff all four Dini numbers of f at x are equal.

The stage is now set for

Theorem 2.1 (Lebesgue). Let I be an open subinterval of the line and suppose f is a monotone real-valued function defined on I. Then f' exists a.e. on I.

Proof. We may assume that f is nondecreasing, and consider first the case when I is bounded. We will be done once we show that

$$D_- f \leq D^- f \leq D_+ f \leq D^+ f \leq D_- f , \quad \text{a.e. on } I , \tag{2.1}$$

for then all the Dini numbers of f are equal at those x's where (2.1) holds, and f' exists a.e. on I.

As noted above, the first and third inequalities in (2.1) are always true, so we only need to establish the second and fourth inequalities there. Now, this amounts to showing that the (bad) sets

$$B = \{x \in I : D^+f(x) > D_-f(x)\} \text{ and } B' = \{x \in I : D^-f(x) > D_+f(x)\}$$

are null. Since the proof for both sets follows along similar lines we only consider B.

First observe that all Dini numbers are nonnegative, and if for rational numbers $u > v > 0$ we put

$$B_{u,v} = \{x \in I : D^+f(x) > u > v > D_-f(x)\} ,$$

then we have $B = \bigcup_{u,v} B_{u,v}$. Thus the desired conclusion will follow once we show that each of the $B_{u,v}$'s is null. So we suppose that $|B_{u,v}|_e = \eta$, and show that $\eta = 0$.

The idea of the proof is to approximate $B_{u,v}$ by a simpler set consisting of pairwise disjoint intervals (here we use the fact that $D_-f < v$ and the Vitali covering lemma) and then to further approximate the part of $B_{u,v}$ which lies within those intervals by another family of intervals (here we use the fact that $D^+f > u$ and Vitali's covering lemma again).

First observe that since $B_{u,v} \subseteq I$ we have $\eta < \infty$, and, by (1.8) in Chapter V, given $\varepsilon > 0$, there exists an open set $\mathcal{O} \supseteq B_{u,v}$ such that

$$|\mathcal{O}| \le |B_{u,v}|_e + \varepsilon = \eta + \varepsilon .$$

Moreover, since for each x in $B_{u,v}$ we have $D_-f(x) < v$, there exists a sequence $h_{x,n} > 0$ approaching 0 such that the intervals $[x - h_{x,n}, x] \subset \mathcal{O}$ and

$$f(x) - f(x - h_{x,n}) < v h_{x,n} \quad \text{all } n . \tag{2.2}$$

Clearly $\mathcal{V} = \{[x - h_{x,n}, x]\}$ is a covering of $B_{u,v}$ in the sense of Vitali and consequently, by Theorem 1.1 there is a finite collection $I_1 = [x_1, x_1 - h_1]$, $\ldots, I_n = [x_n, x_n - h_n]$, say, of pairwise disjoint intervals of $\mathcal{V}$ such that

$$\left| B_{u,v} \setminus \bigcup_{j=1}^n I_j \right|_e < \varepsilon . \tag{2.3}$$

Next let $\tilde{I}_j$ denote the interior of $I_j, 1 \le j \le n$, and observe that since for each x in $B'_{u,v} = B_{u,v} \cap \left(\bigcup_{j=1}^n \tilde{I}_j \right)$ we have $D^+f(x) > u$, there is a sequence $k_{x,m} > 0$ tending to 0 such that $[x,x + k_{x,m}] \subseteq I_j$ for some $1 \le j \le n$, and

$$f(x + k_{x,m}) - f(x) > uk_{x,m}\,, \quad \text{all } m\,. \tag{2.4}$$

Since $\mathcal{V}_1 = \{[x,x + k_{x,m}]\}$ is a covering of $B'_{u,v}$ in the sense of Vitali, we can find a finite collection $J_1 = [x'_1,x'_1 + k_1], \ldots, J_m = [x'_m,x'_m + k_m]$, say, of pairwise disjoint intervals of $\mathcal{V}_1$ with the property that

$$\left|B'_{u,v} \setminus \bigcup_{i=1}^{n} J_i\right|_e < \varepsilon\,. \tag{2.5}$$

Let now

$$0 \le \Delta_J = \sum_{i=1}^{m} (f(x'_i + k_i) - f(x'_i))$$

denote the increase of f along the J_i's, and, similarly, let Δ_I denote the increase of f along the I_j's. It is not hard to check that

$$\Delta_J \le \Delta_I\,. \tag{2.6}$$

Indeed, suppose that $J_1, \ldots, J_{m_1}$ are ordered from left to right and are contained in I_1; that $J_{m_1+1}, \ldots, J_{m_2}$ are ordered from left to right and are contained in I_2, and so on. Since $\{J_i\}_{i=1}^{m_1}$ is a pairwise disjoint collection of intervals contained in I_1 it readily follows that

$$\begin{aligned} &\sum_{i=1}^{m_1} (f(x'_i + k_i) - f(x'_i)) \\ &\qquad = f\left(x'_{m_1} + k_{m_1}\right) - \cdots - \left(f(x'_2) - f(x'_1 + k_1)\right) - f(x_1) \\ &\qquad \le f(x'_{m_1} + k_{m_1}) - f(x'_1) \le f(x_1) - f(x_1 - h_1)\,. \end{aligned}$$

Whence, by adding up the increase of f along these blocks of J_i's it follows that (2.6) holds.

Now, by (2.2), and since the I_j's are all contained in $\mathcal{O}$, it is clear that

$$\Delta_I < \sum_{j=1}^{n} v|I_j| = v|\mathcal{O}| \le v(\eta + \varepsilon)\,. \tag{2.7}$$

On the other hand, by (2.4) we get

$$\Delta_J > u \sum_{i=1}^{m} |J_i|\,, \tag{2.8}$$

and consequently, we need a lower bound for the right-hand side of (2.8). Since $B_{u,v} = \left(B_{u,v} \setminus \bigcup_{j=1}^{n} I_j\right) \cup B'_{u,v} \cup N$, where $N = \{$endpoints of the I_j's$\}$ is a finite set, it is clear that

$$B_{u,v} \subseteq \left(B_{u,v} \setminus \bigcup_{j=1}^{n} I_j\right) \cup (B'_{u,v} \setminus \bigcup_{i=1}^{m} J_i) \cup (\bigcup_{i=1}^{m} J_i) \cup N\,.$$

Thus, on account of (2.3) and (2.5) it readily follows that

$$\eta = |B_{u,v}|_e \le \varepsilon + \varepsilon + \sum_{i=1}^{m} |J_i|\,, \quad \text{or} \quad \sum_{i=1}^{m} |J_i| \ge \eta - 2\varepsilon\,.$$

Substituting this estimate in (2.8), and combining it with (2.6) and (2.7), we see that

$$(\eta - 2\varepsilon)u < \Delta_J \le \Delta_I < v(\eta + \varepsilon)\,, \quad \varepsilon > 0. \tag{2.9}$$

Since ε in (2.9) is arbitrary, we also have $\eta u \le \eta v$, and since $0 < v < u$, this can only hold if, as asserted, $\eta = 0$. This completes the proof when I is bounded.

On the other hand, if the interval I is unbounded, observe that

$$I = \bigcup_{k=1}^{\infty}(I \cap (-k,k)) = \bigcup_{k=1}^{\infty} I_k\,,$$

say, where I_k is a bounded open interval, $k = 1, 2, \ldots$ Thus f' exists a.e. on each I_k, and consequently also a.e. on I. ■

Not only does the derivative of a monotone function exist a.e., but it also is integrable. More precisely, we have Lebesgue's theorem,

Theorem 2.2. Suppose f is a nondecreasing real-valued function defined on $I = (a,b)$. Then $f' \in L(I)$ and

$$\int_I f'\,dx \le f(b^-) - f(a^+)\,. \tag{2.10}$$

Proof. Extend f to R by setting $f(x) = f(a^+)$ if $x \le a$, and $f(x) = f(b^-)$ if $x \ge b$. Put now

$$f_n(x) = n(f(x + 1/n) - f(x))\,, \quad x \in R\,, n = 1, 2, \ldots\,,$$

and observe that since by Theorem 2.1 f' exists a.e., we have

$$\lim_{n\to\infty} f_n = f' \quad \text{a.e. on } I\,.$$

Thus, by Fatou's Lemma, it follows that

$$\int_I f'\,dx \le \liminf \int_I f_n\,dx\,. \tag{2.11}$$

It is rather straightforward to compute the integral on the right-hand side of (2.11). Indeed, by 4.6 in Chapter VII, and with n sufficiently large, we have

$$\begin{aligned}\int_I f_n\,dx &= n\int_I f(x+1/n)\,dx - n\int_I f\,dx\\ &= n\int_{[b,b+1/n)} f\,dx - n\int_{(a,a+1/n]} f\,dx = A+B\,,\end{aligned}$$

say. Clearly $A = f(b^-)$, and $B \geq f(a^+)$. Whence, for all sufficiently large n the integral in question is dominated by $f(b^-) - f(a^+)$, and (2.10) holds. ■

Corollary 2.3. Suppose f is BV on a bounded interval $I = [a,b]$. Then $f' \in L(I)$ and

$$\int_I |f'|\,dx \leq V(f;a,b). \tag{2.12}$$

Our aim now is to discover when equality holds in (2.10); for this we need the concept of absolutely continuous functions.

3. ABSOLUTELY CONTINUOUS FUNCTIONS

Absolutely continuous, or AC, functions were introduced in 3.6 in Chapter VIII. They are continuous functions whose increment along any collection of pairwise disjoint intervals of sufficiently small total length is arbitrarily small. This concept excludes the Cantor-Lebesgue function f which, although being locally constant on a subset of [0,1] of full measure, it nevertheless increases from 0 to 1 there. To see this we cover the Cantor set C by a union $\bigcup_n(a_n,b_n)$, say, of pairwise disjoint open intervals with $\sum_n(b_n-a_n)$ is arbitrarily small. Extend f so that $f(x)=0$ for $x<0$ and $f(x)=1$ for $x>1$. Then it is not hard to verify that $\sum_n(f(b_n)-f(a_n)) = 1$, and consequently, $\sum_{n=1}^N(f(b_n)-f(a_n)) > 1/2$ for sufficiently large N, while at the same time, $\sum_{n=1}^N(b_n-a_n)$ is arbitrarily small.

So, which among the continuous functions are AC, and what properties do AC functions satisfy?

Proposition 3.1. Let $I = [a,b]$ and suppose f is AC on I. Then f is BV on I, and consequently, by Corollary 2.3, f' exists a.e. and it is integrable there.

Proof. Let δ be the real number that corresponds to the choice $\varepsilon = 1$ in the AC definition of f, and let the integer $N > (b-a)/\delta$. Note that, in particular, we have

$$V(f;x,x+\eta) \leq 1\,, \quad \text{any } x \in I\,, 0 < \eta \leq \delta\,. \tag{3.1}$$

The idea now is to use (3.1) to put together the estimates along the partitions of I. So, let $\mathcal{P} = \{a = x_0 < \ldots < x_n = b\}$ be a partition of I and let $\mathcal{P}'$ be the partition of I obtained by adjoining the points $a + (b-a)/N$, $a + 2(b-a)/N, \ldots, b$ to $\mathcal{P}$. Since $\mathcal{P}'$ is finer than $\mathcal{P}$ it readily follows that

$$\sum_{\text{over } \mathcal{P}} |\Delta_k f| \leq \sum_{\text{over } \mathcal{P}'} |\Delta_k f|\,. \tag{3.2}$$

It is not hard to estimate the right-hand side of (3.2); indeed, by (3.1) it does not exceed

$$V(f;a,a+(b-a)/N) + \cdots + V(f;a+(N-1)(b-a)/N,b) \leq N\,.$$

Since $\mathcal{P}$ is arbitrary, by (3.2) it follows that $V(f;a,b) \leq N$ and f is BV on I. ■

So, AC functions are continuous and BV, but is the converse to this statement true? It is partially true, and to discuss it we need some preliminary results.

Lemma 3.2. Suppose $A \subseteq I = [a,b]$, and let f be a real-valued function defined on I such that

$$|f'(x)| \leq M\,, \quad x \in A\,.$$

Then

$$|f(A)|_e \leq M|A|_e\,. \tag{3.3}$$

Proof. Given $\varepsilon > 0$, let $\mathcal{O}$ be an open set such that

$$|\mathcal{O}| \leq |A|_e + \varepsilon\,, \quad \mathcal{O} \supseteq A\,. \tag{3.4}$$

Break up A into two disjoint parts, $A_1 = \{x \in A : f \text{ is constant in a neighbourhood of } x\}$, and $A_2 = \{x \in A : f \text{ is not constant in any neighbourhood of } x\}$, say. We construct a covering $\mathcal{V}$ of $f(A)$ in the sense of Vitali as follows: If $f(x) \in f(A)$ and $x \in A_1$, then there is an interval $J \subseteq \mathcal{O}$ such that $f(x) = f(y)$ for all $y \in J$. Now, if $I(w,w')$ denotes the closed interval with endpoints w and w' (note that $w' < w$ is possible), then we can find h so that $I(x,x+Mh) \subset \mathcal{O}$ and f is constant on $I(x,x+Mh)$. Such values $f(x)$ are then assigned the intervals $I(f(x),f(x)+Mb)$, where b satisfies $I(x,x+Mb) \subseteq I(x,x+Mh)$.

On the other hand, if $f(x) \in f(A)$ and $x \in A_2$, then there is a sequence $h_{x,n} \neq 0$, $n = 1,2,\ldots$, converging to 0, such that $I(x,x+h_{x,n}) \subset \mathcal{O}$ for all n, and

$$0 < |f(x+h_{x,n}) - f(x)| \leq (M+\varepsilon)|h_{x,n}|\,, \quad n = 1,2,\ldots$$

To these values $f(x)$ we assign the intervals $I(f(x),f(x+h_{x,n}))$, $n = 1,2,\ldots$

Now, the collection $\mathcal{V}$ of all the intervals introduced above is a covering of $f(A)$ in the sense of Vitali and consequently, there is an at most countable family consisting pairwise disjoint intervals $I_1,\ldots,I_k\ldots$, say, such that

$$|f(A) \setminus \cup_k I_k|_e = 0\,.$$

Whence, we also have

$$|f(A)|_e \leq |\cup_k I_k| = \sum_{k=1}^{m} |I_k|\,. \tag{3.5}$$

To estimate the right-hand side of (3.5) we separate the I_k's into two families: Those that correspond to $f(x)$ with $x \in A_1$, call them I_k^1's, and those corresponding to $f(x)$ with $x \in A_2$, call them I_k^2's. Furthermore, if $I_k^1 = I(f(x_k),f(x_k)+Mb_k)$, let $J_k^1 = I(x_k,x_k+b_k)$, and if $I_k^2 = I(f(x_k),f(x_k+h_k))$, let $J_k^2 = I(x_k,x_k+h_k)$.

Since the I_k's are pairwise disjoint, so are the J_k's, and they are also contained in $\mathcal{O}$. Consequently, by (3.4) we have

$$\begin{aligned}\sum |I_k| &= \sum |I_k^1| + \sum |I_k^2| \leq \sum M|b_k| + \sum (M+\varepsilon)|h_k| \\ &= M\sum |J_k^1| + (M+\varepsilon)\sum |J_k^2| \leq (M+\varepsilon)\sum |J_k| \\ &\leq (M+\varepsilon)|\mathcal{O}| \leq (M+\varepsilon)(|A|_e+\varepsilon)\,,\end{aligned}$$

which substituted into (3.5) gives $|f(A)|_e \leq (M+\varepsilon)(|A|_e+\varepsilon)$. Moreover, since ε is arbitrary, the above inequality is also true with $\varepsilon = 0$, and (3.3) holds. ■

An interesting consequence of Lemma 3.2 is

Lemma 3.3. Suppose f is a real-valued function defined on $I = [a,b]$, and let A be a measurable subset of I so that $f'(x)$ exists everywhere on A and is measurable there. Then

$$|f(A)|_e \leq \int_A |f'|\, dx\,. \tag{3.6}$$

Proof. First suppose that for some integer M we have $|f'(x)| \leq M$ on A and consider the level sets

$$A_{k,n} = \{x \in A : (k-1)/2^n \leq |f'(x)| < k/2^n\}, k = 1, \ldots, M2^n, n = 1, 2, \ldots$$

Since for each n we have $A = \bigcup_k A_{k,n}$, by Lemma 3.2 and Chebychev's inequality it readily follows that

$$\begin{aligned}
|f(A)|_e &\leq \sum_k |f(A_{k,n})|_e \leq \sum_k (k/2^n)|A_{k,n}| \\
&= \sum_k ((k-1)/2^n)|A_{k,n}| + \frac{1}{2^n}\sum_k |A_{k,n}| \\
&\leq \sum_k \int_{A_{k,n}} |f'|\, dx + \frac{1}{2^n}|A| \leq \int_A |f'|\, dx + \frac{1}{2^n}|A|\,.
\end{aligned}$$

Since this estimate holds for every n, (3.6) is true in this case. As for the general case, note that

$$A = \bigcup_{k=1}^{\infty}\{x \in A : k-1 \leq |f'| < k\} = \bigcup_{k=1}^{\infty} A_k\,,$$

say, where the A_k's are pairwise disjoint and $|f'|$ exists and is bounded and measurable on each A_k. Then, by the first part of the proof we have

$$|f(A)|_e \leq \sum_k |f(A_k)|_e \leq \sum_k \int_{A_k} |f'|\, dx = \int_A |f'|\, dx\,,$$

and (3.6) holds. ■

We are now ready for

Theorem 3.4 (Banach-Zarecki). Suppose f is a continuous, BV, real-valued function defined on $I = [a,b]$. Then f is AC on I iff f maps null sets into null sets, i.e.,

$$|A| = 0 \quad \text{implies} \quad |f(A)| = 0\,.$$

Proof. We show the necessity first: It is enough to prove that given a null set $A \subseteq (a,b)$ and $\varepsilon > 0$, we have

$$|f(A)|_e \leq \varepsilon \,. \tag{3.7}$$

We invoke (iii) in 4.14 below: From the hypothesis of AC there exists $\delta > 0$ such that no matter what finite pairwise disjoint family $\{I_k = (a_k,b_k)\}$ of subintervals of (a,b) we take, with the notation $\omega(f,J) = \sup_J f - \inf_J f$, we have

$$\sum (b_k - a_k) < \delta \quad \text{implies} \quad \sum \omega(f, I_k) < \varepsilon \,. \tag{3.8}$$

Also observe that since $|f(I_k)|_e \leq \omega(f, I_k)$ for each k, by (3.8) it readily follows that

$$\left|f\left(\bigcup_k I_k\right)\right|_e \leq \sum |f(I_k)|_e \leq \sum \omega(f, I_k) \leq \varepsilon \,.$$

Choose now an open set $\mathcal{O}$ with $|\mathcal{O}| < \delta$ so that $A \subset \mathcal{O} = \bigcup_{k=1}^\infty (a_k,b_k) \subseteq (a,b)$, where the (a_k,b_k)'s are pairwise disjoint, and note that the above estimate implies that

$$\left|f\left(\bigcup_{k=1}^n (a_k,b_k)\right)\right|_e < \varepsilon \,, \quad \text{for } n = 1,2,\ldots$$

(3.7) follows at once from this.

As for the sufficiency, suppose that $\varepsilon > 0$ is given, and let $\{(a_k,b_k)\}$ be a finite pairwise disjoint family of subintervals of I. Then, if $A_k = \{x \in [a_k,b_k] : f'(x) \text{ exists }\}$, we have $|[a_k,b_k] \setminus A_k| = 0$ for all k. Furthermore, since f is continuous we also have

$$|f(b_k) - f(a_k)| \leq |f([a_k,b_k])|_e \,, \quad \text{all } k \,.$$

Whence, combining these remarks, and by our assumption and Lemma 3.3, we obtain

$$\begin{aligned} \sum_k |f(b_k) - f(a_k)| &\leq \sum_k |f([a_k,b_k])|_e \\ &\leq \sum_k |f([a_k,b_k] \setminus A_k)|_e + \sum_k |f(A_k|_e \\ &\leq \sum_k \int_{A_k} |f'| \, dx = \int_{\bigcup A_k} |f'| \, dx \,. \end{aligned} \tag{3.9}$$

Now, since f is BV on I, by Corollary 2.3 f' is integrable on I, and we are in a position to invoke 3.7 in Chapter VIII: Choose $\delta > 0$ so that the conclusion there holds for the ε we fixed at the beginning of the argument, and observe that since $\bigcup_k A_k \subseteq \bigcup_k [a_k, b_k]$, we also have

$$|\bigcup_k A_k| \le \delta \quad \text{whenever} \quad \sum_k (b_k - a_k) \le \delta\,.$$

Therefore, by (3.9) it follows at once that

$$\sum_k |f(b_k) - f(a_k)| \le \varepsilon \quad \text{whenever} \quad \sum_k (b_k - a_k) \le \delta\,,$$

and f is AC on I. ■

That the assumption that f is BV is necessary for the validity of Theorem 3.4 follows from a construction which is reminiscent of the discussion preceding (1.5) in Chapter III. Consider $I = [0,1]$ and a Cantor-like subset K of I; the measure of K may be positive or not. Write the set $I \setminus K = \bigcup_n (a_n, b_n)$ as the at most countable pairwise disjoint union of open intervals, and let c_n denote the midpoint of (a_n, b_n). If d_n is a sequence of positive numbers with limit 0, define f on I as follows: $f(x) = 0$ for $x \in K$, $f(c_n) = d_n$ for all n, and f is linear in $[a_n, c_n]$ and $[c_n, b_n]$. Then f is continuous, and $V(f; 0, 1) = 2\sum_{n=1}^{\infty} d_n$. To see that f maps null sets into null sets, consider a null subset A of I. By 5.8 in Chapter I, we have $f(A) = f(A \cap K) \cup \bigcup_n f(A \cap (a_n, b_n))$. Since f is linear in $[a_n, c_n]$ and in $[c_n, b_n]$, it readily follows that $|f(A \cap (a_n, b_n)| = 0$ for all n, and so $|f(A) \le |\{0\}| + \sum_n |f(A \cap (a_n, b_n)| = 0$. If $\sum_n d_n = \infty$, then f fails to be AC on I since it is not BV there.

In order to establish further properties of AC functions we introduce the following definition: Suppose f is a real-valued a.e. differentiable function on an interval I. We then say that f is singular if $f' = 0$ a.e. on I.

How do AC singular functions look?

Proposition 3.5. Suppose f is an AC singular function defined on an interval I. Then f is constant.

Proof. Let A be a subset of I of full measure so that $f'(x) = 0$ for $x \in A$. By Lemma 3.3 we have

$$|f(A)|_e \le \int_A |f'|\,dx = 0\,. \tag{3.10}$$

Further, since $|I \setminus A| = 0$, by the necessity of Theorem 3.4 it follows that

$$|f(I \setminus A)| = 0\,. \tag{3.11}$$

Whence combining (3.10) and (3.11) we get

$$|f(I)|_e \le |f(A)|_e + |f(I \setminus A)|_e = 0\,. \tag{3.12}$$

Now, since f is continuous, unless f is constant, $f(I)$ contains an interval and (3.12) does not hold. Therefore f must be constant. ∎

We are now ready to characterize, following Lebesgue, the class of functions which may be reconstructed by integrating their derivatives.

Theorem 3.6. Suppose f is a real-valued function defined on $I = [a,b]$. Then, f' exists a.e. in (a,b), it is integrable there, and

$$f(x) - f(a) = \int_{[a,x]} f'(t)\,dt\,, \quad a \le x \le b\,, \tag{3.13}$$

iff f is AC on I.

Proof. We do the sufficiency first: Since f is AC on I, f' exists a.e. on I and it is integrable, therefore we only need to show that (3.13) holds. For this purpose put $F(x) = \int_{[a,x]} f'(t)\,dt$, $a \le x \le b$, and observe that by 3.6 in Chapter VIII also F is AC on I, and, by the Lebesgue differentiation theorem, $F' = f$ a.e. Let $g = F - f$; g is AC and singular on I, and, by Proposition 2.6, g is constant there. More precisely, $F(x) - f(x) = F(a) - f(a)$, $a \le x \le b$, and since $F(a) = 0$, it readily follows that

$$F(x) = \int_{[a,x]} f'(t)\,dt = f(x) - f(a)\,, \quad a \le x \le b\,,$$

as we wanted to show.

Conversely, since

$$f(x) = f(a) + \int_{[a,x]} f'(t)\,dt\,, \quad a \le x \le b\,,$$

the sufficiency follows from Theorem 2.2 in Chapter VIII. ∎

Implicit in the proof of Theorem 3.6 is the following important result concerning BV functions.

Theorem 3.7 (Lebesgue). Suppose f is BV on $I = [a,b]$. Then there exist an AC function g and a singular function h such that

$$f(x) = g(x) + h(x)\,, \quad x \in I\,. \tag{3.14}$$

Up to constants, the decomposition in (3.14) is unique.

Proof. Since f is BV on I, f' exists a.e. there, and it is integrable. Let $g(x) = \int_{[a,x]} f'\,dt$, $a \le x \le b$, and set

$$h(x) = f(x) - g(x)\,, \quad a \le x \le b\,.$$

Then g is AC on I and by the Lebesgue differentiation theorem, $h' = f' - g' = 0$ a.e. on I. Thus $f = g + h$ is a desired decomposition.

As for the uniqueness (modulo constants), suppose that also $f = g_1 + h_1$, where g_1 is AC on I and h_1 is singular. We then have

$$g - g_1 = h_1 - h\,, \tag{3.15}$$

where the expression on the left-hand side of (3.15) is AC and that on the right-hand side is singular. By Proposition 3.5 it readily follows that this function is constant, c say. Thus $g = g_1 + c$, and $h = h_1 - c$. ∎

4. PROBLEMS AND QUESTIONS

4.1 Show that in Vitali's covering lemma we may also demand that given $\varepsilon > 0$, $\sum_k |I_k| \le (1+\varepsilon)|E|_e$.

4.2 State and prove Vitali's covering lemma, including the conclusion of 4.1, for subsets E of R^n with $|E|_e < \infty$, and covered in the sense of Vitali by closed n-dimensional intervals.

4.3 Let E be a subset of R^n that is the union of sets, each being an open interval together with any of its edges. Prove that E is Lebesgue measurable.

4.4 A measure μ on $(R^n, \mathcal{L})$ is said to be doubling provided there exists an absolute constant c such that

$$\mu(I(x,2r)) \le c\,\mu(I(x,r))\,, \quad \text{all } x \in R^n\,, r > 0\,.$$

Given a doubling measure μ, a family $\mathcal{V}$ of closed intervals of R^n is said to be a covering of $E \in \mathcal{L}$ in the sense of Vitali if for any $x \in E$ and $\varepsilon > 0$, there is an interval $I \in \mathcal{V}$ which contains x and so that $\mu(I) \le \varepsilon$. Show that if $\mathcal{V}$ is a covering of E in the sense of Vitali, and $\mu(E) < \infty$, then there exists an at most countable family $\{I_k\}$ of pairwise disjoint intervals of $\mathcal{V}$ such that $\mu\left(E \setminus \bigcup_k I_k\right) = 0$.

4.5 Suppose f is a real-valued function defined on an interval I, and that all the Dini numbers of f for x in I lie between $-k$ and k, where k is some positive constant. Must f be Lipschitz on I? If so, what is the relation between the Lipschitz constant of f and k?

4.6 Let f be a real-valued function defined on $I = [a,b]$ and suppose there exist real constants u, v such that

$$u \leq D^+ f(x) \leq v\,, \quad \text{all } x \in I\,.$$

Is it then true that for all $a \leq x < x + h \leq b$ we have

$$uh \leq f(x+h) - f(x) \leq vh\,?$$

4.7 Suppose f is a continuous real-valued function defined on $I = [a,b]$, $A \subset I$ is at most countable, and $D^+ f(x) \geq 0$ on $I \setminus A$. Prove that f is nondecreasing on I.

4.8 Compute the Dini numbers of the Cantor-Lebesgue function at each point x of $[0,1]$.

4.9 (Fubini's Lemma) Let $\{f_k\}$ be a sequence of nondecreasing functions defined on an interval I of the line. Show that if the series $f(x) = \sum_k f_k$ converges to a finite limit on I, then $f' = \sum_k f'_k$ a.e. on I.

4.10 Let (a_n) be a sequence of distinct points in an interval I of the line, and suppose (u_n), (v_n) are sequences of real numbers such that $\sum |u_n|, \sum |v_n| < \infty$. Put

$$f_n = \begin{cases} 0 & \text{if } x < a_n \\ u_n & \text{if } x = a_n \\ v_n & \text{if } x > a_n\,, \end{cases}$$

and show that $s = \sum f_n$ has a finite derivative a.e. on I and that $s' = 0$ a.e.

4.11 Let E be any subset of R. Show that for almost all x in E we have

$$\lim_{h \to 0} \frac{|E \cap [x-h, x+h]|_e}{2h} = 1\,.$$

4.12 Does there exist a strictly increasing function f defined on an interval I so that $f' = 0$ a.e. on I?

4.13 Suppose f is a real-valued function defined on I. Show that if f is not constant and $f' = 0$ a.e., then f cannot be Lipschitz on I.

4.14 Let f be a real-valued defined on $I = [a,b]$. Prove that the following statements are equivalent:

(i) f is AC on I.

(ii) Given $\varepsilon > 0$, there exists $\delta > 0$ such that for any finite collection $\{[a_i,b_i]\}$ of nonoverlapping subintervals of $[a,b]$ we have $|\sum_i (f(b_i) - f(a_i))| < \varepsilon$, whenever $\sum_i (b_i - a_i) < \delta$.

(iii) If

$$\omega(f,J) = \sup_J f - \inf_J f\,,$$

then for each $\varepsilon > 0$, there is $\delta > 0$ such that for any finite collection $\{I_k = [a_k,b_k]\}$ of nonoverlapping subintervals of I we have $\sum_k \omega(f,I_k) \leq \varepsilon$, whenever $\sum_k (b_k - a_k) < \delta$.

4.15 Let f be a real-valued continuous function defined on $I = [a,b]$, and suppose that f is AC on $[a,d]$, for any $d < b$. Show, then, that f is AC on I.
Is this result true if the assumption that f is continuous on I is dropped?

4.16 If the functions f, g are AC on an interval I, show that their difference, sum and product are also AC on I. If g vanishes nowhere on I, then the quotient f/g is also AC on I.

4.17 Show that the composition of AC functions need not be AC, or even BV for that matter.

4.18 Let f be AC on I, and $f(I) \subseteq J$. If $\phi\colon J \to R$ is Lipschitz, show that $\phi \circ f$ if AC on I.

4.19 Suppose f is a nondecreasing, AC function on I, and $f(I) \subseteq J$. Show that if ϕ is AC on J, then $\phi \circ f$ is AC on I.

4.20 Suppose f is a monotone function defined on $I = [0,1]$, and let $E = \{x \in I : D^+ f(x) = \infty\}$. Prove that f is AC on I iff $|f(E)| = 0$.

4.21 Suppose f is a real-valued continuous function defined on I, and let A be an F_σ subset of I. Prove that $f(A)$ is F_σ too, and show that as a consequence of this, if f is AC on I, then it maps Lebesgue measurable sets into Lebesgue measurable sets.

4.22 Let N be a Lebesgue null subset of $I = [0,1]$. Show that there is a real-valued function f defined on I which is AC there, and such that $f'(x) = \infty$ for each $x \in N$.

4.23 Show that there is an AC function f defined on $[0,1]$ which is monotone on no subinterval of I.

4.24 Let f be a real-valued Lebesgue measurable function defined on $I = [a,b]$, and suppose $\varepsilon, \eta > 0$ are given. Show that there exist a

Lebesgue measurable set $B \subset I$, and an AC function F defined on I such that $\int_{I\setminus B} |f - F| < \varepsilon$ and $|B| < \eta$.

4.25 Let E be a bounded Lebesgue measurable set in the line and let

$$f_n(x) = n \int_{[x,x+1/n]} \chi_E \, dy\,, \quad n = 1,2,\ldots$$

Show that each f_n is AC on every bounded interval of the line, that $0 \le f_n \le 1$, that $\lim_{n\to\infty} f_n = \chi_E$ a.e., and that $\lim_{n\to\infty} d\,(f_n, \chi_E) = 0$.
Is this result sufficient to prove that AC functions are dense in the metric of $L(R)$?

4.26 Let g be a continuous function defined on $I = [a,b]$ and suppose f is AC there. Prove that $\int_a^b g\, df = \int_I g f'\, dy$.

4.27 Let $I = [a,b]$, and suppose f is BV on I. Prove that for each Borel set $E \subseteq I$ we have $\int_E |f'|\, dy \le |V(f)(E)|_e$, and that there is equality here provided that f is AC on I.
A related result is the following: Suppose f is a strictly monotone AC function defined on I, and let $f(I) = J$. Show that for every Borel subset E of J we have $\int_{f^{-1}(E)} f'(y)\, dy = |E|$.

4.28 (Change of Variable) Let $I = [a,b]$, and $g\colon I \to R$, $g(I) \subset J \subset R$, be continuous there. Furthermore, if $J = [c,d]$ and $f\colon J \to R$ is integrable, put $F(x) = \int_{[c,x]} f\, dy$, $c \le x \le d$. Now, suppose that g and $F \circ g$ are a.e. differentiable on their domains of definition, and prove that the relation $(F \circ g)' = (f \circ g)\, g'$ holds a.e. on I. Finally, show that $F \circ g$ is AC on I iff

(i) $(f \circ g)\, g' \in L([a,b])$.

(ii) For each subinterval $I' = [a', b']$ of I we have

$$\int_{g(a')}^{g(b')} f\, dy = \int_{I'} (f \circ g)\, g'\, dy\,.$$

4.29 (Integration by Parts) Let $I = [a,b]$ be a bounded interval, and suppose $F, g \in L(I)$. Show that if

$$F(x) = \int_{[a,x]} f\, dy\,, \quad G(x) = \int_{[a,x]} g\, dy\,, \quad a \le x \le b\,,$$

then $fG, gF \in L(I)$ and

$$\int_I fG\, dy + \int_I Fg\, dy = F(b)G(b) - F(a)G(a)\,.$$

Also, if f, g are AC on I, we have

$$\int_I f g'\,dy + \int_I f' g\,dy = f(b)g(b) - f(a)g(a)\,.$$

4.30 Suppose $f, f' \in L(R)$. Prove that $\int_R f'\,dy = 0$.

4.31 Let $I = [a,b]$, and suppose f is a continuous real-valued function defined on I. The estimate (1.5) in Chapter III implies that if $V(x)$ is AC on I, then also f is AC on I. Discuss whether the converse is true, to wit, does the assumption that f is AC on I imply that $V(x)$ is AC on I?

4.32 Let $\{f_n\}$ be a sequence of AC functions defined on $I = [0,1]$ such that $f_n(0) = 0$ for all n. Assume that the sequence of derivatives $\{f_n'\}$ is Cauchy in $L(I)$, i.e., $\lim_{n,m\to\infty} \int_I |f_n'(x) - f_m'(x)|\,dx = 0$. Prove that $\{f_n\}$ converges uniformly to a function f, and that f is AC in I.

4.33 Prove that a real-valued function f defined on R is of the form

$$f(x) = \int_{(-\infty,x]} \phi\,dy\,, \quad \text{where } \phi \in L(R)\,,$$

iff (a) f is AC on $[-n,n]$ for all n, (b) $V(f;-n,n) \le k < \infty$ for all n, and, (c) $\lim_{|x|\to\infty} f(x) = 0$. Prove it.

4.34 Let $I = [a,b]$, and suppose f is a continuous function defined on I. Show that f is AC on I iff there exists a sequence $\{f_n\}$ of Lipschitz functions defined on I such that $\lim_{n\to\infty} V(f - f_n; a, b) = 0$.

4.35 Suppose f, f_n are BV on $I = [a,b]$, $n = 1, 2, \ldots$, and

$$V(f_n - f;\, x) \to 0\,, \quad \text{for some } a < x < b\,.$$

Prove that there exists a subsequence $n_k \to \infty$ such that

$$f_{n_k}'(x) \to f'(x)\,.$$

CHAPTER XI

Signed Measures

In this chapter we consider σ-additive set functions of arbitrary sign, or signed measures, establish their basic properties and describe, in the Theorems of Lebesgue and Radon-Nikodým, their basic form. I learned the proof of these theorems from R. Bradley.

1. ABSOLUTE CONTINUITY

In Chapter IV we dealt briefly with additive set functions of arbitrary sign; these functions appear quite naturally in applications. We consider here extended real-valued σ-additive set functions; we motivate our interest in them with two simple examples.

Let μ_1, μ_2 be measures on $(X, \mathcal{M})$, and let ν be the set function defined on $\mathcal{M}$ by

$$\nu(E) = \mu_1(E) - \mu_2(E)\,, \quad E \in \mathcal{M}\,. \tag{1.1}$$

Although ν is not necessarily nonnegative, it satisfies many of the properties of a measure provided, of course, that the right-hand side of (1.1) is defined.

Similar considerations apply to the set function

$$\nu(E) = \int_E f\,d\mu_1\,, \quad E \in \mathcal{M}\,, \tag{1.2}$$

where f is an extended real-valued measurable function defined on X for which the integral in (1.2) exists.

To consider the general setting we introduce the following definition: Given a set X and a σ-algebra $\mathcal{M}$ of subsets of X, we say that a set function ν defined on $\mathcal{M}$ is a signed measure provided the following three properties hold:

(i) $\nu: \mathcal{M} \to [-\infty,\infty]$, and ν assumes, at most, one of the values $-\infty$ or ∞.

(ii) $\nu(\emptyset) = 0$.

(iii) If $\{E_k\}_{k=1}^{\infty} \subseteq \mathcal{M}$ is a sequence of pairwise disjoint sets, then

$$\nu\left(\bigcup_{k=1}^{\infty} E_k\right) = \sum_{k=1}^{\infty} \nu(E_k)\,. \tag{1.3}$$

The equality in (1.3) means, in particular, that if $|\nu(\bigcup_k E_k)| < \infty$, then the series on the right-hand side of (1.3) converges absolutely and unconditionally, and that it diverges properly to $\pm\infty$ otherwise.

Also note that property (ii) rules out the possibility that ν is identically $-\infty$ or identically ∞.

The usual properties of a measure are true in this more general setting as well. For instance, properly interpreted, the results concerning limits discussed in Proposition 3.1 of Chapter IV hold; it is incumbent upon the reader to verify that this is the case.

How do signed measures look, and how can they be represented? Referring to 4.8-4.12 in Chapter IV, given a signed measure ν on $(X, \mathcal{M})$, let ν_+, ν_- and $|\nu|$ denote the positive variation, the negative variation, and the total variation of ν, respectively. By 4.14 in Chapter IV, all the variations are actually measures on $(X, \mathcal{M})$ and, by 4.11 in that Chapter, the Jordan decomposition

$$\nu(E) = \nu_+(E) - \nu_-(E)\,, \quad E \in \mathcal{M}\,,$$

obtains. Thus, a general representation in the spirit of (1.1) holds for arbitrary signed measures. We note in passing that the Jordan decomposition is not unique: If μ is any finite measure on $(X, \mathcal{M})$ and $\nu_1 = \nu_+ + \mu$ and $\nu_2 = \nu_- + \mu$, then we also have

$$\nu(E) = \nu_1(E) - \nu_2(E)\,, \quad E \in \mathcal{M}\,.$$

However, the Jordan decomposition satisfies a "minimality" condition, cf. 3.8 below.

As for (1.2), it also leads to an interesting theory; the observations in 3.7 in Chapter VIII are relevant here. In fact, motivated by those considerations we introduce the following definition: Suppose μ is a measure and ν is a signed measure on $(X, \mathcal{M})$; we say that ν is absolutely continuous with respect to μ, and denote this by $\nu \ll \mu$, if $\nu(A) = 0$ for any $A \in \mathcal{M}$ with $\mu(A) = 0$. Informally, each μ-null set has ν measure 0.

Our first result explains how this nomenclature is derived from our usual understanding of absolute continuity.

Proposition 1.1. Let μ_F be the Borel measure induced by the distribution function $F \in \mathcal{D}$. Then μ_F is absolutely continuous with respect to the Lebesgue measure iff F is AC on every bounded interval of R.

Proof. We show the necessity first: If F is not AC on every bounded interval of R, then there exist an interval $I = [a,b]$ and $\varepsilon > 0$, such that for every $\delta > 0$ there is a family $\{[a_k,b_k]\}$ of nonoverlapping subintervals of I so that

$$\sum_k |F(b_k) - F(a_k)| \geq \varepsilon\,, \quad \sum_k (b_k - a_k) \leq \delta\,. \tag{1.4}$$

Now, since as is readily seen, cf. Theorem 2.1 in Chapter IX,

$$|F(b_k) - F(a_k)| \leq \mu_F([a_k, b_k])\,, \quad \text{all } k\,,$$

(1.4) implies that the set $B = \bigcup_k [a_k,b_k]$ verifies

$$\mu_F(B) > \varepsilon\,, \quad |B| \leq \delta\,. \tag{1.5}$$

We now invoke (1.5) with $\delta = 1/2^n, n = 1,2,\ldots$, and construct a sequence of (bad) sets $\{B_n\}$ so that

$$\mu_F(B_n) > \varepsilon\,, \quad |B_n| \leq 1/2^n\,, \quad B_n \subseteq I\,, \quad \text{all } n\,.$$

Thus, on the one hand by the Borel-Cantelli Lemma it follows that $|\limsup B_n| = 0$, and, on the other hand, by 4.25 in Chapter IV, we have $\mu_F(\limsup B_n) > 0$, contradicting the fact that μ_F is absolutely continuous with respect to the Lebesgue measure.

As for the sufficiency, suppose that $|A| = 0$, and given $\varepsilon > 0$, let $\delta > 0$ be the number that corresponds to the choice of ε in the AC definition of F on $[-2n,2n], n > 0$. Observe that since $|A \cap [-n,n]| = 0$, there exists an open set $\mathcal{O} = \bigcup_k (a_k,b_k) \subseteq [-2n,2n]$ such that

$$A \cap [-n,n] \subset \mathcal{O}\,, \quad |\mathcal{O}| < \delta\,.$$

Also note that by (2.4) in Theorem 2.1 in Chapter IX,

$$\mu_F((a_k,b_k)) \leq \mu_F((a_k,b_k]) = F(b_k) - F(a_k)\,, \quad \text{all } k\,,$$

and consequently, since $\sum_{k=1}^N (b_k - a_k) \leq |\mathcal{O}| \leq \delta$, we get

$$\mu_F\left(\bigcup_{k=1}^N (a_k,b_k)\right) \leq \sum_{k=1}^N (F(b_k) - F(a_k)) \leq \varepsilon\,, \quad \text{all } N\,.$$

Since this inequality holds for all N, it follows that $\mu_F(\mathcal{O}) \le \varepsilon$, and since ε is arbitrary and μ_F is regular, we have

$$\mu_F(A \cap [-n,n]) = 0\,, \quad \text{all } n\,.$$

But this can only be true if $\mu_F(A) = 0$, and consequently, μ_F is absolutely continuous with respect to the Lebesgue measure. ■

Also corresponding to the ε-δ definition of AC functions, there is the following result.

Proposition 1.2. Suppose μ is a measure and ν is a signed measure on $(X, \mathcal{M})$ so that

$$|\nu(E)| < \infty \quad \text{whenever} \quad \mu(E) < \infty\,. \tag{1.6}$$

Then, $\nu \ll \mu$ iff given $\varepsilon > 0$, there exists $\delta > 0$, such that

$$|\nu(E)| < \varepsilon \quad \text{whenever} \quad \mu(E) < \delta\,. \tag{1.7}$$

Proof. To show that the condition is sufficient observe that if $\mu(E) = 0$, then $\mu(E) < \delta$ for all $\delta > 0$, and (1.7) gives that $\nu(E) = 0$.

As for the necessity, suppose that $\nu \ll \mu$ and that (1.7) is false. Then there exist a sequence $\{B_n\} \subseteq \mathcal{M}$ and $\varepsilon > 0$, such that

$$|\nu(B_n)| > \varepsilon \quad \text{and} \quad \mu(B_n) \le 1/2^n\,, \quad n = 1, 2, \ldots$$

Pick now a subsequence $\{B_{n_k}\}$, say, so that all the $\nu(B_{n_k})$'s are of the same sign, and observe that since $\mu(\bigcup_k B_{n_k}) < \infty$, by (1.6) it also follows that $|\nu(\bigcup_k B_{n_k})| < \infty$. The proof may now be finished in a stroke: By 4.25 in Chapter IV, $|\nu(\limsup B_{n_k})| > 0$, and by the Borel-Cantelli Lemma, $\mu(\limsup B_{n_k}) = 0$; this contradicts the fact that $\nu \ll \mu$. ■

Observe that even if ν is a measure, (1.6) is necessary for Proposition 1.2 to hold. Consider, for example,

$$\nu(E) = \int_E x^2\, dx\,, \quad E \in \mathcal{L}\,.$$

Then $|E| = 0$ implies $\nu(E) = 0$, but since for $|x|$ large the set $E = (x, x+\eta)$ has $|E| = \eta$ and $\nu(E)$ is large, (1.7) fails.

Next suppose that μ is a probability measure and ν is a signed measure defined on $(X, \mathcal{M})$, and that

$$|\nu(E)| \le \mu(E)\,, \quad E \in \mathcal{M}\,. \tag{1.8}$$

Clearly $\nu \ll \mu$; the question is, then, how to go about constructing a measurable function f so that

$$\nu(E) = \int_E f \, d\mu \,, \quad E \in \mathcal{M} \,. \tag{1.9}$$

Now, for $A \in \mathcal{M}$, the function

$$f = (\nu(A)/\mu(A))\,\chi_A + (\nu(X \setminus A)/\mu(X \setminus A))\,\chi_{X\setminus A} \,, \tag{1.10}$$

satisfies (1.9) for $E = A, X \setminus A$. f is a first and crude approximation to the solution to our problem, and it is well-defined if the convention $0/0 = 0$ is in force; in fact, by (1.8), it is natural to adopt this convention.

Observe that f is measurable and $|f| \le 1$. We may think of f as $f_{\mathcal{P}}$, namely, as a function associated to the measurable partition $\mathcal{P} = \{A, X \setminus A\}$ of X. Intuitively, the more refined the partition $\mathcal{P}$ is, the more spread out the function $f_{\mathcal{P}}$ associated to $\mathcal{P}$ by a formula similar to (1.10) will be. We remind the reader that we only consider measurable partitions of X, and that, given partitions $\mathcal{P}$ and $\mathcal{P}'$ of X, we say that $\mathcal{P}'$ is finer than $\mathcal{P}$ if given $A' \in \mathcal{P}'$, there exists $A \in \mathcal{P}$ such that $A' \subseteq A$.

Now, on a probability measure space a natural measure of the amount of spread of a measurable function is its "variance"; in the process described in (1.10) above the finer the partition $\mathcal{P}$ becomes, the greater the variance of the associated function $f_{\mathcal{P}}$ is. We expect, then, that the function f that verifies (1.9) will emerge as the limit function of the $f_{\mathcal{P}}$'s as the partitions $\mathcal{P}$ get finer, or as the variance of the $f_{\mathcal{P}}$'s approaches a supremum. The "expected value" of the $f_{\mathcal{P}}$'s is the same, and, as in the case of the function defined by (1.10), it equals $\int_X f \, d\mu = \nu(X)$. Thus by the well-known formula

$$\text{variance} = (\text{second moment}) - (\text{expected value})^2 \,,$$

it is clear that maximizing the variance of the $f_{\mathcal{P}}$'s is equivalent to finding the maximum of the second moments of these functions. The proof of (1.9) we present below follows along these lines. More precisely, we construct the function f by maximizing the second moments of an appropriate family of functions.

We begin by formalizing the relation between the partitions $\mathcal{P}$ of X and the functions $f_{\mathcal{P}}$ associated with them. Specifically, given a partition $\mathcal{P} = \{A_1, \ldots, A_n\}$ of X, let $f_{\mathcal{P}} \colon X \to [-1,1]$ be defined by the expression

$$f_{\mathcal{P}} = \sum_{k=1}^{n} (\nu(A_k)/\mu(A_k))\,\chi_{A_k} \,, \tag{1.11}$$

where the convention $0/0 = 0$ is in force.

Clearly, if $A \in \mathcal{M}$ is any set of the form $A = \bigcup_{j=1}^{m} A_{k_j}, 1 \le k_1 < \ldots < k_m \le n$, we have

$$\nu(A) = \int_A f_{\mathcal{P}}\, d\mu\,,$$

and, in particular,

$$\nu(X) = \int_X f_{\mathcal{P}}\, d\mu\,, \quad \text{all partitions } \mathcal{P}\,.$$

The following properties of partitions are essential to carry out the verification of (1.9).

Lemma 1.3. Let $(X, \mathcal{M}, \mu)$ be a probability measure space, and suppose $\mathcal{P}$ and $\mathcal{P}'$ are partitions of X, with $\mathcal{P}'$ finer than $\mathcal{P}$. If ν is a signed measure on $(X, \mathcal{M})$ which satisfies (1.8), $A \in \mathcal{P}$, and $f_{\mathcal{P}}$ is given by (1.11), the following five properties hold:

$$\nu(A) = \int_A f_{\mathcal{P}}\, d\mu = \int_A f_{\mathcal{P}'}\, d\mu\,, \tag{1.12}$$

$$\int_A f_{\mathcal{P}} f_{\mathcal{P}'}\, d\mu = \int_A f_{\mathcal{P}}^2\, d\mu\,, \tag{1.13}$$

$$\int_A f_{\mathcal{P}}\,(f_{\mathcal{P}'} - f_{\mathcal{P}})\, d\mu = 0\,, \tag{1.14}$$

$$\int_A f_{\mathcal{P}'}^2\, d\mu = \int_A (f_{\mathcal{P}'} - f_{\mathcal{P}})^2\, d\mu + \int_A f_{\mathcal{P}}^2\, d\mu\,, \tag{1.15}$$

and

$$\int_A f_{\mathcal{P}}^2\, d\mu \le \int_A f_{\mathcal{P}'}^2\, d\mu\,. \tag{1.16}$$

By the additivity of the integral it is clear that the above properties also hold for any set $E \in \mathcal{M}$ of the form $E = \bigcup_k A_k$, where $A_k \in \mathcal{P}$ for each k. In particular, they are true for $E = X$.

Proof. Since $\mathcal{P}'$ is finer than $\mathcal{P}$, each $A \in \mathcal{P}$ may be written as a finite union $A = \bigcup_k A_k$, $A_k \in \mathcal{P}'$. Furthermore, since

$$f_{\mathcal{P}'}(x) = \nu(A_k)/\mu(A_k)\,, \quad x \in A_k\,,$$

it readily follows that

$$\int_A f_{\mathcal{P}'}\, d\mu = \sum_k \int_{A_k} f_{\mathcal{P}'}\, d\mu = \sum_k (\nu(A_k)/\mu(A_k))\, \mu(A_k) = \nu(A)\,.$$

Similarly, since

$$f_{\mathcal{P}}(x) = \nu(A)/\mu(A)\,, \quad x \in A\,, \tag{1.17}$$

we also have

$$\int_A f_{\mathcal{P}}\, d\mu = (\nu(A)/\mu(A))\, \mu(A) = \nu(A)\,,$$

and (1.12) holds.

The verification of (1.13) is also simple: If $A \in \mathcal{P}$ is partitioned into A_k's $\in \mathcal{P}'$ as before, we have

$$f_{\mathcal{P}}(x) f_{\mathcal{P}'}(x) = (\nu(A)/\mu(A)) \sum_k (\nu(A_k)/\mu(A_k))\, \chi_{A_k}(x)\,, \quad x \in A\,,$$

and consequently, the left-hand side of (1.13) is equal to

$$\begin{aligned} \int_A f_{\mathcal{P}} f_{\mathcal{P}'}\, d\mu &= (\nu(A)/\mu(A)) \sum_k (\nu(A_k)/\mu(A_k)) \int_{A_k} d\mu \\ &= (\nu(A)/\mu(A))\, \nu(A)\,. \end{aligned}$$

As for the right-hand side of (1.13), on account of (1.17) it equals

$$\int_A (\nu(A)/\mu(A))^2\, d\mu \;=\; \nu(A)^2/\mu(A)\,,$$

and (1.13) holds. (1.14) is a simple rewriting of (1.13).

Next we consider (1.15). Since $f_{\mathcal{P}'} = (f_{\mathcal{P}'} - f_{\mathcal{P}}) + f_{\mathcal{P}}$, it readily follows that

$$f_{\mathcal{P}'}^2 = (f_{\mathcal{P}'} - f_{\mathcal{P}})^2 + 2\,(f_{\mathcal{P}'} - f_{\mathcal{P}})\, f_{\mathcal{P}} + f_{\mathcal{P}}^2\,. \tag{1.18}$$

Whence, integrating the identity (1.18) over A, and invoking (1.14), we get at once that (1.15) holds. (1.16) follows from (1.15), and we have finished. ■

We are now ready to prove

Theorem 1.4. Suppose $(X, \mathcal{M}, \mu)$ is a probability measure space and ν is a signed measure defined on $(X, \mathcal{M})$ such that

$$|\nu(E)| \le \mu(E)\,, \quad \text{all } E \in \mathcal{M}\,. \tag{1.19}$$

Then there exists a unique measurable function $f\colon X \to [-1,1]$ such that

$$\nu(E) = \int_E f\, d\mu\,, \quad \text{all } E \text{ in } \mathcal{M}\,. \tag{1.20}$$

Uniqueness is understood in the following sense: If f_1 is another measurable function defined on X for which (1.20) holds, then $f = f_1$ μ-a.e.

Proof. Consider the collection of all the finite measurable partitions $\mathcal{P}$ of X, to each associate the function $f_{\mathcal{P}}$ given by (1.11), and set

$$\eta = \sup_{\mathcal{P}} \int_X f_{\mathcal{P}}^2 \, d\mu \, .$$

Clearly $0 \leq \eta \leq 1$. Let now $\{\mathcal{P}_n\}$ be a sequence of partitions of X with the property that

$$\int_X f_{\mathcal{P}_n}^2 \, d\mu \geq \eta - 1/4^n \, , \quad n = 1, 2, \ldots \tag{1.21}$$

It is more efficient, however, to work with the sequence $\{\mathcal{P}'_n\}$ consisting of the common refinement of the $\mathcal{P}_n$'s; specifically, let $\mathcal{P}'_1 = \mathcal{P}_1$, $\mathcal{P}'_2$ be the common refinement of $\mathcal{P}_1$ and $\mathcal{P}_2$, and so on. Denote $f_{\mathcal{P}'_n}$ by f_n, and observe that since $\mathcal{P}'_n$ refines $\mathcal{P}_n$ for all n, by (1.16), (1.21), and the definition of η, we have

$$\eta - 1/4^n \leq \int_X f_{\mathcal{P}_n}^2 \, d\mu \leq \int_X f_n^2 \, d\mu \leq \eta \, . \tag{1.22}$$

Next we construct a "maximal" element from the f_n's. Since for each n $\mathcal{P}'_{n+1}$ refines $\mathcal{P}'_n$, by (1.15) and (1.22) it follows that

$$\begin{aligned} \int_X (f_{n+1} - f_n)^2 \, d\mu &= \int_X f_{n+1}^2 \, d\mu - \int_X f_n^2 \, d\mu \\ &\leq \eta - (\eta - 1/4^n) = 1/4^n \, , \quad n = 1, 2, \ldots \end{aligned}$$

Thus, by the Cauchy-Schwarz inequality, cf. (3.1) in Chapter VIII, we have

$$\begin{aligned} \int_X |f_{n+1} - f_n| \, d\mu &\leq \left(\int_X (f_{n+1} - f_n)^2 \, d\mu \right)^{1/2} (\mu(X))^{1/2} \\ &\leq 1/2^n \, , \quad n = 1, 2, \ldots \end{aligned}$$

and consequently, since

$$\int_X \sum_{n=1}^{\infty} |f_{n+1} - f_n| \, d\mu = \sum_{n=1}^{\infty} \int_X |f_{n+1} - f_n| \, d\mu \leq \sum_{n=1}^{\infty} 1/2^n < \infty \, ,$$

we may invoke 4.7 in Chapter VII and obtain that $\sum_n (f_{n+1} - f_n)$ converges absolutely, and pointwise, μ-a.e. on X. Put now

$$g = \lim_{N \to \infty} \sum_{n=1}^{N} (f_{n+1} - f_n) = \lim_{N \to \infty} f_{N+1} - f_1 \, ;$$

g is well-defined μ-a.e. on X.

Next we show that the function $f = g + f_1$ gets the job done; in other words, (1.20) is true for f and any E in $\mathcal{M}$. Given $E \in \mathcal{M}$, let $\{\mathcal{Q}_n\}$ be the sequence of partitions consisting of the common refinement of the partitions $\mathcal{P}'_n$ and $\{E, X \setminus E\}$. Clearly $\mathcal{Q}_n$ is finer than $\mathcal{P}'_n$, $n = 1, 2, \ldots$, and $\mathcal{Q}_{n+1}$ is finer than $\mathcal{Q}_n$, $n = 1, 2, \ldots$ Thus, if $h_n = h_{\mathcal{Q}_n}$, $n = 1, 2, \ldots$, by (1.22) and (1.16) it readily follows that

$$\eta - 1/4^n \le \int_X f_n^2 \, d\mu \le \int_X h_n^2 \, d\mu \le \eta \,, \quad \text{all } n \,,$$

and, by (1.15) and (1.22), we also get

$$\begin{aligned} \int_X (h_n - f_n)^2 \, d\mu &= \int_X h_n^2 \, d\mu - \int_X f_n^2 \, d\mu \\ &\le \eta - (\eta - 1/4^n) = 1/4^n \,, \quad \text{all } n. \end{aligned} \tag{1.23}$$

Consequently, once again by the Cauchy-Schwarz inequality and (1.23), we have

$$\begin{aligned} \left| \int_E (h_n - f_n) \, d\mu \right| &\le \left(\int_E (h_n - f_n)^2 \, d\mu \right)^{1/2} \mu(E)^{1/2} \\ &\le \left(\int_X (h_n - f_n)^2 \, d\mu \right)^{1/2} \le 1/2^n \,, \quad \text{all } n. \end{aligned} \tag{1.24}$$

We are now ready to compute $\nu(E)$. Since $\mathcal{Q}_n$ is finer than $\{E, X \setminus E\}$ for all n, by (1.12) we have

$$\nu(E) = \int_E h_n \, d\mu = \int_E (h_n - f_n) \, d\mu + \int_E f_n \, d\mu = A_n + B_n \,,$$

say. By (1.24), $\lim_{n \to \infty} A_n = 0$. As for the B_n's, first observe that

$$\lim_{n \to \infty} f_n = f \ \mu\text{-a.e.} \,, \quad \text{and} \quad |f_n| \le 1 \quad \text{all } n.$$

Whence, by LDCT it follows that

$$\lim_{n \to \infty} B_n = \int_E f \, d\mu \,,$$

and (1.20) holds. Further, if f_1 is another measurable function for which (1.20) is true, then $\int_E (f - f_1) \, d\mu = 0$ for all $E \in \mathcal{M}$, and consequently, $f - f_1 = 0$ μ-a.e. ■

The applications of Theorem 1.4 are numerous and interesting; we begin by showing how a signed measure may be decomposed into "positive" and "negative" parts. Theorem 1.5, and the remarks that follow it, are known as Hahn's decomposition theorem. This decomposition is true in greater generality, cf. 3.6 below, but the result presented here is sufficient for the applications we have in mind.

Theorem 1.5. Let ν be a signed measure defined on $(X, \mathcal{M})$, and suppose that its variation $|\nu|$ is a probability measure on $(X, \mathcal{M})$. Then there exist two disjoint, measurable sets A and B, $A \cup B = X$, so that
(i) $\nu(E \cap A) \geq 0$, all $E \in \mathcal{M}$,
(ii) $\nu(E \cap B) \leq 0$, all $E \in \mathcal{M}$.

Proof. Since

$$|\nu(E)| \leq |\nu|(E), \quad \text{all } E \in \mathcal{M},$$

Theorem 1.4 applies with $\mu = |\nu|$. In particular, there is a measurable function f, $|f| \leq 1$, that satisfies

$$\nu(E) = \int_E f \, d|\nu|, \quad \text{all } E \in \mathcal{M}.$$

We claim that $|f| = 1$ μ-a.e. First note that since $|\nu|(X) = 1$, by 4.12 in Chapter IV, given $\varepsilon > 0$, there exists a partition $\mathcal{P} = \{A_1, \ldots, A_m\}$ of X such that

$$1 - \varepsilon \leq \sum_{k=1}^{m} |\nu(A_k)| \leq |\nu|(X) = 1.$$

Thus, by (1.12), it readily follows that

$$\begin{aligned} 1 - \varepsilon &\leq \sum_{k=1}^{m} |\nu(A_k)| \\ &= \sum_{k=1}^{m} |\text{value of } f_{\mathcal{P}} \text{ on } A_k| \, |\nu|(A_k) = \int_X |f_{\mathcal{P}}| \, d|\nu|. \end{aligned}$$

This estimate, together with the Cauchy-Schwarz inequality, and with the meaning of η introduced in Theorem 1.4, give

$$(1 - \varepsilon)^2 \leq \int_X f_{\mathcal{P}}^2 \, d|\nu| \leq \eta \leq 1.$$

Since ε is arbitrary, it follows at once that $\eta = 1$. Referring once again to the proof of Theorem 1.4, by LDCT we obtain

$$1 = \eta = \lim_{n \to \infty} \int_X f_n^2 \, d|\nu| = \int_X f^2 \, d|\nu|,$$

and consequently, since $|\nu|$ is a probability measure, we have $f^2 = 1$ $|\nu|$-a.e., and $|f| = 1$ $|\nu|$-a.e., as claimed.

Let now $A = \{x \in X : f(x) = 1\}$, and $B = X \setminus A$; $\{A, B\}$ is a disjoint partition of X. Since $f = -1$ $|\nu|$-a.e. on B, for any $E \in \mathcal{M}$ we have

$$\nu(E \cap A) = \int_{E \cap A} f \, d|\nu| = |\nu|(E \cap A) \geq 0 ,$$

and

$$\nu(E \cap B) = \int_{E \cap B} f \, d|\nu| = -|\nu|(E \cap B) \leq 0 . \quad \blacksquare$$

Two remarks concerning this result: First, the decomposition $X = A \cup B$ is not unique, and second, the result is true in greater generality. For instance, if $|\nu|$ is σ-finite, then we have $X = \bigcup_k X_k$, where the X_k's are pairwise disjoint and $|\nu|(X_k) < \infty$ for all k. By rescaling if necessary we may assume that $|\nu|(X_k) = 1$ for all k, and apply Theorem 1.4 to each X_k, thus obtaining sequences $\{A_k\}$, $\{B_k\}$ as in that theorem; then the sets $A = \bigcup_k A_k$ and $B = \bigcup_k B_k$ correspond to the "positive" and "negative" parts of ν respectively.

Another interesting consequence of Theorem 1.4 is

Theorem 1.6. Suppose that λ, μ are σ-finite measures on $(X, \mathcal{M})$ and that

$$\lambda(E) \leq \mu(E), \quad \text{all } E \in \mathcal{M} .$$

Then there exists a unique (in the μ-a.e. sense as explained above) nonnegative measurable function $f : X \to [0,1]$ such that

$$\lambda(E) = \int_E f \, d\mu , \quad \text{all } E \in \mathcal{M} . \tag{1.25}$$

Furthermore, if g is a measurable extended real-valued function defined on X, then

$$\int_X g \, d\lambda = \int_X gf \, d\mu . \tag{1.26}$$

The equality in (1.26) is understood as follows: If the integral on either side of (1.26) exists, then the integral on the other side also exists and they are equal.

Proof. Write $X = \bigcup_k X_k$, where the union is pairwise disjoint and $\mu(X_k) < \infty$ for all k. By rescaling if necessary, assume that $\mu(X_k) = 1$, and invoke Theorem 1.4 for the measures λ_k, μ_k on $(X, \mathcal{M})$ given by

$$\lambda_k(E) = \lambda(E \cap X_k) , \quad \mu_k(E) = \mu(E \cap X_k) , \quad k = 1, 2, \ldots$$

The function f in (1.25) is obtained as $f = \sum_k f_k \chi_{X_k}$, where the f_k's are the (unique) functions that satisfy (1.20) for the measures λ_k and μ_k.

As for (1.26), suppose first that $g \geq 0$, and let $\{\phi_n\}$ be a sequence of simple functions that increases to g μ-a.e. and consequently, also λ-a.e. Note that if $\phi_n = \sum_j a_j \chi_{E_j}$, from (1.25) we get that

$$\begin{aligned}\int_X \phi_n \, d\lambda &= \sum_j a_j \lambda(E_j) = \sum_j a_j \int_{E_j} f \, d\mu \\ &= \int_X \Big(\sum\nolimits_j a_j \chi_{E_j}\Big) f \, d\mu = \int_X \phi_n f \, d\mu \,,\end{aligned}$$

and, by MCT, it follows that (1.26) holds in this case. As for a general function g, note that $g = g^+ - g^-$, and apply the above result to g^+ and g^- separately. ■

2. THE LEBESGUE AND RADON-NIKODÝM THEOREMS

Let μ and ν be signed measures on $(X, \mathcal{M})$. We say that μ and ν are mutually singular, and denote this by $\mu \perp \nu$, if there exists a disjoint partition $\{A, B\}$ of X such that $|\mu|(A) = |\nu|(B) = 0$. In this case we also say that μ is singular with respect to ν, or symmetrically, that ν is singular with respect to μ.

For instance, in Theorem 1.5 the measures

$$\nu_1(E) = \nu(E \cap A)\,, \quad \nu_2(E) = \nu(E \cap B)\,, \quad E \in \mathcal{M}\,,$$

are mutually singular. Another interesting example we have encountered is that of $\mu = \mu_F$, the Borel measure induced by the extension F to R of the Cantor-Lebesgue function, and ν the Lebesgue measure on the line.

To show that this last example is part of a general state of affairs, and to elucidate the notion of singularity, we present a preliminary result.

Lemma 2.1. Suppose μ_F is a finite Borel measure induced by a distribution $F \in \mathcal{D}$. If A is a Borel set on the line such that F' exists on A and $M, m \geq 0$, then we have

(i) If $F'(x) \leq M$ for $x \in A$, then $\mu_F(A) \leq M|A|$.

(ii) If $F'(x) \geq m$ for $x \in A$, then $\mu_F(A) \geq m|A|$.

Proof. Given $\varepsilon > 0$, let $\mathcal{I}_n$ denote the collection of intervals of the form $(u,v]$ that satisfy the following two properties:

(a) u, v are rational, and $0 < v - u < 1/n$.

(b) $F(v) - F(u) \le (M + \varepsilon)(v - u)$.

Observe that the sets

$$A_n = \bigcup_{(u,v] \in \mathcal{I}_n} (A \cap (u,v]) , \quad n = 1, 2 \ldots$$

are Borel sets, and that under the hypothesis of (i), they increase to A.

Let $\{I_{n,k}\}$ be a sequence of nonoverlapping intervals, open on the left and closed on the right, such that

$$A_n \subseteq \bigcup_k I_{n,k} , \quad \text{and} \quad \sum_k |I_{n,k}| \le |A_n| + \varepsilon .$$

By working with the $I_{n,k}$'s it is possible to assume that each $I_{n,k}$ meets A_n, has rational endpoints, and that $|I_{n,k}| < 1/n$ for all k. Then, (a) and (b) above apply to each $I_{n,k} = (u_{n,k}, v_{n,k}]$, and it readily follows that

$$\begin{aligned} \mu_F(A_n) &\le \sum_k \mu_F(I_{n,k}) = \sum_k (F(v_{n,k}) - F(u_{n,k})) \\ &\le (M + \varepsilon) \sum_k (v_{n,k} - u_{n,k}) = (M + \varepsilon) \sum_k |I_{n,k}| \\ &\le (M + \varepsilon)(|A_n| + \varepsilon) . \end{aligned}$$

Since ε is arbitrary the above inequality implies that

$$\mu_F(A_n) \le M|A_n| , \quad \text{all } n ,$$

and since the A_n's increase to A, a similar inequality holds with A in place of A_n above, and (i) is true. The proof of (ii) follows along similar lines, a Vitali covering argument also works in this case, and is therefore left to reader. ■

We are now ready to prove

Proposition 2.2. Suppose μ_F is a finite Borel measure induced by a distribution $F \in \mathcal{D}$. Then μ_F is singular with respect to the Lebesgue measure on the line iff F is singular, i.e., $F' = 0$ a.e.

Proof. We first show that the condition is sufficient. Given $\eta > 0$, by (i) in Lemma 2.1 we get

$$\mu_F(\{x : |x| \leq n, F' < \eta\}) \leq 2n\eta\,,$$

and consequently, by first letting $\eta \to 0$ and then $n \to \infty$, it readily follows that

$$\mu_F(\{F' = 0\}) = 0\,.$$

Put now $A = \{F' = 0\}$ and $B = R \setminus A$. Since F is singular we have

$$\mu_F(A) = |B| = 0\,, \quad A \cap B = \emptyset\,, \quad A \cup B = R\,, \tag{2.1}$$

and consequently, μ_F is singular with respect to the Lebesgue measure.

On the other hand, if (2.1) holds for Borel sets A and B, by the other half of Lemma 2.1, it readily follows that for $\eta > 0$ we have

$$\begin{aligned}\eta|\{x : F' \geq \eta\}| &= \eta|\{x \in A : F' \geq \eta\}| + \eta|\{x \in B : F' \geq \eta\}| \\ &= \eta|\{x \in A : F' \geq \eta\}| \leq \mu_F(A) = 0\,.\end{aligned}$$

So, for each $\eta > 0$ we have $|\{F' \geq \eta\}| = 0$, and by letting $\eta \to 0$ we get that $F' = 0$ except on a null Lebesgue set. ∎

Propositions 1.1 and 2.2 suggest that it may be possible to decompose signed measures in terms of absolutely continuous and singular measures. This indeed is the case; we begin by discussing a preliminary result in this direction.

Theorem 2.3. Suppose λ and μ are finite measures on $(X, \mathcal{M})$. Then there exist finite measures λ_a and λ_s which satisfy the following properties:

(i) $\lambda_a \ll \mu\,, \lambda_s \perp \mu$.

(ii) $\lambda = \lambda_a + \lambda_s\,.$

Furthermore, the measures λ_a, λ_s are unique.

Proof. The set function

$$(\lambda + \mu)(E) = \lambda(E) + \mu(E)\,, \quad E \in \mathcal{M}\,,$$

is a measure on $(X, \mathcal{M})$ which satisfies

$$\lambda(E) \leq (\lambda + \mu)(E)\,, \quad \text{all } E \text{ in } \mathcal{M}\,.$$

Whence, by (1.25) in Theorem 1.6, there exists a measurable function $f: X \to [0,1]$ such that

$$\lambda(E) = \int_E f\, d(\lambda + \mu) = \int_E f\, d\lambda + \int_E f\, d\mu\,, \quad E \in \mathcal{M}\,. \tag{2.2}$$

Next, let $g = f\chi_E$, and observe that by (1.26), with $\mu = \lambda + \mu$ there, we have

$$\int_X f\chi_E\, d\lambda = \int_X f\chi_E f\, d(\lambda + \mu)\,,$$

and consequently, (2.2) may be rewritten as

$$\lambda(E) = \int_E f^2\, d\lambda + \int_E (f + f^2)\, d\mu\,, \quad E \in \mathcal{M}\,. \tag{2.3}$$

Now, this procedure may be iterated with $g = f^2\chi_E$, $g = f^3\chi_E$, and so on, and (2.3) becomes

$$\lambda(E) = \int_E f^n\, d\lambda + \int_E (f + \cdots + f^n)\, d\mu\,, \quad n = 1, 2, \ldots \tag{2.4}$$

Let $B = \{f = 1\}$, put

$$\lambda_s(E) = \lambda(E \cap B)\,, \quad E \in \mathcal{M}\,,$$

and observe that (2.4) becomes

$$\begin{aligned} \lambda(E) &= \lambda_s(E) + \int_{E\cap(X\setminus B)} f^n\, d\lambda + \int_E (f + \cdots + f^n)\, d\mu \\ &= I + J + K\,, \end{aligned} \tag{2.5}$$

say. Clearly, by LDCT, $\lim_{n\to\infty} J = 0$.

As for K, note that by MCT and (2.4) it follows that

$$\lim_{n\to\infty} (f + \cdots + f^n) = \frac{f}{1-f} \in L(\mu) \tag{2.6}$$

and consequently, $\mu(B) = 0$, and

$$\lim_{n\to\infty} K = \int_E \frac{f}{1-f}\, d\mu\,.$$

Now, by (2.6) we obtain that the measure

$$\lambda_a(A) = \int_A \frac{f}{1-f}\, d\mu\,, \quad A \in \mathcal{M}\,,$$

is absolutely continuous with respect to μ. Thus, returning to (2.5), we get

$$\lambda(E) = \lim_{n\to\infty}(I + J + K) = \lambda_s(E) + \lambda_a(E)\,,$$

and (ii) holds.

Next we show that $\lambda_s \perp \mu$; since $\mu(B) = 0$, this reduces to checking that

$$\lambda_s(X \setminus B) = \int_{(X\setminus B)\cap B} d\lambda = 0\,,$$

which is obviously true.

Finally we show that the decomposition is unique; this is not hard. Suppose $\lambda = \lambda'_a + \lambda'_s$ is a decomposition of λ that satisfies (ii) above, and let

$$\lambda'_s(X \setminus B') = \mu(B') = 0\,, \quad B' \in \mathcal{M}\,.$$

We claim that

$$\lambda_a(E) = \lambda'_a(E)\,, \quad \text{all } E \in \mathcal{M}\,. \tag{2.7}$$

Indeed, since $\mu(B \cup B') = 0$, by the absolute continuity of λ_a and λ'_a we have

$$\lambda_a(E \cap (B \cup B')) = \lambda'_a(E \cap (B \cup B')) = 0\,, \quad \text{all } E \in \mathcal{M}\,. \tag{2.8}$$

Moreover, since $E \cap (X \setminus (B \cup B'))$ is a subset of both $X \setminus B$ and $X \setminus B'$, we also have

$$\lambda_s(E \cap (X \setminus (B \cup B'))) = \lambda'_s(E \cap (X \setminus (B \cup B'))) = 0\,, \quad E \in \mathcal{M}\,.$$

Whence, since $\lambda_a + \lambda_s = \lambda'_a + \lambda'_s$, it readily follows that

$$\lambda_a(E \cap (X \setminus (B \cup B'))) = \lambda'_a(E \cap (X \setminus (B \cup B')))\,, \quad E \in \mathcal{M}\,. \tag{2.9}$$

Finally, combining (2.8) and (2.9), for each $E \in \mathcal{M}$ we get

$$\begin{aligned}\lambda_a(E) &= \lambda_a(E \cap (B \cup B')) + \lambda_a(E \cap (X \setminus (B \cup B')))\\ &= \lambda_a(E \cap (X \setminus (B \cup B'))) = \lambda'_a(E \cap (X \setminus (B \cup B')))\\ &= \lambda'_a(E \cap (B \cup B'))) + \lambda'_a(E \cap (X \setminus (B \cup B'))) = \lambda'_a(E)\,,\end{aligned}$$

and (2.7) holds. Further, since $\lambda_a = \lambda'_a$ and since all the measures involved are finite, then $\lambda_s = \lambda'_s$ also holds. ∎

We are now ready to prove a result in the spirit of Theorem 3.7 in Chapter X, it is appropriately known as the Lebesgue Decomposition Theorem.

Theorem 2.4. Suppose μ and λ are σ-finite measures defined on $(X, \mathcal{M})$. Then there exist σ-finite measures λ_a, λ_s defined on $(X, \mathcal{M})$ such that:

(i) $\lambda_a \ll \mu\,, \quad \lambda_s \perp \mu\,.$

(ii) $\lambda = \lambda_a + \lambda_s$.

Furthermore, the measures λ_a and λ_s are unique.

Proof. The idea of the proof is to reduce the general hypothesis to the special case when the measures involved are finite, and then to invoke Theorem 2.3. First note that since μ and λ are σ-finite, we can write X as a pairwise disjoint union

$$X = \bigcup_k X_k\,, \quad \text{with} \quad \mu(X_k)\,, \lambda(X_k) < \infty\,, \quad \text{all } k\,.$$

Next we localize the problem at hand by introducing the finite measures μ_k and λ_k defined on $(X, \mathcal{M})$ by

$$\mu_k(E) = \mu(E \cap X_k)\,, \quad \lambda_k(E) = \lambda(E \cap X_k)\,, \quad k = 1, 2, \ldots$$

For each $k \geq 1$, let $\lambda_k = \lambda_{k,a} + \lambda_{k,s}$ be the unique decomposition of λ_k obtained in Theorem 2.3 with the property that

$$\lambda_{k,a} \ll \mu_k\,, \quad \text{and} \quad \lambda_{k,s} \perp \mu_k\,.$$

Moreover, since the X_k's are pairwise disjoint it is also simple to verify that if we put

$$\lambda_a = \sum_k \lambda_{k,a}\,, \quad \text{and} \quad \lambda_s = \sum_k \lambda_{k,s}\,,$$

then $\lambda = \lambda_a + \lambda_s$ is the unique decomposition of λ for which (i) above holds. ■

It is not hard to extend this result to signed measures. Indeed, we have

Theorem 2.5. Suppose μ is a σ-finite measure and ν is a signed measure defined on $(X, \mathcal{M})$. If $|\nu|$ is σ-finite, then there exist signed measures ν_a, ν_s defined on $(X, \mathcal{M}$ such that:

(i) $\nu_a \ll \mu\,, \quad$ and $\quad \nu_s \perp \mu$.

(ii) $\nu = \nu_a + \nu_s$.

Furthermore, the signed measures ν_a and ν_s are unique.

Proof. By the remarks following the Hahn decomposition theorem we can find σ-finite measures λ, λ' on $(X, \mathcal{M})$ such that $\nu = \lambda - \lambda'$. Let now

$$\lambda = \lambda_a + \lambda_s\,, \quad \lambda' = \lambda'_a + \lambda'_s$$

be the unique decomposition of λ and λ' obtained in Theorem 2.4 with the property that

$$\lambda_a, \lambda'_a \ll \mu\,, \quad \text{and} \quad \lambda_s, \lambda'_s \perp \mu\,.$$

It is not hard to check that if we put

$$\nu_a = \lambda_a - \lambda'_a \quad \text{and} \quad \nu_s = \lambda_s - \lambda'_s\,,$$

then $\nu = \nu_a + \nu_s$ is the (unique) decomposition of ν that does the job. ■

These interesting results still do not address the question raised in considering (1.2) above. Now, in that example we had $\nu \ll \mu$, and under this assumption we show that (a preliminary version of) the Radon-Nikodým Theorem holds.

Theorem 2.6. Suppose μ and λ are finite measures defined on $(X, \mathcal{M})$, and that $\lambda \ll \mu$. Then there exists a nonnegative integrable function h such that

$$\lambda(E) = \int_E h\,d\mu\,, \quad \text{all } E \in \mathcal{M}. \tag{2.10}$$

h is called the Radon-Nikodým derivative of λ with respect to μ, and one writes

$$h = d\lambda/d\mu\,, \quad \text{or} \quad d\lambda = h\,d\mu\,.$$

Furthermore, h is unique.

Proof. The proof is identical to that of Theorem 2.3. First observe that since in the notation of Theorem 2.3 $\mu(B) = 0$, and since $\lambda \ll \mu$, we also have $\lambda(B) = 0$. But this readily implies that $\lambda_s(E) = 0$ for all $E \in \mathcal{M}$, and consequently for any E in $\mathcal{M}$ we have

$$\lambda(E) = \int_E h\,d\mu\,, \quad h = \frac{f}{1-f} \in L(\mu)\,. \quad ■$$

It is clear that we cannot expect (2.10) to hold in general. For instance, if λ is the Lebesgue measure on R and $\mu = \delta$ is the Dirac delta measure at 0, it is obvious that (2.10) cannot hold for any $h \in L(\mu)$; thus the assumption $\lambda \ll \mu$ is necessary. On the other hand, if μ is the counting measure on [0,1], then $\lambda \ll \mu$ but still (2.10) is not true for any nonnegative measurable function h defined on [0,1]. The difficulty here is that the integral on the right-hand side of (2.10) is finite only when the set $\{h \neq 0\} \cap E$ is at most countable, and [0,1] is an uncountable set of finite Lebesgue measure.

A moment's thought will convince the reader that these are the only difficulties in extending Theorem 2.6 to a more general setting. Indeed, we have

Theorem 2.7 (Radon-Nikodým Theorem). Let μ be a σ-finite measure, and ν a signed measure defined on $(X, \mathcal{M})$. If $|\nu|$ is σ-finite and $\nu \ll \mu$, then there exists an extended real-valued measurable function h defined on X such that if $E \in \mathcal{M}$ and $|\nu|(E) < \infty$,

$$\nu(E) = \int_E h \, d\mu . \tag{2.11}$$

h is called the Radon-Nikodým derivative of ν with respect to μ, and one writes

$$h = d\nu / d\mu , \quad \text{or} \quad d\nu = h d\mu .$$

Also h is unique, in the sense that if h' is another extended real-valued measurable function defined on X for which (2.11) holds, then $h = h'$ μ-a.e.

Proof. The proof follows along the lines to that of Theorem 2.5; first we localize the problem at hand and reduce it to the particular case of finite measures. Write X as a pairwise disjoint union

$$X = \bigcup_k X_k , \quad \text{with} \quad \mu(X_k), |\nu|(X_k) < \infty , \quad \text{all } k .$$

Consider now the finite measures μ_k and signed measures ν_k defined on $(X, \mathcal{M})$ by

$$\mu_k(E) = \mu(E \cap X_k) , \quad \text{and} \quad \nu_k(E) = \nu(E \cap X_k) , \quad k = 1, 2, \ldots$$

Further, write $\nu_k = \lambda_k - \lambda'_k$, where the λ_k's and λ'_k's are finite measures on $(X, \mathcal{M})$ such that $\nu_k = \lambda_k - \lambda'_k$ for $k = 1, 2, \ldots$ It is also easy to check that under our assumptions we have

$$\lambda_k, \lambda'_k \ll \mu_k , \quad \text{and} \quad \lambda_{k,s} - \lambda'_{k,s} = 0 , \quad \text{all } k .$$

Now, by Theorem 2.6 there exist (unique) nonnegative functions h_k, h'_k in $L(\mu_k)$ with the property that

$$\lambda_k(E) = \int_E h_k\, d\mu_k\,, \quad \lambda'_k(E) = \int_E h'_k\, d\mu_k\,, \quad \text{all } E \in \mathcal{M}\,,$$

and consequently for each $k = 1, 2, \ldots$, we have

$$\nu_k(E) = \int_E (h_k - h'_k)\, d\mu_k\,, \quad E \in \mathcal{M}\,.$$

It is now readily seen that the function $h = \sum_k (h_k - h'_k)$ has all the desired properties. ■

The assumption that μ is σ-finite is essential for the validity of Theorem 2.7. For, suppose X is an uncountable set, and let $\mathcal{M}$ be the σ-algebra of those subsets of X which are either at most countable or so that their complements are at most countable. For $E \in \mathcal{M}$ put $\mu(E) =$ the number of elements of E if E is finite and $\mu(E) = \infty$ otherwise, and $\nu(E) = 0$ or 1 according as to whether E is at most countable or not. Then $\nu \ll \mu$, but no integral representation such as the one given in (2.11) is possible. Clearly μ is not σ-finite.

The theorems of Lebesgue and Radon-Nikodým have many interesting applications, we discuss one next.

If ν is a signed Borel measure, we say that ν is differentiable at $x \in R^n$, provided that

$$D\nu(x) = \lim_{r \to 0} \frac{\nu(I(x,r))}{|I(x,r)|} \quad \text{exists}\,.$$

Then the following is true.

Proposition 2.8. Suppose ν is a signed Borel measure so that $|\nu|$ is finite on bounded sets of R^n. Then ν is differentiable a.e.

More precisely, if $\nu = \nu_a + \nu_s$ is the Lebesgue decomposition of ν with respect to the Lebesgue measure μ, $\nu_a \ll \mu$, $\nu_s \perp \mu$, we have

$$D\nu_s = 0 \ \text{ a.e.} \quad \text{and} \quad D\nu_a = d\nu_a / d\mu \ \text{a.e.}$$

Proof. Since $|\nu|$ is σ-finite, Theorem 2.4 applies. Let h denote the Radon-Nikodým derivative of ν_a with respect to the Lebesgue measure and observe that

$$\begin{aligned}\frac{\nu(I(x,r))}{|I(x,r)|} &= \frac{\nu_s(I(x,r))}{|I(x,r)|} + \frac{1}{|I(x,r))|} \int_{I(x,r)} h\, dy \\ &= A + B\,,\end{aligned}$$

say. Since by the Lebesgue Differentiation Theorem we have $\lim_{r\to 0} B = h$ a.e., we are reduced to showing that $\lim_{r\to 0} A = 0$ a.e.; this is not hard. First observe that since ν_s is the difference of two Borel measures whose variations are finite on bounded sets, also $|\nu_s|$ enjoys this property. The proof now follows along the lines of Proposition 2.2. Since

$$|\nu_s(I(x,r))| \le |\nu_s|(I(x,r)),$$

it clearly suffices to show that

$$D|\nu_s| = 0 \quad \text{a.e.} \tag{2.12}$$

Let B be a Borel set such that $|\nu_s|(B) = |R^n \backslash B| = 0$, and for $k = 1,2,\ldots,$ let

$$F_k = \left\{ x \in R^n : \limsup_{r\to 0} \frac{|\nu_s|(I(x,r))}{|I(x,r)|} > \frac{1}{k} \right\}.$$

Since $|F_k \cap (R^n \setminus B)| = 0$, in order to prove (2.12) it is enough to show that

$$|F_k \cap B| = 0, \quad \text{all } k. \tag{2.13}$$

Now, since $|\nu_s|$ is a regular Borel measure, given $\varepsilon > 0$, there exists an open set $\mathcal{O} \supseteq B$ such that $|\nu_s|(\mathcal{O}) \le \varepsilon$. Further, to each $x \in F_k \cap B$ associate an interval $I(x,r)$ with the property that

$$|\nu_s|(I(x,r_x)) > |I(x,r_x)|/k, \quad I(x,r) \subseteq \mathcal{O}. \tag{2.14}$$

Observe that if K is an arbitrary compact subset of $F_k \cap B$, by a covering argument similar to that in the proof of the Hardy-Littlewood maximal theorem, specifically estimate (2.15) there, there exists a finite family $I(x_1,r_1),\ldots,I(x_m,r_m)$ of pairwise disjoint subintervals of $\mathcal{O}$ such that

$$|K| \le \sum |I(x_i,3r_i)| = 3^n \sum |I(x_i,r_i)|. \tag{2.15}$$

But the intervals that appear on the right-hand side of (2.15) are special. In addition to being pairwise disjoint, they all satisfy (2.14). Therefore the sum on the right-hand side of (2.15) can be estimated by

$$3^n k \sum |\nu_s|(I(x_i,r_i)) \le 3^n k |\nu_s|(\mathcal{O}) \le 3^n k\varepsilon.$$

Since ε is arbitrary, this means that $|K| = 0$. By the regularity of the Lebesgue measure it follows that (2.13) holds, and we have finished. ■

3. PROBLEMS AND QUESTIONS

In what follows, all measures are assumed to be defined on a σ-algebra $\mathcal{M}$ of subsets of a space X, even when this is not explicitly stated.

3.1 Referring to the notation of 4.8 in Chapter IV and 3.10 in Chapter IX, if ν is a signed measure, describe $\nu^+ \wedge \nu^-$, and $\nu^+ \vee \nu^-$.

3.2 Let μ be a measure and f an extended real-valued measurable function defined on X. If ν is the signed measure given by

$$\nu(E) = \int_E f \, d\mu , \quad E \in \mathcal{M} ,$$

describe ν^+, ν^- and $|\nu|$ in terms of f.
What if μ is a signed measure and f has constant sign? What if both μ and f are allowed to have variable sign?

3.3 Let $\mathcal{F}$ denote the class of all measures on $(X, \mathcal{M})$. Show that the relation $\lambda \prec \mu$ iff

$$\lambda(E) \leq \mu(E) , \quad \text{all } E \in \mathcal{M} , \tag{3.1}$$

is a partial ordering on $\mathcal{F}$.

3.4 Show that if (3.1) holds then $L^1(\mu) \subseteq L^1(\lambda)$. Is the converse true?

3.5 A subset $A \in \mathcal{M}$ is said to be positive with respect to a signed measure ν if for each $E \subseteq A$, $E \in \mathcal{M}$, we have $\nu(E) \geq 0$. Similarly, a measurable set C is said to be negative with respect to ν if for each $E \subseteq C$, $E \in \mathcal{M}$, we have $\nu(E) \leq 0$. Finally, a set $N \in \mathcal{M}$ is said to be null with respect to ν if it is simultaneously positive and negative with respect to ν.
Show that there may be sets of measure 0 which are not null, as there may be sets of positive measure which are not positive, and sets of negative measure which are not negative, all with respect to ν. Also, investigate some of the properties of these classes of sets. For instance, show that every measurable subset of a positive set is positive, etc.

3.6 If ν is a signed measure which does not assume the value ∞, and if $\nu(E) > -\infty$, show that E contains a measurable subset A such that: (a) $\nu(A) \geq \nu(E)$, and, (b) A is positive with respect to ν.

3.7 If ν is a signed measure, prove that E is null with respect to ν iff $|\nu|(E) = 0$.

3.8 If ν is a signed measure and λ, ν are measures such that $\nu = \lambda - \mu$, show that

$$\lambda(E) \geq \nu^+(E), \quad \text{and} \quad \nu(E) \geq \nu^-(E), \quad \text{all } E \in \mathcal{M}.$$

3.9 Suppose μ_F is the probability measure on R induced by the distribution function F, and that $F(x) = \mu_F((-\infty, x])$. If λ denotes the Lebesgue measure on R and $d\mu_F/d\lambda = f$, show that

$$F(x) = \int_{(-\infty,x]} f(y)\, d\mu_F(y), \quad -\infty < x < \infty.$$

The nonnegative function f is known as the probability density of F.
Compute the probability density that corresponds to the distribution function

$$F(x) = \begin{cases} 0 & \text{if } x \leq a \\ (x-a)/(b-a) & \text{if } a < x < b \\ 1 & \text{if } x \geq b. \end{cases}$$

3.10 Suppose

$$F(x) = \begin{cases} 0 & \text{if } x \leq -1 \\ 1 - x^2 & \text{if } x > -1, \end{cases}$$

and let μ be the signed measure on $(R, \mathcal{B})$ that satisfies $\mu((x,y]) = F(y) - F(x)$. Find the Hahn decomposition of μ and for an arbitrary interval $I = (x,y]$ of R find explicit formulas for $\mu_+(I)$, $\mu_-(I)$ and $|\mu|(I)$.

3.11 Does there exist an increasing distribution function F on R such that the induced Borel measure μ_F is not absolutely continuous with respect to the Lebesgue measure on R?

3.12 If ν is a finite measure and μ a measure, show that the following are equivalent: (a) $\nu \ll \mu$, and, (b) If the sequence $\{E_n\} \subseteq \mathcal{M}$ has the property that $\lim_{n\to\infty} \mu(E_n) = 0$, then $\lim_{n\to\infty} \nu(E_n) = 0$.

3.13 Show that $\nu \ll \mu$ iff $\nu^+, \nu^- \ll \mu$ iff $|\nu| \ll \mu$.

3.14 If ν_1, ν_2 are signed measures, μ is a measure and $\nu_1, \nu_2 \ll \mu$, prove that $\nu_1 + \nu_2 \ll \mu$ and $d(\nu_1 + \nu_2)/d\mu = d\nu_1/d\mu + d\nu_2/d\mu$.

3.15 Given measures λ, μ, show that

$$d\lambda/d(\lambda + \mu) + d\mu/d(\lambda + \mu) = 1 \ (\lambda + \mu)\text{-a.e.}$$

3.16 If λ, μ, ν are measures such that $\lambda \ll \mu$ and $\mu \ll \nu$, show that $\lambda \ll \nu$ and

$$d\lambda/d\nu = (d\lambda/d\mu) \cdot (d\mu/d\nu) \ \lambda\text{-a.e.}$$

3.17 Suppose μ, ν are σ-finite measures so that $\mu \ll \nu$ and $\nu \ll \mu$. Show that

$$d\nu/d\mu = 1/(d\mu/d\nu) \ \mu\text{-a.e.}$$

3.18 Let μ, ν be σ-finite measures, and suppose $\mu - \nu$ is a measure so that $\nu \ll \mu - \nu$. Show that $\mu(\{d\nu/d\mu = 1\}) = 0$.

3.19 Let μ, ν be σ-finite measures, and suppose $\nu \ll \mu$. Show that $\nu(\{d\nu/d\mu = 0\}) = 0$.

3.20 Referring to 4.14 in Chapter VII, if μ is a finite measure and λ is a σ-finite measure on $(Y, \mathcal{N})$, then show that there exists a function $h \in L^1(\lambda)$ such that $\int_X f \circ \tau \, d\mu = \int_Y f \, d\nu = \int_Y fh \, d\lambda$ for all f in $L^1(\nu)$.

3.21 Let μ, ν be σ-finite measures and assume $\nu \ll \mu$. Show that for all $f \in L^1(\nu)$, $f(d\nu/d\mu) \in L(\mu)$ and $\int_X f \, d\nu = \int_X f(d\nu/d\mu) d\mu$.

3.22 Consider the relation defined on classes of measures by $\lambda \sim \mu$ iff $\lambda \ll \mu$ and $\mu \ll \lambda$. Show that $\sim$ is an equivalence relation and describe the relation between $L^1(\lambda)$ and $L^1(\mu)$.

3.23 Given a measure μ, put

$$\nu(E) = \begin{cases} 0 & \text{if } \mu(E) = 0 \\ \infty & \text{if } \mu(E) > 0. \end{cases}$$

Show that ν is measure on $(X, \mathcal{M})$, and that $\nu \ll \mu$. Also find $d\nu/d\mu$.

3.24 Let $(X, \mathcal{M}, \mu)$ be a measure space, $\mathcal{N} \subset \mathcal{M}$ a σ-algebra of subsets of X, and $\nu = \mu|\mathcal{N}$ the restriction of μ to $\mathcal{N}$. Show that given $f \in L(\mu)$, there exists a uniquely determined, ν-a.e. that is, $f_1 \in L(\nu)$ such that $\int_A f \, d\mu = \int_A f_1 \, d\nu$ for every $A \in \mathcal{N}$. f_1 is called the conditional expectation of f with respect to $\mathcal{N}$.

3.25 Show that the Radon-Nikodým Theorem remains true if μ is a σ-finite signed measure.

3.26 Let μ, ν be regular Borel measures. Show that if $\operatorname{supp} \mu \cap \operatorname{supp} \nu = \emptyset$, then $\mu \perp \nu$.

3.27 Show that if $\nu_1, \nu_2 \perp \mu$, then $\nu_1 + \nu_2 \perp \mu$.

3.28 Show that $\mu \perp \nu$ iff $|\mu| \perp |\nu|$.

3.29 Let μ, ν be σ-finite measures. Show that $\mu \perp \nu$ iff there exists no nonzero measure λ such that $\lambda \ll \mu$ and $\lambda \ll \nu$.

3.30 Suppose λ and μ are finite measures. Prove there is a (measurable) partition $\{A_1, A_2, A_3\}$ of X such that:
(i) $\lambda(A_1) = 0$,
(ii) $\mu(A_2) = 0$,
and
(iii) On A_3, $\lambda \ll \mu$ and $\mu \ll \lambda$.
Also, show there exists a finite positive measurable function h on A_3 such that for every nonnegative measurable function f on A_3,

$$\int_{A_3} f\, d\lambda = \int_{A_3} fh\, d\mu \quad \text{and} \quad \int_{A_3} f\, d\mu = \int_{A_3} (f/h)\, d\lambda\,.$$

3.31 Suppose ν is a signed Borel measure with the property that $|\nu|$ is finite on bounded sets on R^n. Referring to 3.25 in Chapter VIII, show that if $\mathcal{R} = \{R\}$ is a regular family, then

$$\lim_{\operatorname{diam}(R)\to 0} \frac{\nu(\{x+R\})}{|R|} \quad \text{exists a.e.}$$

3.32 Suppose ν is as in 3.31 and that A is a Borel subset of R^n so that $D\nu(x) \geq \lambda$ for all $x \in A$. Prove that $\nu(A) \geq \lambda|A|$.

3.33 Assume μ, μ_m are finite Borel measures, $m = 1, 2, \ldots$, such that $\lim_{m\to\infty} \mu_m(A) = \mu(A)$ for all $A \in \mathcal{B}_n$, monotonically (either nondecreasing or nonincreasing). Prove that

$$\lim_{m\to\infty} D\mu_m = D\mu \quad \text{a.e.}$$

3.34 A complex-valued set function ν of the form

$$\nu(E) = \nu_1(E) + i\nu_2(E)\,, \quad E \in \mathcal{M}\,,$$

where ν_1 and ν_2 are signed measures is called a complex measure. ν_1 and ν_2 are called the real and imaginary parts of ν, respectively. This is an open ended question: Discuss the properties of complex measures. For instance, show that if $|\nu|$ is given by 4.12 in Chapter IV, then the set function $|\nu|$ is a measure called the (total) variation of ν. Also prove that if $\eta = \sup\{|\nu(A)| : A \in \mathcal{M}, A \subset E\}$, then

$$\eta \leq |\nu|(E) \leq 4\eta\,, \quad \text{all } E \in \mathcal{M}\,.$$

3.35 Let μ and ν be complex measures. We say that ν is absolutely continuous with respect to μ, and we write $\nu \ll \mu$, if $|\mu|(E) = 0$ implies $\nu(E) = 0$, for $E \in \mathcal{M}$.
Discuss properties of absolutely continuous measures. For instance, prove that if $\nu = \nu_1 + i\nu_2$, then the following are equivalent: (i) $\nu \ll \mu$, (ii) $\nu_1^+, \nu_1^-, \nu_2^+, \nu_2^- \ll \mu$, and, (iii)$|\nu| \ll \mu$.

3.36 Let μ and ν be complex measures. We say that μ and ν are singular, and we write $\mu \perp \nu$, if there exists a set $A \in \mathcal{M}$ such that $|\mu|(A) = |\nu|(X \setminus A) = 0$. Discuss properties of singular measures.

3.37 Can you think of a Lebesgue decomposition theorem in case $(X, \mathcal{M}, \mu)$ is a σ-finite measure space and ν is a complex measure on $(X, \mathcal{M})$?

CHAPTER XII

L^p Spaces

In this chapter we introduce the Lebesgue spaces of p-integrable functions and study their basic properties. In considering the various results discussed here, the reader should keep in mind the three basic examples: The L^p spaces of Lebesgue measurable functions on the line, or R^n, the L^p spaces of Lebesgue measurable functions on a bounded interval, and the sequence ℓ^p spaces.

1. THE LEBESGUE L^p SPACES

Let $(X, \mathcal{M}, \mu)$ be a measure space and f an extended real-valued measurable function defined on X. Then, for $0 < p < \infty$, $|f|^p$ is also measurable and the expression

$$\|f\|_p = \left(\int_X |f|^p \, d\mu\right)^{1/p}, \quad 0 < p < \infty, \tag{1.1}$$

whether finite or not, is well-defined and is called the "p norm" of f. The case $p = 1$ has been studied in Chapter VIII, and the natural question to consider is to what extent those results can be extended to values of p other than 1.

First a definition. The Lebesgue class $L^p(X, \mu)$, or plainly $L^p(X)$ or $L^p(\mu)$, is defined as

$$L^p(X, \mu) = \{f \text{ measurable} : \|f\|_p < \infty\}, \quad 0 < p < \infty. \tag{1.2}$$

Our immediate goals are to show that $L^p(\mu)$ is a linear class and to introduce a metric that will turn $L^p(\mu)$ into a complete metric space.

It is not hard to verify that $L^p(\mu)$ is a linear class. Indeed, given $f, g \in L^p(\mu)$ and a real scalar λ, first note that $f + \lambda g$ is measurable. Furthermore, since for nonnegative real numbers a, b we have

$$(a+b)^p \le 2^p(a^p + b^p), \quad 0 < p < \infty, \tag{1.3}$$

it follows that $|f(x) + \lambda g(x)|^p \le 2^p(|f(x)|^p + |\lambda|^p|g(x)|^p), x \in X$, and consequently, we also have

$$\|f + \lambda g\|_p^p \le 2^p \left(\|f\|_p^p + |\lambda|^p \|g\|_p^p\right) < \infty .$$

As for the metric, inspired by (1.1) in Chapter VIII, a natural choice is

$$d_p(f,g) = \|f - g\|_p, \quad 0 < p < \infty .$$

Since $d_p(f,g) = 0$ implies $f = g$ μ-a.e, strictly speaking, the elements of $L^p(\mu)$ are equivalence classes of functions defined on X, where we agree that $f = g$ means that $f = g$ μ-a.e.

Now, since the cases $0 < p < 1$ and $1 < p < \infty$ are essentially different, we treat them separately; we do the former case first. So, fix $0 < p < 1$, and note that (1.3) can be improved to

$$(a+b)^p \le a^p + b^p, \quad a, b \ge 0, \tag{1.4}$$

with equality occurring in (1.4) iff $ab = 0$.

To see this observe that if $ab = 0$ there is nothing to prove. On the other hand, if $ab \ne 0$, and since (1.4) is to hold for any $a, b > 0$, we may replace a by ab there and consider the equivalent inequality

$$\phi(t) = (1+t)^p - 1 - t^p \le 0, \quad \text{all } t \ge 0, \tag{1.5}$$

with $\phi(t) = 0$ iff $t = 0$. But this is easily checked: Indeed, since $0 < p < 1$, we have $\phi'(t) = p(1+t)^{p-1} - pt^{p-1} < 0$ for $t > 0$, and consequently, $\phi(t) < \phi(0) = 0$, which gives (1.5), including the remark concerning equality.

With this observation out of the way, let A, B be disjoint measurable subsets of X neither of which is null, and let $f = \chi_A$, $g = \chi_B$. Then, putting $a = \mu(A)^{1/p}$ and $b = \mu(B)^{1/p}$ in (1.4) we obtain

$$\begin{aligned} d_p(f,g) &= \left(\int_X |f-g|^p \, d\mu\right)^{1/p} = (\mu(A) + \mu(B))^{1/p} \\ &= \left((\mu(A)^{1/p})^p + (\mu(B)^{1/p})^p\right)^{1/p} \ge \mu(A)^{1/p} + \mu(B)^{1/p}, \end{aligned}$$

or, in other words,

$$d_p(f,g) \geq d_p(f,0) + d_p(0,g)\,.$$

Thus, the inequality opposite to the triangle inequality holds, and d_p cannot be a distance function on $L^p(\mu)$. This is not, however, a serious difficulty. Indeed, by (1.4), the expression

$$d_p^p(f,g) = \int_X |f-g|^p\,d\mu$$

satisfies the requirements of a metric, and essentially all the results discussed in Chapter VIII are true: Endowed with this metric $L^p(\mu)$ is a complete metric space and $C_0(R^n)$ is dense in $L^p(R^n)$. It is interesting to point out that L^p integrable functions are not necessarily integrable, or locally integrable for that matter. For instance, when $0 < p < 1$, by 4.46 in Chapter VII, the function $|x|^{-n} \in L^p(I(0,1))$, but it is not locally integrable in any neighbourhood of the origin.

In order to deal with the case $1 < p < \infty$ we need a preliminary result which is essential in what follows. We have already encountered a particular instance of this result, the Cauchy-Schwarz inequality, in 3.1 in Chapter VIII.

We say that $1 < p, q < \infty$ are conjugate indices provided that

$$1/p + 1/q = 1\,.$$

Conjugate indices are also related by the expressions $p + q = pq$ and $q = p/(p-1)$.

An important property that the conjugate indices $p = 1/\eta$, $q = 1/(1-\eta)$, $0 < \eta < 1$, satisfy, is this: For any $a, b \geq 0$ we have

$$a^\eta b^{1-\eta} \leq \eta a + (1-\eta)b\,. \tag{1.6}$$

Furthermore, equality holds in (1.6) iff $a = b$.

To see this note that if $ab = 0$ there is nothing to prove. Otherwise, if $ab \neq 0$, and since (1.6) is to hold for any $a, b > 0$, we may replace a by ab there and consider the equivalent assertion

$$\psi(t) = t^\eta - \eta t - (1-\eta) \leq 0\,, \quad \text{all } t \geq 0\,,$$

with $\psi(t) = 0$ iff $t = 1$. Note that since $0 < \eta < 1$, we have

$$\psi'(t) = \eta t^{\eta-1} - \eta = \begin{cases} > 0 & \text{if } 0 < t < 1 \\ = 0 & \text{if } t = 1 \\ < 0 & \text{if } t > 1. \end{cases}$$

Thus $\psi(t)$ decreases to 0 as t increases to 1, and then increases to ∞ as $t \to \infty$, and $\psi(t) = 0$ iff $t = 1$. This givs (1.6), including the remark concerning equality.

A useful reformulation of (1.6), known as Young's inequality, is this: If $a, b > 0$ and p, q are conjugate indices, then

$$ab \leq \frac{a^p}{p} + \frac{b^q}{q}, \tag{1.7}$$

and equality holds in (1.7) iff $a^p = b^q$.

We are now ready to prove Hölder's inequality.

Theorem 1.1. Let $(X, \mathcal{M}, \mu)$ be a measure space, p, q conjugate indices, and suppose $f \in L^p(\mu)$, $g \in L^q(\mu)$. Then fg is integrable, and

$$\int_X |fg|\, d\mu \leq \|f\|_p \|g\|_q \,. \tag{1.8}$$

Moreover, equality holds in (1.8) iff there exist nonnegative constants A, B, $AB \neq 0$, such that

$$A|f|^p = B|g|^q \;\; \mu\text{-a.e.} \tag{1.9}$$

Proof. If the right-hand side of (1.8) is 0, then either $f = 0$ μ-a.e. or else $g = 0$ μ-a.e., and we have equality in (1.8). If, on the other hand, $\|f\|_p \|g\|_q \neq 0$, by replacing f and g by $f/\|f\|_p$ and $g/\|g\|_q$ if necessary, we may assume that $\|f\|_p = \|g\|_q = 1$. Now, since f and g are finite μ-a.e., from (1.7) it follows that

$$|f||g| \leq \frac{|f|^p}{p} + \frac{|g|^q}{q}, \quad \mu\text{-a.e.}, \tag{1.10}$$

and consequently, integrating over X we obtain

$$\int_X |fg|\, d\mu \leq \frac{1}{p} \int_X |f|^p \, d\mu + \frac{1}{q} \int_X |g|^q \, d\mu = 1\,,$$

which is precisely (1.8).

A moment's thought will convince the reader that equality can occur in (1.8) iff it occurs in (1.10) μ-a.e. But, by (1.7), this is true iff $|f|^p = |g|^q$ μ-a.e., and this gives (1.9) when $\|f\|_p = \|g\|_q = 1$. As for arbitrary functions f and g, we normalize them as above and note that equality holds iff $|f|^p/\|f\|_p^p = |g|^q/\|g\|_q^q$ μ-a.e. Whence a possible choice for the (nonunique) constants is $A = \|g\|_q^q$ and $B = \|f\|_p^p$. ∎

We are now in a position to prove one of the essential results in the theory of L^p spaces, which is due to Minkowski (1864–1909), namely, Minkowski's inequality.

Theorem 1.2. Suppose $(X, \mathcal{M}, \mu)$ is a measure space and $f, g \in L^p(\mu)$, $1 \le p < \infty$. Then

$$\|f+g\|_p \le \|f\|_p + \|g\|_p . \tag{1.11}$$

As for equality in (1.11) there are two separate cases, depending on whether $p = 1$ or not. If $1 < p < \infty$ the condition is: There exist constants A, B, $AB \ne 0$, such that $Af = Bg$ μ-a.e. On the other hand, if $p = 1$ the condition is: There exists a nonnegative measurable function h such that $fh = g$ μ-a.e. on the set $\{fg \ne 0\}$.

Proof. (1.11), in case $p = 1$, was already established in Chapter VIII. As for $p > 1$, since f, g are finite μ-a.e., so is $f + g$, and we have that $|f+g|^p$ is less than or equal to

$$|f+g|\,|f+g|^{p-1} \le |f|\,|f+g|^{p-1} + |g|\,|f+g|^{p-1} \quad \mu\text{-a.e.} \tag{1.12}$$

Whence integrating (1.12) over X we get

$$\begin{aligned} \|f+g\|_p^p &\le \int_X |f||f+g|^{p-1}\, d\mu + \int_X |g||f+g|^{p-1}\, d\mu \\ &= I + J\,, \end{aligned} \tag{1.13}$$

say. To estimate I and J we apply Hölder's inequality with indices p and its conjugate $q = p/(p-1)$, and note that

$$I \le \|f\|_p \left(\int_X |f+g|^{(p-1)\frac{p}{p-1}}\, d\mu \right)^{(p-1)/p} = \|f\|_p \|f+g\|_p^{p-1}, \tag{1.14}$$

and similarly,

$$J \le \|g\|_p \|f+g\|_p^{p-1} . \tag{1.15}$$

Whence substituting (1.14) and (1.15) into (1.13) it follows that

$$\|f+g\|_p^p \le \|f+g\|_p^{p-1} \left(\|f\|_p + \|g\|_p \right),$$

which is equivalent to (1.11).

Finally, when $1 < p < \infty$, it is clear that we only have equality in (1.11) provided we have equality in (1.12) and in the estimates of I and J in (1.14) and (1.15), respectively. Now, equality holds in (1.12) if f and g are of the same sign μ-a.e. Also, by (1.9), equality holds in the estimates of I and J if for some constants A, B we have $A|f|^p = |f+g|^q = B|g|^p$ μ-a.e. Whence combining these remarks it follows that equality holds if, as asserted, $Af = Bg$ μ-a.e.

When $p = 1$ we must have equality in $|f+g| \le |f| + |g|$ μ-a.e., and this occurs when f and g are of the same sign μ-a.e. Then the function $h = g/f$ will do the job. ■

An interesting consequence of Minkowski's inequality is that d_p is a metric on $L^p(\mu), 1 < p < \infty$. The only property that offers any difficulty is the triangle inequality, and it is obtained as follows: If $f, g, h \in L^p(\mu)$, we have

$$\begin{aligned} d_p(f,g) &= \|f-g\|_p = \|(f-h)+(h-g)\|_p \\ &\le \|f-h\|_p + \|h-g\|_p = d_p(f,h) + d_p(h,g)\,. \end{aligned}$$

Furthermore,

Theorem 1.3 (F.Riesz-Fischer). Let $(X, \mathcal{M}, \mu)$ be a measure space. Then, the distance function d_p turns $L^p(\mu)$ into a complete metric space, $1 < p < \infty$.

Proof. We fix $1 < p < \infty$, and assume that $\{f_n\}$ is a Cauchy sequence of functions in $L^p(\mu)$. We must show that there is a function $f \in L^p(\mu)$ so that $\lim_{n\to\infty} d_p(f_n, f) = 0$. First observe that since $\{f_n\}$ is Cauchy we can find an increasing sequence $n_{k+1} > n_k$ such that

$$d_p(f_n, f_{n_k}) \le 1/2^k\,, \quad \text{all } n \ge n_k, \quad k = 1, 2, \ldots$$

Put now $f_{n_0} = 0$, and let

$$g_m = \sum_{k=1}^{m} |f_{n_k} - f_{n_{k-1}}|\,, \quad m = 1, 2, \ldots$$

Clearly the sequence $\{g_m\}$ is nondecreasing, let g denote its limit. Furthermore, since by (a simple extension of) Minkowski's inequality it follows that $\|g_m\|_p \le 1$ for all m, by MCT we get that $\int_X g^p \, d\mu \le 1$.Thus, in particular, g is finite μ-a.e., and the series with terms $f_{n_k} - f_{n_{k-1}}$, $k = 1, 2, \ldots,$ converges absolutely to a finite sum μ-a.e. Let then

$$f = \lim_{m\to\infty} \sum_{k=1}^{m} (f_{n_k} - f_{n_{k-1}})\,;$$

f is measurable and finite μ-a.e. Moreover, since the sum on the right-hand side above telescopes to f_{n_m}, it readily follows that

$$\lim_{k\to\infty} f_{n_k} = f \quad \mu\text{-a.e.}$$

We want to show that the convergence is also in the metric of $L^p(\mu)$. First observe that since

$$|f| \le g\,, \quad \text{and} \quad |f_{n_k}| \le g_{n_k} \le g\,, \quad \text{all } k,$$

and since $g \in L^p(\mu)$, by LDCT we get

$$\lim_{k\to\infty} \int_X |f_{n_k} - f|^p \, d\mu = \lim_{k\to\infty} d_p^p(f_{n_k}, f) = 0\,.$$

To complete the proof we invoke the well-known fact that if a Cauchy sequence in a metric space has a convergent subsequence, then the sequence itself converges to the same limit. ∎

As in the case of integrable functions, the metric structure of $L^p(X,\mu)$ permits us to establish the following properties:

(i) Simple functions are dense in $L^p(X,\mu)\,, 1 < p < \infty$.

(ii) $C_0(R^n), C_0^\infty(R^n)$ are dense in $L^p(R^n), 1 < p < \infty$.

(iii) $C_0^k(I) = \{f \in C_0^k(R^n) : f \text{ vanishes off } I\}$ is dense in $L^p(I)$, $1 < p < \infty, 0 \le k \le \infty$.

(iv) The class of sequences $\{c_n\}$ which eventually vanish is dense in ℓ^p, $0 < p < \infty$.

(v) The translates of $L^p(R^n)$ functions are continuous in the norm, i.e., if $f \in L^p(R^n)$, then $\lim_{|h|\to 0} \|f(\cdot + h) - f(\cdot)\|_p = 0$, $1 < p < \infty$.

Since the proof of these statements follows along the lines of that of the corresponding results for $p = 1$, we leave their verification to the reader.

Finally, in the scale of Lebesgue spaces there are two limiting cases left to consider: The case $p = 0$ and the case $p = \infty$. The idea here is to let $p \to 0$ and $p \to \infty$ in the expressions corresponding to $\|f\|_p^p$ and $\|f\|_p$ respectively, and study what happens.

In the limiting case $p = 0$, it is clear that the limit exists and it equals $\|f\|_0 = \int_{\{f\ne 0\}} d\mu$. The class $L^0(X,\mu)$, consisting of those extended real-valued finite μ-a.e. measurable functions f defined on X so that $\mu(\{f \ne 0\}) < \infty$, enjoys many of the properties of the $L^p(X,\mu)$ spaces and is very useful in applications; we discuss it no further here.

As for the case $p = \infty$, to fix ideas consider $I = [0,1]$ and a measurable function f defined on I. If $1 < p < q < \infty$ are given, observe that $r = q/p > 1$ and $r' = r/(r-1) = q/(q-p)$ are conjugate indices, and consequently, by Hölder's inequality, and assuming that the quantities involved are finite, we have

$$\int_I |f|^p \, dx \le \left(\int_I |f|^{pr} \, dx\right)^{1/r} \left(\int_I dx\right)^{1/r'} = \left(\int_I |f|^q \, dx\right)^{q/p},$$

or equivalently,

$$\|f\|_p \le \|f\|_q\,, \quad 1 < p < q < \infty\,.$$

Whence, the p norms of f are nondecreasing and $\lim_{q\to\infty}\|f\|_q$ exists. Suppose this limit is finite, this is the case of interest to us, and call it L. By Chebychev's inequality we have

$$\lambda|\{|f| > \lambda\}|^{1/q} \le \|f\|_q \le L\,, \quad \text{all } \lambda > 0\,. \tag{1.16}$$

Moreover, since

$$\lim_{q\to\infty} |\{|f| > \lambda\}|^{1/q} = 1 \quad \text{whenever} \quad |\{|f| > \lambda\}| \ne 0\,,$$

by (1.16) it readily follows that

$$|\{|f| > \lambda\}| = 0\,, \quad \text{all } \lambda > L\,. \tag{1.17}$$

Functions that satisfy (1.17) are called "essentially bounded". In contrast to bounded functions, essentially bounded functions may assume infinite values, but only on null sets.

These observations motivate our next definition. Let $(X, \mathcal{M}, \mu)$ be a measure space and f an extended real-valued measurable function defined on X. The expression

$$\|f\|_\infty = \inf\{\lambda > 0 : \mu(\{|f| > \lambda\}) = 0\}\,, \tag{1.18}$$

whether finite or not, is well-defined and is called the μ-essential sup of $|f|$, or the "∞ norm" of f. The Lebesgue class $L^\infty(X,\mu)$ consists of those measurable functions f with $\|f\|_\infty < \infty$, and it is also denoted by $L^\infty(\mu)$ or $L^\infty(X)$. Note that since

$$\mu(\{|f| > \|f\|_\infty + 1/n\}) = 0\,, \quad n = 1, 2, \ldots\,,$$

we also have

$$\mu(\{|f| > \|f\|_\infty\}) = 0\,, \tag{1.19}$$

and the infimum in (1.18) is attained.

The next step in the study of the L^∞ spaces is to establish which properties of the L^p spaces remain valid in this setting, and which do not. For instance, since convergence in the L^∞ norm corresponds to uniform convergence, if $I = [0,1]$, only continuous functions can be approximated in $L^\infty(I)$ by continuous functions, and consequently, $C_0(I)$ is not dense in $L^\infty(I)$.

On the other hand, on a positive note we have

Theorem 1.4. Endowed with the metric

$$d_\infty(f,g) = \|f-g\|_\infty\,, \tag{1.20}$$

$L^\infty(X,\mu)$ becomes a complete metric space.

Proof. To show that d_∞ is a metric the only property that offers any difficulty is the triangle inequality. The proof of this depends on the following variant of Minkowski's inequality:

$$\|f+g\|_\infty \le \|f\|_\infty + \|g\|_\infty\,. \tag{1.21}$$

To see this observe that

$$\{|f+g| > \|f\|_\infty + \|g\|_\infty\} \subseteq \{|f| > \|f\|_\infty\} \cup \{|g| > \|g\|_\infty\}\,,$$

and since by (1.19) the sets on the right-hand side above are null, so is the set on the left-hand side, and (1.21) follows.

The completeness of $L^\infty(\mu)$ with the metric d_∞ follows along the lines of Theorem 1.3 and is therefore left for the reader to verify. ■

Another important property of $L^\infty(\mu)$ is that 1 and ∞ are conjugate indices, in the sense that $1/1+1/\infty = 1$. A justification for this convention is given by the following extension of Hölder's inequality.

Proposition 1.5. Let $(X,\mathcal{M},\mu)$ be a measure space, and suppose $f \in L(\mu)$, $g \in L^\infty(\mu)$. Then $fg \in L(\mu)$, and we have

$$\int_X |fg|\,d\mu \le \|f\|_1\|g\|_\infty\,. \tag{1.22}$$

Equality holds in (1.22) iff $|g| = \|g\|_\infty$ μ-a.e.

Proof. By (1.19) we may assume the integral in (1.22) to be extended over the set $\{\,|g| \le \|g\|_\infty\}$, and in this case (1.22) holds trivially. As for the case of equality in (1.22), observe that since

$$|g|/\|g\|_\infty \le 1 \quad \mu\text{-a.e.},$$

the relation

$$\int_X |f|(|g|/\|g\|_\infty)\,d\mu = \int_X |f|\,d\mu\,,$$

can only be true if, as asserted, $|g|/\|g\|_\infty = 1$ μ-a.e. ■

2. FUNCTIONALS ON L^p

The next topic we consider is that of mappings defined on $L^p(X,\mu)$, and the simplest case is that of scalar-valued mappings. For instance, point-evaluation is a well-defined mapping on $C_0(R^n)$, and since this class is dense in $L^p(R^n)$, $0 < p < \infty$, it is natural to consider whether point-evaluation may be extended, in some natural way, to $L^p(R^n)$. Since functions in these classes need only be defined μ-a.e. it is not intuitively clear how to construct such an extension, or whether, in fact, one such extension exists. The answer to these questions is postponed to Chapter XIV, where the Hahn-Banach Theorem is discussed.

In this chapter we take a different approach, one suggested by Hölder's inequality. For example, referring to (1.22), given $g \in L^\infty(\mu)$, let L_g be the mapping on $L(\mu)$ given by

$$L_g f = \int_X fg\,d\mu\,, \quad f \in L(\mu)\,. \tag{2.1}$$

Clearly L_g is well-defined, and it satisfies the following properties: (a) For $f, h \in L(\mu)$ and a scalar λ we have $L_g(f + \lambda h) = L_g f + \lambda L_g h$, (b) $|L_g f| \le \|g\|_\infty \|f\|_1$ for all $f \in L(\mu)$, and, (c) If $\lim_{n\to\infty} d_1(f_n, f) = 0$, then $\lim_{n\to\infty} L_g f_n = L_g f$.

In fact, L_g is a prototype of those mappings L on the Lebesgue L^p spaces which satisfy the following properties:

(i) L is a well-defined scalar-valued mapping.

(ii) (Linearity) For each scalar λ and $f, g \in L^p(\mu)$,

$$L(f + \lambda g) = Lf + \lambda Lg\,.$$

(iii) (Continuity) If $\lim_{n\to\infty} d_p(f_n, f) = 0$, then $\lim_{n\to\infty} Lf_n = Lf$.

(iv) (Boundedness) There is a constant k such that

$$|Lf| \le k\|f\|_p\,, \quad \text{all } f \in L^p(\mu)\,.$$

More precisely, a mapping L which satisfies (i) above is called a functional. L is said to be a linear, continuous or bounded functional if (ii), (iii) or (iv), respectively, hold.

Although not apparent at a first glance, the concepts of continuity and boundedness are equivalent.

Proposition 2.1. Suppose L is a linear functional on $L^p(\mu)$. Then L is continuous iff L is bounded.

Proof. To prove the necessity, suppose that L is continuous but unbounded. Then, for each n, there is $f_n \in L^p(\mu)$ such that

$$|Lf_n| \ge n\|f_n\|_p\,, \quad \text{or} \quad |L(f_n/n\|f_n\|_p)| \ge 1\,, \quad n = 1,2,\ldots$$

Now, the sequence $g_n = (1/n\|f_n\|_p)f_n, n = 1,2,\ldots$, satisfies

$$\lim_{n\to\infty} d_p(g_n, 0) = 0\,, \quad \text{and} \quad |Lg_n| \ge 1\,, \quad \text{all } n\,,$$

thus contradicting the continuity of L.

As for the sufficiency, observe that if $f, f_n \in L^p(\mu), n = 1,2,\ldots$, by the linearity and boundedness of L we have

$$|Lf_n - Lf| = |L(f_n - f)| \le k\,\|f_n - f\|_p\,.$$

Whence, if $d_p(f_n, f) \to 0$, the right-hand side in the above inequality tends to 0 with n, as does the left-hand side there, and the desired conclusion follows. ■

It is rather straightforward to construct bounded linear functionals on l^p, $0 < p \le 1$. Indeed, if $(m_n) \in \ell^\infty$, the mapping

$$L(c_n) = \sum_n m_n c_n\,, \quad (c_n) \in \ell^p\,,$$

is readily seen to be such a functional. In fact, by (1.4) it is clear that

$$|L(c_n)| \le \|(m_n)\|_\infty \|(c_n)\|_p\,,$$

and (iv) above holds with $k = \|(m_n)\|_\infty$ there.

The following result, then, is a bit surprising.

Theorem 2.2 (M.M. Day). Let $I = [0,1]$, and suppose L is a continuous linear functional on $L^p(I)$, $0 < p < 1$. Then L is the zero functional, i.e., for every $f \in L^p(I)$ we have $Lf = 0$.

Proof. Suppose, to the contrary, that L is not the zero functional. Then, by rescaling if necessary, we may assume that there is a function $f \in L^p(\mu)$ such that

$$Lf = 1\,, \quad \|f\|_p \ne 0\,. \tag{2.2}$$

Observe that as a function of x in I, $f\chi_{[0,x]}$ is continuous in the metric of $L^p(\mu)$, and consequently, since L is continous we have that

$$\phi(x) = L\left(f\chi_{[0,x]}\right)\,, \quad x \in I\,,$$

is a continuous real-valued function defined on I which satisfies $\phi(0) = 0$, and $\phi(1) = Lf = 1$. Since ϕ is continuous there is $x \in (0,1)$ such that $\phi(x) = 1/2$; further, since $f = f\chi_{[0,x]} + f\chi_{[x,1]}$ (in $L^p(\mu)$), by the linearity of L it readily follows that

$$L\left(f\chi_{[0,x]}\right) = L\left(f\chi_{[x,1]}\right) = 1/2\,. \tag{2.3}$$

Moreover, since

$$\|f\chi_{[0,x]}\|_p^p + \|f\chi_{[x,1]}\|_p^p = \|f\|_p^p\,, \tag{2.4}$$

one of the summands on the left-hand side of (2.4) does not exceed $\|f\|_p^p/2$. Call g_1 one of the functions in (2.4), either $f\chi_{[0,x]}$ or $f\chi_{[x,1]}$, such that $\|g_1\|_p^p \le \|f\|_p^p/2$, and rewrite this estimate

$$\|2g_1\|_p^p \le 2^{(p-1)}\|f\|_p^p\,. \tag{2.5}$$

Put now $f_1 = 2g_1$ and observe that, by combining (2.3) and (2.5), we get

$$Lf_1 = 1\,, \quad \text{and} \quad \|f_1\|_p^p \le 2^{(p-1)}\|f\|_p^p\,.$$

Repeating the above argument with f_1 in place of f above we obtain a function f_2, say, such that

$$Lf_2 = 1\,, \quad \text{and} \quad \|f_2\|_p^p \le 2^{(p-1)}\|f_1\|_p^p \le 2^{2(p-1)}\|f\|_p^p\,.$$

It is now apparent that iterating this inequality we get a sequence $\{f_n\} \subseteq L^p(I)$ which satisfies

$$Lf_n = 1\,, \quad \text{and} \quad \|f_n\|_p^p \le 2^{n(p-1)}\|f\|_p^p\,. \tag{2.6}$$

Since $0 < p < 1$, (2.6) implies that the f_n's satisfy

$$Lf_n = 1\,, \quad \lim_{n\to\infty} d_p^p(f_n, 0) = 0\,,$$

which is impossible if L is continuous. This contradiction was derived from (2.2), and hence L is the zero functional. ■

The situation is quite different when $p \ge 1$: Not only are there plenty of functionals on $L^p(\mu)$, but it is also possible to characterize them. We begin with a definition.

The norm $\|L\|$ of a bounded linear functional L on $L^p(\mu)$, $1 \le p \le \infty$, is defined by the quantity

$$\|L\| = \sup_{\|f\|_p \ne 0} \frac{|Lf|}{\|f\|_p}\,. \tag{2.7}$$

For instance, as pointed out in (ii) above, the functional L_g on $L(\mu)$ given by (2.1) satisfies $\|L_g\| \le \|g\|_\infty$. In fact, if μ is σ-finite, or more generally semifinite, we have equality here. To see this, given $\varepsilon > 0$, let $E \in \mathcal{M}$ be a set of positive measure so that

$$|g| > \|g\| - \varepsilon\,, \quad \mu\text{-a.e. on } E\,.$$

Moreover, since μ is σ-finite, or semifinite, we may also assume that $\mu(E) < \infty$ and consequently, the function $f = (\operatorname{sgn} g)\chi_E \in L(\mu)$. Finally, since

$$\begin{aligned} L_g f &= \int_E (\operatorname{sgn} g) g \, d\mu = \int_E |g| \, d\mu \\ &\ge (\|g\|_\infty - \varepsilon)\,\mu(E) = (\|g\|_\infty - \varepsilon)\,\|f\|_1\,, \end{aligned}$$

it follows that

$$\|L_g\| \ge |L_g f|/\|f\|_1 \ge \|g\|_\infty - \varepsilon\,,$$

which gives the desired conclusion since ε is arbitrary.

It is natural to pose the analogous question for L_g, $0 \ne g \in L^q(\mu)$, considered as a bounded linear functional on $L^p(\mu)$, $1/p + 1/q = 1$. First observe that by Hölder's inequality we have

$$|L_g f| \le \|g\|_q \|f\|_p\,, \tag{2.8}$$

and indeed L_g is a well-defined continuous linear functional on $L^p(\mu)$ with $\|L_g\| \le \|g\|_q$. To see that we actually have equality in the norms, if $1 < p < \infty$, put $f = (\operatorname{sgn} g)|g|^{q/p}$, and note that $\|f\|_p^p = \|g\|_q^q < \infty$ and $fg = |g|^{q/p}(\operatorname{sgn} g) = |g|^q$. Thus, for this particular L^p function f we have $L_g f = \int_X |g|^q \, d\mu$, and consequently,

$$\|L_g\| \ge |L_g f|/\|f\|_p = \|g\|_q^q/\|g\|_q^{q/p} = \|g\|_q\,,$$

as we wanted to show.

The case $p = \infty$ is even simpler, as the limiting process gives the right answer: If $0 \ne g \in L(\mu)$ now, put $f = \operatorname{sgn} g$, and observe that $\|f\|_\infty = 1$ and

$$L_g f = \int_X (\operatorname{sgn} g) g \, d\mu = \|g\|_1\,.$$

Whence, as asserted, $\|L_g\| \ge L_g f = \|g\|_1$.

A natural question to consider is whether every bounded linear functional L on $L^p(\mu)$ is of the form $L = L_q$, for some $g \in L^q(\mu)$, $1/p+1/q = 1$. In Chapter XIV we will see that this is not the case for $p = \infty$, but how about for finite p's?

In order to address this question we consider a converse to Hölder's inequality.

Proposition 2.3. Let $(X, \mathcal{M}, \mu)$ be a measure space, $1 \le p \le \infty$, $1/p + 1/q = 1$, and suppose $f \in L^p(\mu)$. Then if $1 \le p < \infty$, we have

$$\|f\|_p = \sup_{\|g\|_q \le 1} |L_g f| . \tag{2.9}$$

If μ is σ-finite it is also true that

$$\|f\|_\infty = \sup_{\|g\|_1 \le 1} |L_g f| . \tag{2.10}$$

Proof. We may assume that $f \ne 0$ on a set of μ positive measure, for otherwise there is nothing to prove. Now, if $1 \le p < \infty$, by Hölder's inequality it follows that the sup on the right-hand side of (2.9) is less than or equal to $\|f\|_p$. Furthermore, putting $g = (1/\|f\|_p^{p-1})(\operatorname{sgn} f)|f|^{p-1}$, we see at once that $\|g\|_q = 1$, and that

$$L_g f = \frac{1}{\|f\|_p^{p-1}} \int_X |f|^{p-1} (\operatorname{sgn} f) f \, d\mu = \|f\|_p ,$$

and (2.9) holds.

As for (2.10), again by Hölder's inequality it suffices to show that the sup on the right-hand side there is at least $\|f\|_\infty$. Now, since μ is σ-finite, given $\varepsilon > 0$, we can find a set $E \in \mathcal{M}$ such that

$$|f| > \|f\|_\infty - \varepsilon , \quad \mu\text{-a.e. on } E, \quad 0 < \mu(E) < \infty .$$

Then the function $g = (1/\mu(E))(\operatorname{sgn} f)\chi_E$ is integrable, has norm 1, and satisfies

$$L_g f = \frac{1}{\mu(E)} \int_E f (\operatorname{sgn} f) \, d\mu \ge \|f\|_\infty - \varepsilon .$$

Since ε is arbitrary, (2.10) holds. ∎

An interesting and useful variant of this result is

Theorem 2.4. Let $(X, \mathcal{M}, \mu)$ be a measure space, $1 \le p \le \infty$, $1/p + 1/q = 1$, and let f be an extended real-valued measurable function defined on X with the property that for every simple function ϕ defined on X we have

$$|L_\phi f| \le k \|\phi\|_q . \tag{2.11}$$

Then, if μ is σ-finite, it follows that $f \in L^p(\mu)$ and $\|f\|_p \le k$, where k is the constant in (2.11).

Proof. Write $X = \bigcup_n X_n$ as an increasing union of sets of finite measure, and begin by constructing a sequence $\{\psi_n\}$ of simple functions such that

$$|\psi_n| \le |f|, \quad \text{and} \quad \lim_{n\to\infty} \psi_n = f \quad \text{everywhere.} \tag{2.12}$$

Further, let $\phi_n = \psi_n \chi_{X_n}, n = 1,2,\ldots$, and note that also $\{\phi_n\}$ is a sequence of simple functions, and that (2.12) is also true with the ψ_n's replaced by the ϕ_n's there.

We distinguish three cases, to wit, $p = 1$, $1 < p < \infty$, and $p = \infty$. If $p = 1$ put $f_n = (\operatorname{sgn} f)\chi_{X_n}$, $n = 1,2,\ldots$, and observe that by MCT and (2.11) it follows that

$$\int_X |f|\, d\mu = \lim_{n\to\infty} \int_X f f_n \, d\mu \le k\,,$$

as we wished to show.

Next, if $1 < p < \infty$, put $f_n = |\phi_n|^{p-1}(\operatorname{sgn} f)$, $n = 1,2,\ldots$, and note that the f_n's are also simple functions which satisfy, for each n, the following relations:

(i) $|f_n\phi_n| = |\phi_n|^p$.

(ii) $f_n f \ge 0$.

(iii) $\|f_n\|_q^q = \|\phi_n\|_p^p$.

The idea now is to invoke Fatou's Lemma to estimate $\|f\|_p$; we need to obtain a bound for $\|\phi_n\|_p$. First note that on account of (i), (ii), (2.11), and (iii) we have

$$\begin{aligned} \int_X |\phi_n|^p \, d\mu &= \int_X |f_n\phi_n|\, d\mu \le \int_X |f_n f|\, d\mu \\ &= \left| \int_X f f_n \, d\mu \right| \le k\|f_n\|_q \le k\|\phi_n\|_p^{p/q}\,. \end{aligned}$$

It is clear that unless $f = 0$ μ-a.e., and in this case there is nothing to prove, we may also assume that $0 < \|\phi_n\|_p < \infty$ for all n, and consequently the above inequality gives

$$\|\phi_n\|_p^{p-p/q} = \|\phi_n\|_p \le k\,, \quad \text{or} \quad \|\phi_n\|_p^p \le k^p\,.$$

With this computation out of the way, observe that by Fatou's Lemma we have

$$\int_X |f|^p \, d\mu \le \liminf \int_X |\phi_n|^p \, d\mu \le k^p\,,$$

and (2.11) holds.

As for the case $p = \infty$, let $\varepsilon > 0$ be given, and put $E = \{|f| > k + \varepsilon\}$. If $\mu(E) > 0$, since μ is σ-finite, E contains a subset B, say, of positive, finite measure. Setting

$$\phi = \frac{1}{\mu(B)}(\operatorname{sgn} f)\chi_B\,, \quad \|\phi\|_1 = 1\,,$$

by (2.11) we have

$$k \geq \int_X f\phi\, d\mu > \frac{1}{\mu(B)}\int_B (k+\varepsilon)\, d\mu = k + \varepsilon,$$

which is impossible. Thus, we have $\mu(E) = 0$, and $\|f\|_\infty \leq k$. ∎

We are now in a position to describe the bounded linear functionals on $L^p(\mu)$.

Theorem 2.5 (F. Riesz Representation Theorem). Let $(X, \mathcal{M}, \mu)$ be a measure space, $1 \leq p < \infty$, and $1/p + 1/q = 1$. Then, if μ is σ-finite, to each continuous linear functional L on $L^p(\mu)$ there corresponds a unique function $g \in L^q(\mu)$ such that $\|L\| = \|g\|_q$, and

$$Lf = L_g f = \int_X fg\, d\mu\,, \quad \text{all } f \in L^p(\mu)\,.$$

Proof. Suppose first that μ is a finite measure and introduce the set function

$$\nu(E) = L\chi_E\,, \quad E \in \mathcal{M}\,. \tag{2.13}$$

Since L is linear it readily follows that ν is an additive set function on $\mathcal{M}$. We claim that actually ν is a signed measure and that $\nu \ll \mu$.

Clearly $\nu(\emptyset) = 0$. Moreover, since L is bounded, ν only assumes finite values and for each $E \in \mathcal{M}$ we have

$$|\nu(E)| = |L\chi_E| \leq \|L\|\, \mu(E)^{1/p} < \infty. \tag{2.14}$$

Thus, to verify that ν is a signed measure, it only remains to check that ν is σ-additive. Let, then, $\{E_n\}$ be a sequence of pairwise disjoint measurable subsets of X, and put $E = \bigcup_n E_n$. By (2.13) and (2.14), and since $E \setminus \bigcup_{n=1}^m E_n = \bigcup_{n=m+1}^\infty E_n$, we get

$$\begin{aligned} |\nu(E) - \textstyle\sum_{n=1}^m \nu(E_n)| &= |\nu(E \setminus \textstyle\bigcup_{n=1}^m E_n)| = |L\chi_{\bigcup_{n=m+1}^\infty E_n}| \\ &\leq \|L\|\, \mu\big(\textstyle\bigcup_{n=m+1}^\infty E_n\big)^{1/p}\,. \end{aligned} \tag{2.15}$$

Now, since the E_n's are pairwise disjoint and $\mu(E) < \infty$, we have

$$\mu\big(\bigcup_{n=m+1}^{\infty} E_n\big) = \sum_{n=m+1}^{\infty} \mu(E_n) \to 0\,, \quad \text{as } m \to \infty\,,$$

and consequently, the right-hand side of (2.15) tends to 0 as $m \to \infty$. Whence, it readily follows that

$$\nu(E) = \lim_{m\to\infty} \sum_{n=1}^{m} \nu(E_n)\,,$$

and ν is a signed measure on $(X, \mathcal{M})$. Moreover, if $\mu(E) = 0$, from (2.14) we get that $|\nu(E)| \le \|L\|\mu(E)^{1/p} = 0$, and consequently, $\nu \ll \mu$.

We are now in a position to invoke the Radon-Nikodým Theorem. Let $g = d\nu/d\mu$ be the Radon-Nikodým derivative of ν with respect to μ; g is uniquely determined and locally integrable, we want to show that also $g \in L^q(\mu)$ and that $\|g\|_q = \|L\|$. First recall that, in particular, we have

$$\nu(E) = \int_E g\, d\mu\,, \quad E \in \mathcal{M}\,. \tag{2.16}$$

Now, if $f = \sum c_n \chi_{E_n}$ is a simple function defined on X, by (2.13) and (2.16) we get

$$\begin{aligned} Lf &= \sum c_n L\chi_{E_n} = \sum c_n \nu(E_n) = \sum c_n \int_{E_n} g\, d\mu \\ &= \int_X \Big(\sum c_n \chi_{E_n}\Big)\, g\, d\mu = \int_X fg\, d\mu\,. \end{aligned} \tag{2.17}$$

Moreover, since L is bounded, by (2.17) we obtain

$$\left|\int_X fg\, d\mu\right| \le \|L\|\, \|f\|_p\,,$$

and by Theorem 2.4 it follows that $g \in L^q(\mu)$ and $\|g\|_q \le \|L\|$.

In order to show that $L = L_g$, we must still prove that (2.17) is true for arbitrary f's in $L^p(\mu)$. Given $f \in L^p(\mu)$, let $\{f_n\}$ be a sequence of simple functions such that $f_n \to f$ and $|f_n| \le |f|$. By (2.17) we have

$$Lf_n = \int_X f_n g\, d\mu\,, \quad n = 1, 2, \ldots \tag{2.18}$$

and consequently, the limit of the right-hand side of (2.18) and that of the left-hand side there, if they exist, must be equal.

As for the left-hand side, note that since L is bounded, we have

$$|Lf_n - Lf| \leq \|L\| \, \|f_n - f\|_p \to 0\,, \quad \text{as } n \to \infty\,,$$

and consequently,

$$\lim_{n\to\infty} Lf_n = Lf\,. \tag{2.19}$$

On the other hand, by the linearity of the integral and Hölder's inequality, we also have

$$\begin{aligned}\left|\int_X f_n g\, d\mu - \int_X fg\, d\mu\right| &\leq \int_X |f_n - f|\,|g|\, d\mu \\ &\leq \|f_n - f\|_p \|g\|_q \to 0\,, \quad \text{as } n \to \infty\,,\end{aligned}$$

and consequently,

$$\lim_{n\to\infty} \int_X f_n g\, d\mu = \int_X fg\, d\mu\,. \tag{2.20}$$

Whence combining (2.18), (2.19) and (2.20), we get

$$Lf = \int_X fg\, d\mu, \quad \text{all } f \in L^p(\mu)\,. \tag{2.21}$$

To complete the proof in the general case write $X = \bigcup_n X_n$ as an increasing union of sets of finite measure, and consider the restriction L_n of the functional L to $L^p(X_n,\mu)$. Each L_n is a linear functional, and since

$$\|L_n\| = \sup\{|Lf|/\|f\|_p : f \in L^p(X_n,\mu)\} \leq \|L\|, \quad \text{all } n,$$

the L_n's are also bounded. Thus, by the first part of the proof, we can find functions $g_n \in L^q(X_n,\mu)$, $n = 1, 2, \ldots$, such that

$$\|g_n\|_q \leq \|L_n\| \leq \|L\|\,, \quad \text{all n}\,, \tag{2.22}$$

and

$$Lf = \int_X fg_n\, d\mu\,, \quad f \in L^p(X_n,\mu)\,. \tag{2.23}$$

Now, since $X_n \subseteq X_m$ for all $n \leq m$, we also have $L^p(X_n,\mu) \subseteq L^p(X_m,\mu)$. Furthermore, since for each $f \in L^p(X_n,\mu)$ we have $L_n f = L_m f$ for all $m \geq n$, by (2.23) it follows that for such functions we have $\int_X fg_n\, d\mu = \int_X fg_m\, d\mu$, or

$$\int_X f(g_n - g_m)\, d\mu = 0\,, \quad \text{all } m \geq n\,. \tag{2.24}$$

In particular, since $\mu(X_n) < \infty$, the functions $\operatorname{sgn}(g_n - g_m) \in L^p(X_n, \mu)$ for all $m \geq n$, and by (2.24) we get that $|g_n - g_m| = 0$ μ-a.e. on X_n. In other words, $g_n = g_m$ μ-a.e. on X_n for all $m \geq n$, and consequently, the function g on X given by

$$g(x) = g_n(x)\,, \quad x \in X_n\,,$$

is well-defined and it satisfies $|g| = \lim_{n\to\infty} |g_n|$ μ-a.e. Thus, by Fatou's Lemma and (2.22) it follows that

$$\|g\|_q^q \leq \liminf \|g_n\|_q^q \leq \|L\|^q\,, \tag{2.25}$$

and g is an $L^q(\mu)$ function with norm less than or equal to $\|L\|$.

Next note that since for each $f \in L^p(\mu)$ we have

$$f_n = f\chi_{X_n} \to f \;\mu\text{-a.e.}\,, \quad \text{and} \quad |f_n| \leq |f| \;\mu\text{-a.e.}\,,$$

by LDCT it follows that $\lim_{n\to\infty} \|f_n - f\|_p = 0$, and consequently, by the continuity of L we obtain

$$\lim_{n\to\infty} Lf_n = Lf\,.$$

Moreover, since by (2.23) $Lf_n = \int_{X_n} f_n g_n \, d\mu = \int_X f_n g \, d\mu$, by Hölder's inequality we also have $\lim_{n\to\infty} \int_X f_n g \, d\mu = \int_X fg \, d\mu$, and consequently,

$$Lf = \int_X fg \, d\mu\,, \quad \text{all } f \in L^p(\mu)\,,$$

as we wanted to show.

It thus only remains to verify that $\|L\| = \|g\|_q$, and by (2.25) we only need to check that $\|L\| \leq \|g\|_q$. But since $L = L_g$, this is an easy consequence of Hölder's inequality. ■

Two natural questions arise from this result: How can we go about representing the bounded linear functionals on $L^\infty(\mu)$, and, is the assumption concerning the σ-finiteness of μ necessary?

The former question will be addressed in Chapter XIV, and the latter question has two answers, to wit: If $1 < p < \infty$, it is not necessary that μ be σ-finite, and if $p = 1$, it is. We do the case $p = 1$ first.

Let $X = (0,1)$, let $\mathcal{M}$ be the σ-algebra of those subsets E of X which are either countable or such that $X \setminus E$ is countable, and assume μ is the counting measure on $(X, \mathcal{M})$; μ is not σ-finite. If ν denotes the counting measure on $(X, \mathcal{P}(X))$, put

$$Lf = \int_X f\chi_{(0,1/2)}\,d\nu\,, \quad f \in L^1(\mu)\,.$$

Since for $f \in L^1(\mu)$ the set $\{f \neq 0\}$ is μ and ν σ-finite, it is clear that $|Lf| \leq \|f\|_1$, and L is a bounded linear functional on $L^1(\mu)$ of norm less than or equal to 1. It is intuitively clear that if $Lf = L_g f$, then g must be the function $\chi_{(0,1/2)}$, which is measurable with respect to the σ-algebra $\mathcal{P}(X)$, but not measurable with respect to the σ-algebra $\mathcal{M}$; the verification of this observation is left to the reader. On the other hand, the situation is quite different if $1 < p < \infty$, for then $\chi_{(0,1/2)} \notin L^q(\nu)$ for any $q < \infty$.

Finally, to see that in the case $1 < p < \infty$ the σ-finiteness of μ is not needed in the Riesz representation theorem, note that, in the notation of that theorem, given a σ-finite subset E of X, there is a unique function $g = g_E$ vanishing off E so that $g_E \in L^q(E, \mu)$ and

$$Lf = \int_E fg\,d\mu\,, \quad \text{all } f \in L^p(E, \mu)\,.$$

Furthermore, if $L|E$ denotes the restriction of L to $L^p(E, \mu)$, we also have

$$\|g_E\|_q \leq \|L|E\| \leq \|L\|\,, \quad \text{all } \sigma\text{-finite } E.$$

Also (a simple variant of) the argument in (2.24) above gives that if $E_1 \supseteq E$ are σ-finite subsets of X, then we have $g_{E_1} = g_E$ μ-a.e. on E, and $\|g_E\|_q \leq \|g_{E_1}\|_q$. Let now η be the finite quantity

$$\eta = \sup\{\|g_E\|_q : E \text{ is a } \sigma\text{-finite subset of } X\}\,,$$

and let $\{E_n\}$ be a sequence of σ-finite subsets of X with the property that $\lim_{n\to\infty} \|g_{E_n}\|_q = \eta$. Observe that if $E = \bigcup_n E_n$, then E is also a σ-finite subset of X, and since

$$\|g_E\|_q \geq \|g_{E_n}\|_q\,, \quad \text{all } n\,,$$

it readily follows that $\|g_E\|_q = \eta$. Now, life outside E is uneventful. Indeed, let A be a σ-finite subset of X, and put $A_1 = (A \setminus E) \cup E$. Then A_1 is also a σ-finite subset of X, and since $q < \infty$ and

$$\begin{aligned}\int_{A_1} |g_{A_1}|^q\,d\mu &= \int_{A\setminus E} |g_A|^q\,d\mu + \int_E |g_E|^q\,d\mu \\ &= \int_{A\setminus E} |g_A|^q\,d\mu + \eta^q \leq \eta^q\,,\end{aligned}$$

it readily follows that $g_A = 0$ μ-a.e. on $A \setminus E$.

This is all we need to know: If f is an arbitrary function in $L^p(\mu)$, then the set $A = \{f \neq 0\}$ is σ-finite, cf. 4.11 in Chapter VII, and

$$\begin{aligned} Lf &= \int_X f g_A \, d\mu = \int_A f g_A \, d\mu \\ &= \int_{A \setminus E} f g_A \, d\mu + \int_{A \cap E} f g_E \, d\mu = \int_X f g_E \, d\mu \,, \end{aligned}$$

which is the desired representation of L.

3. WEAK CONVERGENCE

Assume $(X, \mathcal{M}, \mu)$ is a measure space, and let $f, f_n \in L^p(\mu)$, $n = 1, 2, \ldots$, $1 \leq p < \infty$. We say that the sequence $\{f_n\}$ converges weakly to f in $L^p(\mu)$, if, with $1/p + 1/q = 1$, we have

$$\lim_{n \to \infty} \int_X f_n g \, d\mu = \int_X f g \, d\mu \,, \quad \text{all } g \in L^q(\mu) \,. \tag{3.1}$$

We now give a few examples to show that there is no connection between weak convergence and any of the other forms of convergence, unless further assumptions are made on either the sequence itself or the measure space involved.

For instance, in ℓ^p, consider the sequence $\{e_n\}$ consisting of those sequences $e_n = (0, \ldots, 1, 0, \ldots)$ with 1 in the nth place and zeroes elsewhere. If $1 < p < \infty$, and $x = (x_1, \ldots, x_n, \ldots) \in \ell^q$, then the functional L_x has the property that

$$L_x e_n = x_n \to 0 \,, \quad \text{as } n \to \infty \,, \tag{3.2}$$

and since by the Riesz representation theorem these are all the functionals on ℓ^p, the sequence $\{e_n\}$ converges weakly to 0. Nevertheless, since $\|e_n - e_m\|_p = 2^{1/p}$ for all $n \neq m$, neither the sequence itself nor any of its subsequences converges to 0. Neither does $\{e_n\}$ converge to 0 in measure, nor uniformly, nor even in the pointwise sense.

Note however that $\{e_n\}$ does not converge weakly to 0 in ℓ^1; this is clear since the sequence $x = (1, 1, \ldots)$ is bounded and $L_x e_n = 1$ for all n. Now, in the case of ℓ^1 we have the following interesting result.

Proposition 3.1 (Schur). If the sequence $\{x_n\}$ converges weakly to x in ℓ^1, then

$$\lim_{n \to \infty} \|x_n - x\|_1 = 0 \,. \tag{3.3}$$

Proof. By considering, if necessary, the sequence $\{x_n - x\}$ we may assume that $\{x_n\}$ converges weakly to 0. Suppose that $\lim_{n\to\infty} \|x_n\|_1 \neq 0$; passing to a subsequence, if needed, we may assume that $\|x_n\|_1 \geq \eta > 0$ for all n. In this case also $(1/\|x_n\|_1)x_n$ converges weakly to 0, and so we may as well assume that $\|x_n\|_1 = 1$ for all n. In addition, if $x_n = (x_{n,1}, \ldots, x_{n,m}, \ldots)$, $n = 1, 2, \ldots$, by the weak convergence it follows that

$$\lim_{n\to\infty} x_{n,m} = 0\,, \quad \text{all } m\,. \tag{3.4}$$

Observe that since $\|x_1\|_1 = 1$, we can find m_1 so that $\sum_{m=1}^{m_1} |x_{1,m}| > 3/4$. Further, by (3.4) it readily follows that there exists an index $n_2 > 1$ such that $\sum_{m=1}^{m_1} |x_{n_2,m}| < 1/4$, and consequently, since $\|x_{n_2}\|_1 = 1$, we can find an index $m_2 > m_1$ so that $\sum_{m=m_1+1}^{m_2} |x_{n_2,m}| > 3/4$. The pattern is now clear: Having chosen sequences $m_0 = 0 < m_1 < m_2 < \ldots < m_k$ and $n_1 = 1 < n_2 < \ldots < n_k$, choose first n_{k+1} with the property that

$$\sum_{m=1}^{m_k} |x_{n_{k+1},m}| < 1/4\,,$$

and then m_{k+1} so that

$$\sum_{m=m_k+1}^{m_{k+1}} |x_{n_{k+1},m}| > 3/4\,.$$

Consider now the sequence $y \in \ell^\infty$ with terms

$$y_m = \operatorname{sgn} x_{n_k,m}\,, \quad m_{k-1} + 1 \leq m \leq m_k\,,\ k = 1, 2, \ldots$$

Since $|y_m| \leq 1$ for all m, the functional L_y on ℓ^1 satisfies

$$\begin{aligned}
|L_y(x_{n_k})| &= \left| \sum_{m=1}^{\infty} x_{n_k,m} y_m \right| \\
&\geq \sum_{m=m_k+1}^{m_{k+1}} |x_{n_k,m}| - \left(\sum_{m=1}^{m_k} + \sum_{m=m_{k+1}+1}^{\infty} \right) |x_{n_k,m} y_m| \\
&= 2 \sum_{m=m_k+1}^{m_{k+1}} |x_{n_k,m}| - \|x_{n_k}\|_1 > 2 \cdot 3/4 - 1 = 1/2\,,\quad k = 1, 2, \ldots
\end{aligned}$$

Thus, $\limsup |L_y(x_n)| \geq 1/2$, which contradicts the weak convergence of $\{x_n\}$ to 0. ■

In a different direction, an interesting example is the sequence $\{f_n\} \subseteq L^p([0,1])$, $1 \le p$, given by $f_n = n\chi_{[0,1/n]}$, which converges to 0 in measure and a.e., and does not converge to 0 weakly in $L^p([0,1])$ for any $p \ge 1$. (Just consider the functional induced by $\chi_{[0,1]}$.)

Finally, the sequence $\{f_n\} \subseteq L^p(R)$, $1 \le p$, given by $f_n = (1/n)\chi_{[1,e^n]}$, converges uniformly to 0, yet it does not converge weakly to 0 in $L^p(R)$ for any $p \ge 1$. (The functional induced by $(1/x)\chi_{[1,\infty)}(x)$ will do.)

There are additional assumptions that we may impose on weakly convergent sequences to ensure they they also are convergent. Again we discuss the ℓ^p case; the result is also true for general $L^p(\mu)$ spaces, but the proof is more complicated.

Proposition 3.2. Suppose the sequence $\{x_n\}$ converges weakly to x in ℓ^p, $1 \le p < \infty$, and that, in addition,

$$\lim_{n\to\infty} \|x_n\|_p = \|x\|_p\,. \tag{3.5}$$

Then we also have

$$\lim_{n\to\infty} \|x_n - x\|_p = 0\,. \tag{3.6}$$

Proof. As before it follows that if

$$x = (x_1, \ldots, x_m, \ldots)\,, \qquad x_n = (x_{n,1}, \ldots, x_{n,m}, \ldots)\,, \qquad n = 1, 2, \ldots$$

then

$$\lim_{n\to\infty} x_{n,m} = x_m\,, \quad \text{all } m\,. \tag{3.7}$$

Also, by (3.5) and (3.7), for each fixed M we have

$$\lim_{n\to\infty} \left(\sum_{m=M}^{\infty} |x_{n,m}|^p \right)^{1/p} = \left(\sum_{m=M}^{\infty} |x_n|^p \right)^{1/p}. \tag{3.8}$$

Whence, for each fixed M we get,

$$\begin{aligned} \|x_n - x\|_p &\le \left(\sum_{m=1}^{M-1} |x_{n,m} - x_m|^p \right)^{1/p} + \left(\sum_{m=M}^{\infty} |x_{n,m} - x_m|^p \right)^{1/p} \\ &= A + B\,, \end{aligned}$$

say. It is not hard to estimate B. By Minkowski's inequality and (3.8) it follows that for all sufficiently large n,

$$B \le \left(\sum_{m=M}^{\infty} |x_{n,m}|^p \right)^{1/p} + \left(\sum_{m=M}^{\infty} |x_m|^p \right)^{1/p} \le 3 \left(\sum_{m=M}^{\infty} |x_n|^p \right)^{1/p},$$

which, since $x \in \ell^p$, can be made arbitrarily small provided M is sufficiently large. Once M is fixed, it is clear that, on account of (3.7), A also can be made arbitrarily small provided n is large enough. Thus (3.6) holds, and we have finished. ∎

As noted above, L^p spaces do not have the Bolzano-Weierstrass property: There are bounded sequences for which no convergent subsequence may be found. The concept of weak convergence is also relevant in this context.

Theorem 3.3. Let $\{f_k\}$ be a bounded sequence in $L^p(R^n)$, $1 < p < \infty$, with bound M. Then there exist a subsequence $k_m \to \infty$ and a function $f \in L^p(R^n)$ with $\|f\|_p \leq M$, such that $\{f_{k_m}\}$ converges weakly to f in $L^p(R^n)$.

Proof. We divide the proof into a number of steps, and begin by showing that there is a subsequence $k_m \to \infty$ with the property that $\lim_{m\to\infty} \int_{R^n} f_{k_m} g\, dx$ exists, provided g is any function in a fixed countable family $\{g_h\} \subseteq L^q(R^n)$, $1/p+1/q = 1$. This is not hard: Let $c_{k,h} = \int_{R^n} f_k g_h$ for $k, h \geq 1$, and note that by Hölder's inequality we have

$$|c_{k,h}| \leq \|f_k\|_p \, \|g_h\|_q \leq M\|g_h\|_q\,, \quad \text{all } k, h\,. \tag{3.9}$$

Fix $h = 1$ now. By (3.9), $(c_{k,1})$ is a bounded sequence and consequently, there is a subsequence $k_1 \to \infty$ such that $\lim_{k_1\to\infty} c_{k_1,1}$ exists. Repeating this argument with $(c_{k_1,1})$ in place of $(c_{k,1})$ above we obtain a new subsequence $k_2 \to \infty$, say, such that $\lim_{k_2\to\infty} c_{k_2,h}$ exists for $h = 1, 2$. These are the first steps of the by now familiar Cantor diagonal process which ensures the existence of a subsequence $k_m \to \infty$ so that $\lim_{k_m\to\infty} c_{k_m,h}$ exists for each h.

We choose now for the g_h's a dense family in $L^q(R^n)$, which, since $L^q(R^n)$ is separable, is clearly possible, and define the functional L on the g_h's by means of the expression $Lg = \lim_{m\to\infty} \int_{R^n} f_{k_m} g\, dx$. Now, L is decidedly linear over these g_h's, i.e., $L(g_{h_1} + \lambda g_{h_2}) = Lg_{h_1} + \lambda L g_{h_2}$ for all scalars λ, and by Hölder's inequality it also satisfies $|Lg| \leq M\|g\|_q$. We claim that L can be extended linearly and continuously to all of $L^q(R^n)$. Indeed, to each $g \in L^q(R^n)$ there corresponds a sequence of g_h's such that $\lim_{h\to\infty} \|g - g_h\|_q = 0$ and $\lim \|g_h\|_q = \|g\|_q$. Now, for these g_h's the sequence of scalars (Lg_h) is Cauchy, and consequently, convergent. Putting $Lg = \lim_{h\to\infty} Lg_h$, L turns out to be a well-defined linear functional on $L^q(R^n)$, and since

$$|Lg| \leq \limsup |Lg - Lg_h| + \limsup |Lg_h| \leq M\|g\|_q\,,$$

L is also bounded and has norm $\|L\| \leq M$. By Theorem 2.5 there exists a function $f \in L^p(R^n)$ with $\|f\|_p \leq M$ such that $Lg = L_f g$; the function f satisfies all the required conditions. ■

The proof of this interesting result relies on the fact that the functional L, originally defined on a subset of $L^q(R^n)$, can be extended to all of $L^q(R^n)$ without an increase of its "norm." A more general setting where this is also true is described in Chapter XIV.

4. PROBLEMS AND QUESTIONS

In what follows $(X, \mathcal{M}, \mu)$ denotes a measure space, and we don't find it necessary to stress this point at each instance.

4.1 Suppose $0 < p < q \leq \infty$. Give examples of functions f defined on R such that $f \in L^r(R)$ iff (a) $p < r < q$, (b) $p \leq r \leq q$, and, (c) $r = p$.

4.2 Let I be a bounded interval of R. By means of an example show that, in general, $\bigcap_{0<p<q} L^p(I) \neq L^q(I)$, $0 < q \leq \infty$.

4.3 The following inequality is often referred to as Hölder's inequality, prove it and identify the cases of equality: If $f \in L^p(\mu)$ and $g \in L^q(\mu), 1 < p, q < \infty, 1/p+1/q = 1$, then $\left|\int_X fg\, d\mu\right| \leq \|f\|_p \|g\|_q$.

4.4 The following is an extension of Hölder's inequality to more than two indices: Suppose $f \in L^p(\mu)$, $g \in L^q(\mu)$, $h \in L^r(\mu)$, $1 < p, q, r < \infty$, $1/p + 1/q + 1/r = 1$. Prove that $fgh \in L(\mu)$ and that

$$\|fgh\|_1 \leq \|f\|_p \|g\|_q \|h\|_r \,.$$

Can you think of a further extension to more than three functions?

4.5 Let $I = [0, \pi]$. Show that $\int_I x^{-1/4} \sin x\, dx \leq \pi^{3/4}$.

4.6 Show that if for some $0 < p < \infty$, $f \in L^p(\mu) \cap L^\infty(\mu)$, then for all $p < q < \infty$, $f \in L^q(\mu)$ and $\|f\|_q \leq \|f\|_p^{p/q} \|f\|_\infty^{1-p/q}$.

4.7 If for some $0 < p, q < \infty$, $f \in L^p(\mu) \cap L^q(\mu)$, then $f \in L^r(\mu)$ for all $p < r < q$, and $\|f\|_r \leq \|f\|_p^{1-\eta} \|f\|_q^{\eta}$, where $0 < \eta < 1$ is given by $1/r = (1 - \eta)/p + \eta/q$.

4.8 Let $I = [0, \pi]$ and $f \in L^2(I)$. Is it possible to have simultaneously $\int_I (f(x) - \sin x)^2\, dx \leq 4/9$ and $\int_I (f(x) - \cos x)^2\, dx \leq 1/9$?

4.9 Suppose an extended real-valued function f defined on R^n satisfies the following two properties: (a) There is a p, $1 \le p < \infty$, such that $f \in L^p(I)$, for every bounded interval I in R^n, and, (b) $|\int_I f|^p \le c|I|^{p-1}\int_I |f|^p$, for every bounded interval I and a constant $0 < c < 1$ independent of I.
Show that $f = 0$ a.e.

4.10 Let $0 < p < q \le \infty$. Prove that $L^p(\mu)$ is not contained in $L^q(\mu)$ iff X contains sets of arbitrarily small, positive, μ measure.

4.11 Let $0 < p < q < \infty$. Show that $L^q(\mu)$ is not contained in $L^p(\mu)$ iff X contains sets of arbitrarily large, finite, μ measure.
What can you say about the case $q = \infty$?

4.12 Prove that if $\lim_{n\to\infty} \|f_n\|_p = 0$, $1 \le p \le \infty$, then there exist a subsequence $\{f_{n_k}\}$ and a nonnegative function $h \in L^p(\mu)$ such that $|f_{n_k}| \le h$ μ-a.e., and $\lim_{k\to\infty} f_{n_k} = 0$ μ-a.e.

4.13 Prove that if $\|f_n - f\|_p \to 0$ and $\|g_n - g\|_q \to 0$, $1 \le p, q \le \infty$, $1/p + 1/q = 1$, then $\|f_n g_n - fg\|_1 \to 0$.

4.14 The $L^p(X, \mu)$ spaces are not, in general, separable when $0 < p \le \infty$. Give an example of a measure space $(X, \mathcal{M}, \mu)$ so that $L^p(\mu)$ is not separable, and prove that the Lebesgue spaces $L^p(R^n)$ are separable for $0 < p < \infty$. On the other hand, $L^\infty(R^n)$ is not separable. When is $L^\infty(\mu)$ separable?

4.15 Suppose $f \in L^p(R^n)$, $0 < p < \infty$, and compute

$$\lim_{|h|\to\infty} \int_{R^n} |f(y+h) + f(y)|^p dy\,.$$

4.16 Suppose $f, f_n \in L^p(\mu)$, $n = 1, 2, \ldots$, satisfy $\lim_{n\to\infty} f_n = f$ μ-a.e., and $\lim_{n\to\infty} \|f_n\|_p = \|f\|_p$, $0 < p < \infty$. Prove that

$$\lim_{n\to\infty} \|f_n - f\|_p = 0\,.$$

4.17 Is the conclusion of 4.16 true if we replace μ-a.e. convergence by convergence in measure?

4.18 Let $(X, \mathcal{M}, \mu)$ be a finite measure space, $0 < r < p$, and $\{f_n\}$ a sequence of $L^p(\mu)$ functions such that $\|f_n\|_p \le k$ for all n and $\lim_{n\to\infty} f_n = f$ μ-a.e. Prove that $\lim_{n\to\infty} \|f_n - f\|_r = 0$, and that the conclusion may fail if either $\mu(X) = \infty$ or $r = p$.

4.19 (Vitali's Convergence Theorem) Let $\{f_n\}$ be a sequence of $L^p(\mu)$ functions, $1 \le p < \infty$, and assume $\lim_{n\to\infty} f_n = f$ μ-a.e. Show that

$f \in L^p(\mu)$ and $\lim_{n\to\infty} \|f_n - f\|_p = 0$ iff: (a) For each $\varepsilon > 0$, there exists a set A_ε such that $\mu(A_\varepsilon) < \infty$ and $\int_{X\setminus A_\varepsilon} |f_n|^p \, d\mu < \varepsilon$ for all n, and, (b) $\lim_{\mu(E)\to 0} \int_E |f_n|^p d\mu = 0$, uniformly in n.

4.20 Verify that for every measurable function f

$$\int_X |f|^p \, d\mu = \int_{[0,\infty)} p\, t^{p-1} \mu(\{|f| > t\}) dt \,, \quad 0 < p < \infty \,.$$

4.21 Suppose the nonnegative functions $f, g \in L^p(\mu)$, $1 < p < \infty$, satisfy the relation $\mu(\{g > \lambda\}) \leq \frac{1}{\lambda} \int_{\{g>\lambda\}} f \, d\mu$ for all $\lambda > 0$. Show that $\|g\|_p \leq p' \|f\|_p$, where $1/p + 1/p' = 1$.

4.22 Suppose $0 < r < p < q \leq \infty$, and let $f \in L^p(\mu)$. Show that f can be written $f = g + h$, where $g \in L^r(\mu)$ and $h \in L^q(\mu)$. Further, given $t > 0$, we can choose g and h so that $\|g\|_r^r \leq t^{r-p} \|f\|_p^p$ and $\|h\|_q^q \leq t^{q-p} \|f\|_p^p$.

4.23 Suppose $f \in$ wk-$L(R^n)$ is such that $|\{f \neq 0\}| < \infty$. Prove that $f \in L^p(R^n)$ for each $0 < p < 1$.
Also, if $f \in$ wk-$L(R^n) \cap L^\infty(R^n)$, show that $f \in L^p(R^n)$ for $1 < p < \infty$.

4.24 The Hardy-Littlewood maximal operator takes $L^p(R^n)$ functions into $L^p(R^n)$ functions, $1 < p < \infty$. More precisely, show there is a constant $c = c_{n,p}$ such that

$$\|Mf\|_p \leq c \|f\|_p \,, \quad \text{all } f \in L^p(R^n) \,.$$

4.25 Show that if $f \in L^p(R^n)$, $1 \leq p \leq \infty$, then the integral of f differentiates to $f(x)$ for almost every x in R^n.

4.26 Given an interval $I = [a,b]$ in the line, show that a necessary and sufficient condition for a function F to be the integral of $f \in L^p(I)$, $1 < p < \infty$, is that the sums

$$\sum_{n=1}^m |F(x_n) - F(x_{n-1})|^p / (x_n - x_{n-1})^{p-1}$$

formed for every partition $\{a \leq x_0 < \ldots < x_n \leq b\}$ be bounded. The sup of these sums is then $\int_I |f(x)|^p \, dx$.

4.27 Suppose μ is a finite measure and $\nu \ll \mu$. Prove that the Radon-Nikodým derivative $h = d\nu/d\mu \in L^p(\mu)$ iff there is a constant c such that for all measurable at most countable partitions $\{E_n\}$ of X, we have $\sum_{n=1}^{\infty} \nu(E_n)^p/\mu(E_n)^{p-1} \le c$. What is the relation between c and $\|h\|_p$?

4.28 If μ is a finite measure, and the sequence $\{f_n\}$ of nonnegative functions satisfies $\limsup \int_X f_n\, d\mu = \eta$ and $\liminf \int_X f_n^p\, d\mu < \infty$ for some $1 < p < \infty$, show that $\mu(\{\limsup f_n \ge \eta\}) > 0$.

4.29 Show that each function $f \in L^p(\mu)$, $0 < p < \infty$, satisfies the following property: $\lim_{\lambda\to\infty} \lambda^p \mu(\{|f| > \lambda\}) = 0$.

4.30 Suppose f is an extended real-valued function defined on a probability measure space $(X, \mathcal{M}, \mu)$. Show that

$$\text{ess sup}\, f = \inf\{\lambda : \mu(\{x \in X : f(x) \le \lambda\}) = 1\}\,.$$

4.31 If f is an extended real-valued function defined on X, define its essential infimum by

$$\text{ess inf}\, f = \sup\{\lambda : \mu(\{x \in X : f(x) < \lambda\}) = 0\}\,.$$

Explore the properties of this quantity. In particular show that if $f \ge 0$ μ-a.e., then $\text{ess inf}\, f = 1/\|1/f\|_\infty$.

4.32 Suppose that $\lim_{n\to\infty} \|f_n - f\|_p = 0$, $0 < p < \infty$, and that g, g_n are uniformly bounded measurable functions, $n = 1, 2, \ldots$, such that $\lim_{n\to\infty} g_n = g$ μ-a.e. Prove that $\lim_{n\to\infty} \|f_n g_n - fg\|_p = 0$.

4.33 Suppose $f, f_n \in L^p(\mu)$, $n = 1, 2, \ldots, 1 < p < \infty$. Show that f_n converges weakly to f in $L^p(\mu)$ iff (a) $\sup \|f_n\|_p < \infty$, and, (b) $\lim_{n\to\infty} \int_E f_n\, d\mu = \int_E f\, d\mu$ for all $E \in \mathcal{M}$.

4.34 Suppose $\{f_n\}$ is a bounded set in $L^p(\mu)$, $1 < p < \infty$. Show that if $\lim_{n\to\infty} f_n = f$ μ-a.e., then also f_n converges weakly to f in $L^p(\mu)$.

4.35 Show that the conclusion of 4.34 holds with the assumption of the pointwise μ-a.e.convergence replaced by convergence in measure.

4.36 Let $(X, \mathcal{M}, \mu)$ be a measure space and $\{f_n\}$ a sequence of $L^p(\mu)$ functions so that $\liminf \|f_n\|_p < \infty$ for some $0 < p < \infty$. Suppose further that $\lim_{n\to\infty} f_n = f$ μ-a.e. What can then be said about $\|f\|_p$? Well, prove that

$$\lim_{n\to\infty} (\|f_n - f\|_p^p - \|f_n\|_p^p - \|f\|_p^p) = 0\,.$$

CHAPTER XIII

Fubini's Theorem

In this chapter we deal with the questions of "exchanging the order of integration" in a double integral, and of evaluating an integral as an iterated one.

1. ITERATED INTEGRALS

The Lebesgue theory of integration developed in Chapter VII is independent of the dimension of the ambient space, and this is an attractive feature. Nevertheless, the computation of an integral in R^n is usually carried out as a succession of n one-dimensional integrals. The following is a familiar situation: Suppose f is a continuous function on a rectangle $I = [a,b] \times [c,d]$ in R^2. Then it is true that

$$\int_I f(x,y)\, dA = \int_a^b \int_c^d f(x,y)\, dy\, dx = \int_c^d \int_a^b f(x,y)\, dx\, dy\,,$$

where all integrals are taken in the sense of Riemann.

Simple examples indicate that, in general, this is not always the case. Consider, for instance, $I = (0,1) \times (0,1)$ and

$$f(x,y) = \frac{x^2 - y^2}{(x^2 + y^2)^2}\,, \quad (x,y) \in I\,.$$

Since as a simple computation shows $\int_0^1 f(x,y)\, dy = 1/(1 + x^2)$, we have $\int_0^1 \int_0^1 f(x,y)\, dy\, dx = \pi/4$. Similarly, $\int_0^1 \int_0^1 f(x,y)\, dx\, dy = -\pi/4$.

Now, a couple of things go wrong with f. First notice that for $0 < x < 1$, we have $\int_0^x f(x,y)\, dy = 1/2x$, and consequently, $\int_0^1 |f(x,y)| dy \geq 1/2x$.

It is then apparent that $\int_0^1 \int_0^1 |f(x,y)|\, dy\, dx = \infty$. Also, since on the region $\{(x,y) \in I : |x| > \sqrt{3}|y|\}$ we have

$$|f(x,y)| \geq \frac{1}{2} \frac{1}{(x^2+y^2)},$$

by x.xx in Cahpter VII, it is clear that $f \notin L(I)$.

Motivated by this example we introduce the following definition: Let $I_1 \subseteq R^n$ and $I_2 \subseteq R^m$ be intervals in R^n and R^m respectively, and consider the "rectangle" $I = I_1 \times I_2 \subseteq R^n \times R^m = R^{n+m}$. An integrable function $f \in L(I)$ is said to satisfy, or have, "property F", provided the following three properties hold:

(i) For almost every $x \in I_1$, in the sense of the n-dimensional Lebesgue measure, $f(x,y)$ is a measurable and integrable function of y on I_2, with respect to the m-dimensional Lebesgue measure.

(ii) As a function of $x \in I_1$, $\int_{I_2} f(x,y)\, dy$ is measurable and Lebesgue integrable.

(iii)
$$\int_I f = \int_{I_1} \int_{I_2} f(x,y)\, dy\, dx\,.$$

What we hope to prove is that every $f \in L(I)$ has property F. This we achieve in a series of lemmas, each involving a basic step required in approximating arbitrary integrable functions.

First an observation: By setting $f = 0$ off I if necessary we may assume that $I_1 = R^n$, $I_2 = R^m$ and $I = R^{n+m}$.

Lemma 1.1. Suppose $\{f_k\}$ is a nondecreasing sequence of functions each of which has property F, and suppose that $\lim_{k\to\infty} f_k = f \in L(R^{n+m})$. Then f also has property F.

Proof. Let N_k be a null set (in R^n) such that $f_k(x,y)$ is measurable and integrable as a function of $y \in R^m$ for $x \notin N_k$, and let $N = \bigcup_k N_k$; clearly $|N| = 0$. Now, for $x \notin N$, $f_k(x,y)$ increases (in the wide sense) to $f(x,y)$, and consequently, for those x's, $f(x,\cdot)$ is measurable. Thus by MCT it follows that

$$\lim_{k\to\infty} \int_{R^m} f_k(x,y)\, dy = \int_{R^m} f(x,y)\, dy\,. \tag{1.1}$$

Moreover, since each f_k has property F, each integral on the left-hand side of (1.1) is a measurable, integrable function of x and consequently,

$\int_{R^m} f(x,y)\,dy$ is also a measurable function of x. Again by MCT, (1.1) gives

$$\lim_{k\to\infty} \int_{R^n} \int_{R^m} f_k(x,y)\,dy\,dx = \int_{R^n} \int_{R^m} f(x,y)\,dy\,dx\,. \tag{1.2}$$

On the other hand, also by MCT,

$$\lim_{k\to\infty} \int_{R^{n+m}} f_k = \int_{R^{n+m}} f\,. \tag{1.3}$$

Now, since the f_k's have property F, for each k the integral on the left-hand side of (1.2) is equal to that on the left-hand side of (1.3) and consequently, we get

$$\int_{R^{n+m}} f = \int_{R^n} \int_{R^m} f(x,y)\,dy\,dx\,.$$

Since $f \in L(R^{n+m})$, f has property F. ■

Corollary 1.2. Suppose $\{f_k\}$ is a nonincreasing sequence of functions each of which has property F, and suppose that $\lim_{k\to\infty} f_k = f \in L(R^{n+m})$. Then f also has property F.

Lemma 1.3. Suppose $E = \bigcap_{k=1}^{\infty} \mathcal{O}_k$, $\mathcal{O}_k$ open, is a G_δ subset of R^{n+m} so that $|\mathcal{O}_k| < \infty$ for some k. Then χ_E satisfies property F.

Proof. We proceed in steps. Note first that if I_1 and I_2 are open, bounded intervals in R^n and R^m respectively, and if $I = I_1 \times I_2$, then $\chi_{I(x,y)} = \chi_{I_1}(x)\chi_{I_2}(y)$, and χ_I has property F.

Also note that if A is any subset of the boundary of I, we then have

$$|\{y \in R^m : (x,y) \in A\}| = 0\,, \quad \text{a.e. } x \text{ in } R^n.$$

Roughly speaking, this relation is true except, possibly, along the "endpoints" of I_1. It then follows that

$$\int_{R^n} \int_{R^m} \chi_A(x,y)\,dy\,dx = \int_{R^n} |\{y \in R^m : (x,y) \in A\}|\,dx = 0\,,$$

which together with $\int_{R^{n+m}} \chi_A = |A| = 0$, implies that χ_A has property F.

Consider next a subset $\tilde{I}$ of R^{n+m} consisting of a rectangle $I = I_1 \times I_2$, I_1, I_2 open in R^n and R^m respectively, plus some portion A of the boundary of I. Since $\chi_{\tilde{I}} = \chi_I + \chi_A$, and since as is readily verified any finite linear combination of functions that satisfy property F also has property F, then by the above results it follows that $\chi_{\tilde{I}}$ has property F.

Now suppose $\mathcal{O}$ is an open set in R^{n+m} with finite measure; we can write $\mathcal{O} = \bigcup_k \tilde{I}_k$ as the pairwise disjoint union of subsets $\tilde{I}_k$ of R^{n+m} consisting of open rectangle I_k and a subset of their boundary.

Thus, $\chi_{\mathcal{O}} = \lim_{n\to\infty} \sum_{k=1}^{n} \chi_{\tilde{I}_k}$, and since each sum on the right-hand side above has property F, by Lemma 1.1 $\chi_{\mathcal{O}}$ also satisfies property F.

We are finally ready to handle χ_E. Under our conditions we may assume that

$$E = \bigcap_{k=1}^{\infty} \mathcal{O}_k\,, \quad |\mathcal{O}_1| < \infty\,.$$

Then $\chi_E = \lim_{N\to\infty} \chi_{\bigcap_{k=1}^{N} \mathcal{O}_k}$, and each of the sets $\bigcap_{k=1}^{N} \mathcal{O}_k$ is open and has finite measure. Therefore, each of the functions on the right-hand side above has property F, and, by Corollary 1.2, χ_E has property F. ■

Lemma 1.4. Let $N \subseteq R^{n+m}$, $|N| = 0$. Then χ_N has property F.

Proof. By Corollary 1.5 in Chapter V there is a G_δ set $E \supseteq N$ such that $|E| = |N| = 0$. Now, χ_E has property F, and, in particular,

$$\int_{R^n} \int_{R^m} \chi_E(x,y)\, dy\, dx = |E| = 0\,.$$

Thus, by Corollary 2.3 in Chapter VII, we get that

$$\int_{R^m} \chi_E(x,y)\, dy = |\{y \in R^m : (x,y) \in E\}| = 0\,, \quad x \text{ a.e. in } R^n,$$

which in turn implies that

$$\int_{R^n} \int_{R^m} \chi_N(x,y)\, dy\, dx = \int |\{y \in R^m : (x,y) \in N\}|\, dx = 0\,.$$

Since also $\int_{R^{n+m}} \chi_N = |N| = 0$, it readily follows that χ_N has property F. ■

Lemma 1.5. Suppose $E \subseteq R^{n+m}$ is Lebesgue measurable, $|E| < \infty$. Then χ_E has property F.

Proof. By 1.5 in Chapter V we have

$$E = H \setminus N\,, \quad H \text{ a } G_\delta \text{ set, and } N \text{ null}\,.$$

where $H = \bigcap_{k=1}^{\infty} \mathcal{O}_k$, and some $\mathcal{O}_k$ has finite measure. Now, since $\chi_E = \chi_H - \chi_N$, and since by Lemmas 1.2 and 1.4 the functions χ_H and χ_N have property F, it follows that χ_E also has property F. ■

The stage is now set for a preliminary version of Fubini's theorem.

Theorem 1.6. Suppose $f \in L(R^{n+m})$. Then f has property F.

Proof. Since $f = f^+ - f^-$, f^+, f^- integrable, we may assume that f is nonnegative. The conclusion follows now by a familiar limiting argument. Indeed, let $\{\phi_k\}$ be a nondecreasing sequence of nonnegative, simple, integrable functions such that

$$0 \le \phi_k \le f\,, \quad \text{and} \quad \lim_{k\to\infty} \phi_k = f \quad \text{a.e.}$$

Now, since each ϕ_k is a finite linear combination of functions that, by Lemma 1.5, satisfy property F, it readily follows that ϕ_k has property F, $k = 1, 2, \ldots$ Whence by Lemma 1.1 f itself has property F. ■

Since the roles of x and y can be interchanged above, we expect a more symmetric statement of Fubini's theorem. First a definition: Given a measurable subset E of R^{n+m}, we consider the "sections" of E given by

$$E_x = \{y \in R^n : (x,y) \in E\}\,, \quad \text{all } x \in R^n\,,$$

and

$$E^y = \{x \in R^n : (x,y) \in E\}\,, \quad \text{all } y \in R^m\,.$$

Although not intuitively apparent, the sections of a measurable subset of R^{n+m} are a.e. measurable in the respective ambient spaces. This is a particular instance of our next result, which extends the following interesting consequence of the preliminary version of Fubini's theorem: If $f(x,y) \in L(R^{n+m})$, then $f(x,y)$ is a Lebesgue measurable function of $y \in R^m$, for almost every $x \in R^n$. We now show that the same is true if f is merely measurable; the result about the sections follows by considering $f = \chi_E$, which is the first step of the proof.

Proposition 1.7. Suppose f is a measurable function defined on R^{n+m}. Then, for almost every $x \in R^n$, $f(x,y)$ is a measurable function of $y \in R^m$. Symmetrically, for almost every $y \in R^m$, $f(x,y)$ is a measurable function of $x \in R^n$.

Proof. Assume first that $f = \chi_E$, where E a measurable subset of R^{n+m}. Then write $E = H \cup N$, where H is of type F_σ in R^{n+m} and $|N| = 0$. Then $E_x = H_x \cup N_x$, where H_x is an F_σ subset of R^m, and by Lemma 1.4, $|N_x| = 0$ for almost every $x \in R^n$. Therefore, E_x is measurable for almost every $x \in R^n$. Similarly for E^y.

If f is any measurable function on R^{n+m} now, given λ real, consider the set $E(\lambda) = \{(x,y) \in R^{n+m} : f(x,y) > \lambda\}$. Since $E(\lambda)$ is measurable in R^{n+m}, the section $E(\lambda)_x$ is measurable in R^m for almost every x in R^n; the exceptional null subset of R^n depends, of course, on λ. The union N of these exceptional sets for all rational λ is also a null set, and the set $\{y \in R^m : f(x,y) > \lambda\}$ is measurable, provided that λ is rational and $x \notin N$. By 4.1 in Chapter VI the same is true for all λ, and consequently $f(x,y)$ is a measurable function of $y \in R^m$, for almost every $x \in R^n$. The other statement is proved in an analogous fashion. ■

We now state Fubini's theorem in its general form.

Theorem 1.8. Let $f(x,y)$ be a measurable function defined on a measurable subset E of R^{n+m}. Then,

(i) For almost every $x \in R^n$, $f(x,y)$ is a measurable function of y on E_x.

(ii) For almost every $y \in R^m$, $f(x,y)$ is a measurable function of x on E^y.

(iii) If $f \in L(E)$, then for almost every x in R^n, $f(x,\cdot) \in L(E_x)$. Moreover, $\int_{E_x} f(x,y)\,dy$ is an integrable function of x and

$$\int_E f = \int_{R^n} \int_{E_x} f(x,y)\,dy\,dx\,.$$

(iv) If $f \in L(E)$, then for almost every y in R^m, $f(\cdot,y) \in L(E^y)$. Moreover, $\int_{E^y} f(x,y)\,dy$ is an integrable function of y and

$$\int_E f = \int_{R^m} \int_{E^y} f(x,y)\,dx\,dy\,.$$

Proof. Let $\tilde{f}$ be the function equal to f on E and equal to 0 elsewhere in R^{n+m}. Since f is measurable on E and E is measurable, $\tilde{f}$ is measurable on R^{n+m}. Therefore, by Proposition 1.7, $\tilde{f}(x,y)$ is a measurable function of y for almost every x in R^n. Moreover, since E_x is measurable for almost every x in R^n, it follows that $f(x,y)$ is measurable as a function of y on E_x, for almost every x. This proves (i); the proof of (ii) is analogous.

Now, if f is integrable on E, then $\tilde{f} \in L(R^{n+m})$ and by Theorem 1.6 we get

$$\int_E f = \int_{R^{n+m}} \tilde{f} = \int_{R^n} \int_{R^m} \tilde{f}(x,y)\, dy\, dx\,. \tag{1.4}$$

On the other hand, E_x is measurable for almost every x in R^n, and for those x's we get

$$\int_{R^m} \tilde{f}(x,y)\, dy = \int_{E_x} f(x,y)\, dy\,. \tag{1.5}$$

Whence integrating (1.5) over R^n, and combining the resulting expression with (1.4), it readily follows that (iii) holds.

The proof of (iv) is analogous. ■

Next we consider whether a converse to Fubini's theorem is true. For instance, suppose f is measurable on R^{n+m}, and the iterated integrals of f exist and are equal. Does it, then, follow that f is integrable?

The answer to this question is no, and the counterexample is constructed along the lines of Figure 5 below.

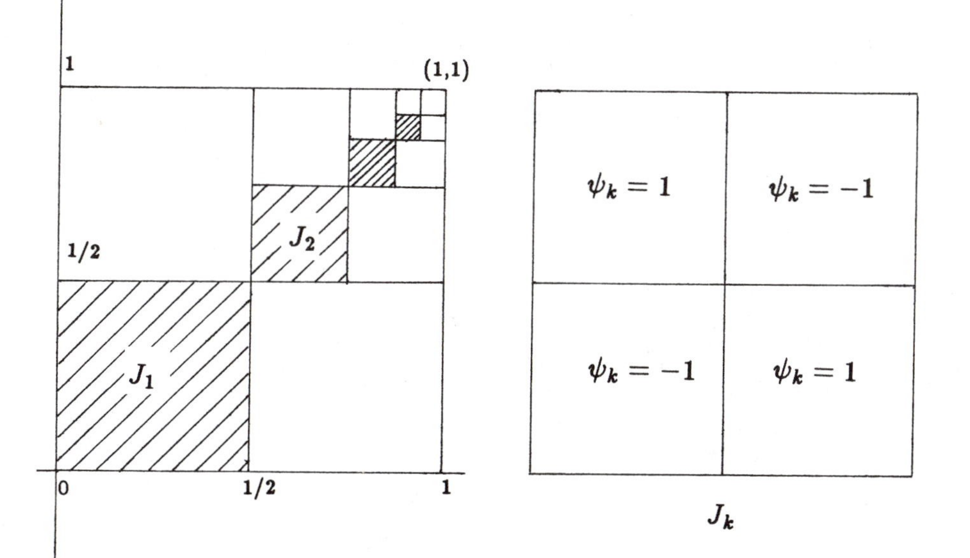

Figure 5

Indeed, let $J = [0,1] \times [0,1]$, divide J into four equal squares, separate the bottom left square, call it J_1, divide the right upper square into four equal squares, call the bottom left square J_2, and so on. On each square J_k, $k = 1,2,\ldots$, define a function ψ_k as follows: Divide each J_k into four equal squares, and let ψ_k equal -1 on the interior of left bottom and right upper squares, equal 1 on interior of the right bottom and left upper squares, and 0 otherwise.

Put now

$$f(x,y) = \sum_{k=1}^{\infty} \frac{1}{|J_k|} \chi_k(x,y) .$$

There is no problem with convergence here since at most one of the terms of the above series is nonzero. Also f, being the sum of measurable functions, is itself measurable.

Now, since for all $x \in [0,1]$ we have $\int_{[0,1]} f(x,y)\,dy = 0$, it is clear that

$$\int_{[0,1]} \int_{[0,1]} f(x,y)\,dy\,dx = 0 .$$

A similar argument gives that the other iterated integral is 0, and so both iterated integrals exist and are equal. Is f integrable, and, if so, is its integral equal to 0? Well, since $|\psi_k(x,y)| = 1$ for $(x,y) \in J_k$, it follows that

$$\int_{[0,1]\times[0,1]} |f| = \sum_k \frac{1}{|J_k|} \int_{J_k} |\psi_k| = \infty ,$$

and the integral of f is not defined on $[0,1] \times [0,1]$.

In deciding what goes wrong with f, we note that it changes signs. In fact, this is the only difficulty.

Theorem 1.9 (Tonelli). Suppose f is a nonnegative measurable function defined on R^{n+m}. Then,

$$\int_{R^{n+m}} f = \int_{R^n} \int_{R^m} f(x,y)\,dy\,dx = \int_{R^m} \int_{R^n} f(x,y)\,dx\,dy . \qquad (1.6)$$

The identity in (1.6) is understood as follows: If any one of the three expressions there is infinite, so are the other two, and if any one is finite, the other two are also finite and all three are then equal.

Proof. Let $\{f_k\}$ be a nondecreasing sequence of integrable functions that converges to f a.e. For instance, if $\chi_k(x,y)$ denotes the characteristic function of $I_1(0,k) \times I_2(0,k)$, then the sequence

$$f_k(x,y) = \min\{f(x,y), k\}\chi_k(x,y) , \quad k = 1,2,\ldots$$

will do. Now we apply Fubini's theorem to each f_k and note that outside of a null subset N_k of R^n, $\int_{R^m} f_k(x,y)\,dy$ is well-defined and

$$\int_{R^{n+m}} f_k = \int_{R^n}\int_{R^m} f_k(x,y)\,dy\,dx \quad k = 1,2,\ldots \tag{1.7}$$

Let $N = \bigcup_k N_k$; N is a null subset of R^n and for $x \in R^n \setminus N$, by MCT, it follows that

$$\lim_{k\to\infty}\int_{R^m} f_k(x,y)\,dy = \int_{R^m} f(x,y)\,dy\,.$$

Thus, again by MCT,

$$\lim_{k\to\infty}\int_{R^n}\int_{R^m} f_k(x,y)\,dy\,dx = \int_{R^n}\int_{R^m} f(x,y)\,dy\,dx\,. \tag{1.8}$$

On the other hand, also by MCT, we get

$$\lim_{k\to\infty}\int_{R^{n+m}} f_k = \int_{R^{n+m}} f\,. \tag{1.9}$$

Whence, combining (1.7), (1.8), and (1.9) we see that

$$\int_{R^{n+m}} f = \int_{R^n}\int_{R^m} f(x,y)\,dy\,dx\,,$$

which is one of the identities in (1.6). The other is obtained in a similar fashion. ■

An interesting application of Tonelli's theorem is the following remark concerning the evaluation of the measure of a set E in R^{n+m} in terms of its sections. Since for each fixed $x \in R^n$, $\chi_E(x,y) \neq 0$ iff $\chi_{E_x}(y) \neq 0$, we have

$$\begin{aligned}|E| &= \int_{R^{n+m}} \chi_E = \int_{R^n}\int_{R^m} \chi_E(x,y)\,dy\,dx \\ &= \int_{R^n}\int_{R^m} \chi_{E_x}(y)\,dy\,dx = \int_{R^n} |E_x|\,dx\,.\end{aligned}$$

A similar argument also gives that $|E| = \int_{R^m} |E^y|\,dy$.

Note that in particular we have that E is null iff $|E_x| = 0$ for x a.e. in R^n iff $|E^y| = 0$ for y a.e. in R^m.

2. CONVOLUTIONS AND APPROXIMATE IDENTITIES

Given integrable functions f, g defined on R^n, we have already noted that the pointwise product fg is not necessarily integrable. In other words, the expression $\int_{R^n} f(y)g(y)\,dy$ is not necessarily finite. It may come as a surprise that a closely related expression, namely the convolution $f*g(x)$ of f and g at x, is nevertheless finite a.e. on R^n. The precise definition is as follows: Suppose $f,g \in L(R^n)$, and consider the integral

$$f*g(x) = \int_{R^n} f(x-y)g(y)\,dy\,, \quad x \in R^n\,. \tag{2.1}$$

Suppose for the moment that $f,g \geq 0$, and note that for any fixed $x \in R^n$, the integrand in (2.1) is a nonnegative measurable function, so that $f*g(x)$ is a well-defined real number or ∞. But, is there any x for which $f*g(x) < \infty$?

A way to go about answering this question is to show that actually the function $f*g$ is integrable, for then it will be finite a.e. As for the integrability, observe that

$$\|f*g\|_1 = \int_{R^n}\int_{R^n} f(x-y)g(y)\,dy\,dx\,,$$

and this expression corresponds to one of the iterated integrals of the function $f(x-y)g(y)$. But it is the other iterated integral, namely,

$$\begin{aligned}\int_{R^n}\int_{R^n} f(x-y)g(y)\,dx\,dy &= \int_{R^n}\int_{R^n} f(x-y)\,dx\,g(y)\,dy\\ &= \|f\|_1\|g\|_1 < \infty\,,\end{aligned}$$

which is easy to handle. To pass from one to the other we wish to invoke Tonelli's theorem, and to do this we need to prove first that the function $h(x,y)$ defined by

$$h(x,y) = f(x-y)g(y)\,, \quad x,y \in R^n\,,$$

is actually measurable in $R^n \times R^n = R^{2n}$. Now, since g is measurable, given a scalar λ, the set

$$\{(x,y) \in R^{2n} : g(y) > \lambda\} = R^n \times \{y \in R^n : g(y) > \lambda\}$$

is a measurable subset of R^{2n}, and g is also measurable when considered as a function defined on R^{2n}.

So, we are reduced to showing that if f is a measurable function defined on R^n, then $h(x,y) = f(x-y)$ is a measurable function on $R^n \times R^n = R^{2n}$. To clarify the ideas behind this result we do the case $n = 1$, where the main difficulties are already apparent; we leave it to the reader to think of the general proof.

We proceed step by step and assume first that $f = \chi_I$, where $I = (a,b)$ is an open interval. In this case the level sets

$$\{(x,y) \in R^2 : f(x-y) > \lambda\}, \quad \lambda \text{ real},$$

are empty if $\lambda \geq 1$, and the open strip $\{(x,y) : a < x-y < b\}$ otherwise. In either case, the level sets are measurable. Next, if $f = \chi_{\mathcal{O}}$, $\mathcal{O}$ open, then write $\mathcal{O} = \bigcup_k I_k$ as the pairwise disjoint union of the open intervals I_k, and note that the levels sets are now $\{f(x-y) > \lambda\} = \bigcup_k \{\chi_{I_k}(x-y) > \lambda\}$, and hence they are also measurable.

If $f = \chi_G$, G a G_δ subset in R, then we have $G = \bigcap_k \mathcal{O}_k$, $\mathcal{O}_k$ open, and consequently $f(x-y)$ is the limit $f(x-y) = \lim_{k\to\infty} \chi_{\mathcal{O}_k}(x-y)$ of measurable functions, and hence measurable. The same is true if $f = \chi_N$, N a null subset of R^n. Indeed, let G be a G_δ subset in R such that $N \subseteq G$, $|G| = 0$, and let $\mathcal{G} = \{(x,y) \in R^2 : x - y \in G\}$. As pointed out above, we have $|\mathcal{G}| = 0$ iff $|\mathcal{G}_x| = 0$ for all $x \in R$. Now,

$$\mathcal{G}_x = \{y \in R : x - y \in G\} = \{y \in R : y = x - g, g \in G\} = x - G,$$

and by the translation invariance of the Lebesgue measure we have $|\mathcal{G}_x| = |x - G| = |G| = 0$ for all $x \in R$. Thus $|\mathcal{G}| = 0$. Consider now

$$\mathcal{N} = \{(x,y) \in R^2 : x - y \in N\} \subseteq \mathcal{G}.$$

By the completeness of the Lebesgue measure it follows that also $|\mathcal{N}| = 0$, and so $\chi_N(x-y)$ is measurable.

Now, if $f = \chi_E$, where E is a measurable subset of R, write $E = G \setminus N$, with G a G_δ subset of R and N a null set, and note that in this case $f(x-y) = \chi_G(x-y) - \chi_N(x-y)$, is also measurable. Whence simple functions ϕ also enjoy the property that $\phi(x-y)$ is measurable, and the same is true of the limits of simple functions, to wit, arbitrary measurable functions. This is precisely what we set to prove.

Returning to the properties of convolutions, we have

Theorem 2.1. Suppose $f, g \in L(R^n)$. Then

$$\int_{R^n} |f(x-y)|\,|g(y)|\,dy < \infty, \quad x \text{ a.e. in } R^n. \tag{2.2}$$

For those x's that satisfy (2.2) we define

$$f * g(x) = \int_{R^n} f(x-y)g(y)\,dy\,. \tag{2.3}$$

Then $f * g \in L(R^n)$, and

$$\|f * g\|_1 \le \|f\|_1 \|g\|_1\,. \tag{2.4}$$

Proof. First observe that the function h on R^{2n} defined by $h(x,y) = |f(x-y)||g(y)|$ is measurable and nonnegative. Thus by Tonelli's theorem we have the equality of the iterated integrals of h, and consequently,

$$\begin{aligned}\int_{R^n}\int_{R^n} |f(x-y)|\,|g(y)|\,dy\,dx &= \int_{R^n}\int_{R^n} |f(x-y)|\,|g(y)|\,dx\,dy \\ &= \int_{R^n}\int_{R^n} |f(x-y)|\,dx\,|g(y)|\,dy\,.\end{aligned}$$

By the translation invariance of the Lebesgue measure it readily follows that the inner-most integral above is equal to $\|f\|_1$, and the right-hand side is $\|g\|_1\|f\|_1$.

Now, since $|f * g(x)| \le \int_{R^n} |f(x-y)|\,|g(y)|\,dy$, (2.4) holds. ■

What more can we say about convolutions? For instance, a natural question to consider is whether the operation of convolution has a unit, i.e., whether there exists an integrable function f such that $f * g = g$ for any $g \in L(R^n)$. Suppose such a function f exists, and given $0 \ne x \in R^n$, let $0 < \eta < |x|$. Put now $g(y) = (1/2\eta)^n \chi_{I(0,\eta)}(y)$ and note that

$$0 = g(x) = f * g(x) = \frac{1}{|I(0,\eta)|}\int_{I(0,\eta)} f(x-y)\,dy\,. \tag{2.5}$$

Now, by the Lebesgue Differentiation Theorem, the limit as $\eta \to 0$ of the above expression is $f(x)$ a.e., and consequently, we get that $f = 0$ a.e. Thus, since $0 * g = 0$, such a function f cannot exist.

Nevertheless, if we put $x = 0$ in (2.5), since $g(0) = 1/|I(0,\eta)|$, again by the Lebesgue Differentiation Theorem it follows that

$$\infty = \lim_{\eta\to 0}\frac{1}{|I(0,\eta)|}\int_{I(0,\eta)} f(-y)\,dy = f(0)\,.$$

Still one more property: Since $f \ge 0$, by taking g to be nonnegative and integrable, we get

$$\int_{R^n} f * g(x)\,dx = \left(\int_{R^n} f\right)\left(\int_{R^n} g\right) = \int_{R^n} g\,,$$

and so we have $\int_{R^n} f = 1$. Combining these remarks we note that the integrable function f, which does not exist, satisfies the following properties: It is 0 a.e. away from the origin, where it assumes the value ∞, and it has integral 1. In other words, f corresponds to the Dirac measure δ_0.

The convolution is also well-defined in the more general context of the Lebesgue L^p-spaces, $1 < p \leq \infty$. More precisely, we have

Theorem 2.2. Suppose $g \in L(R^n)$, $f \in L^p(R^n)$, $1 < p \leq \infty$. Then the convolution $f * g(x)$ is a well-defined $L^p(R^n)$ function that satisfies

$$\|f * g\|_p \leq \|g\|_1 \|f\|_p \,. \tag{2.6}$$

Proof. Since $|f * g(x)| \leq |f| * |g|(x)$ we may assume that f and g are nonnegative. Since the case $p = \infty$ follows at once from the definition of convolution, we assume that $1 < p < \infty$. Let $1 < q < \infty$ be the conjugate index to p, i.e., $1/p + 1/q = 1$, and note that by Hölder's inequality, Tonelli's theorem, and the translation invariance of the integral we have

$$\begin{aligned}
\|f * g\|_p^p &= \int_{R^n} \left(\int_{R^n} f(x-y) g(y)^{1/p} g(y)^{1/q} dy \right)^p dx \\
&\leq \int_{R^n} \left(\int_{R^n} f(x-y)^p g(y)^{p \cdot 1/p} dy \right) \left(\int_{R^n} g(y)^{q \cdot 1/q} \right)^{p/q} dx \\
&= \|g\|_1 \|f\|_p^p \|g\|_1^{p/q} = \|g\|_1^{p/q+1} \|f\|_p^p = \|g\|_1^p \|f\|_p^p \,,
\end{aligned}$$

and, as asserted, (2.6) holds. ■

Thus the convolution inherited the integrability properties of f; next we show that it also inherits the "smoothness" properties of g.

Smooth integrable functions abound; $e^{-|x|^2}$ is one such example. In applications it is important to have at hand a smooth integrable function ϕ that vanishes off $|x| \leq 1$. To construct such a function, let ψ be defined on the line by

$$\psi(t) = \begin{cases} e^{1/t} & t < 0 \\ 0 & t \geq 0. \end{cases}$$

Since all the derivatives of ψ exist when $t \neq 0$ and converge to 0 when $t \to 0$, and since by a simple application of l'Hôpital's rule it follows that all derivatives of ψ vanish when $t = 0$, $\psi \in C^\infty(R)$. The function ϕ is now defined by letting $\phi(x) = \psi(|x|^2 - 1)$.

Before we continue we need a simple observation concerning convolutions: By the translation invariance of the Lebesgue measure it readily follows that at those x's where $f * g(x)$ is defined, we also have

$$f * g(x) = \int_{R^n} f(y)g(x-y)\,dy\,.$$

We are now ready to prove

Theorem 2.3. Suppose $f \in L^P(R^n)$, $1 \le p \le \infty$, and $\phi \in C_0^m(R^n)$, $m \ge 1$. Then $f * \phi \in L^p(R^n) \cap C^m(R^n)$.

Proof. Since, by Theorem 2.2, $f * \phi \in L^p(R^n)$, only the smoothness of the convolution needs to be proved. We first show that $f * \phi$ is continuous. Indeed, given $x,h \in R^n$, note that

$$\begin{aligned} |f*\phi(x+h) - f*\phi(x)| &= \left|\int_{R^n} f(y)(\phi(x+h-y) - \phi(x-y))\,dy\right| \\ &\le \int_{R^n} |f(y)||\phi(x+h-y) - \phi(x-y)|\,dy \\ &\le \|f\|_p \|\phi(x+h-\cdot) - \phi(x-\cdot)\|_q\,, \end{aligned} \tag{2.7}$$

where p, q are conjugate indices. Now, if $q < \infty$, since $\phi \in L^q(R^n)$ and since translations are continuous in L^q, the right-hand side of (2.7) goes to 0 as $|h| \to 0$, and so does the left-hand side; thus $f*\phi(x)$ is continuous. On the other hand, if $q = \infty$, by the first inequality above it readily follows that

$$|f*\phi(x+h) - f*\phi(x)| \le \|f\|_1 \sup|\phi(\cdot + h) - \phi(\cdot)|\,.$$

Now, by the uniform continuity of ϕ we get that the right-hand side of the above inequality goes to 0 with h, and so does the left-hand side there.

As for the smoothness, let $h = (h, 0, \dots, 0)$ denote the vector with scalar $h \ne 0$ in the first position and zeros elsewhere, and put

$$\Phi(x,y,h) = \frac{\phi(x+h-y) - \phi(x-y)}{h} - \left(\frac{\partial \phi}{\partial x_1}\right)(x-y)\,.$$

By the conditions of the theorem it is clear that for each fixed $x \in R^n$, $\Phi(x,y,h) \to 0$, uniformly and boundedly in y as $h \to 0$. Whence

$$\lim_{h\to 0} \int_{R^n} f(y)\Phi(x,y,h)\,dy = 0\,, \quad x \in R^n\,. \tag{2.8}$$

Moreover, since the integral in (2.8) equals

$$\frac{f*\phi(x+h) - f*\phi(x)}{h} - f*\left(\frac{\partial\phi}{\partial x_1}\right)(x)\,,$$

it readily follows that

$$\frac{\partial(f*\phi)}{\partial x_1}(x) = f*\frac{\partial\phi}{\partial x_1}(x)\,, \quad x \in R^n\,.$$

Clearly a similar argument applies to all other derivatives of the convolution, and the statement concerning the smoothness has also been established. ■

Now, the convolution $f*\phi$ of f with a nonnegative continuous kernel ϕ with nonvanishing integral, represents a smoothed out version of the convolution $f*\chi_{I(0,1)}$ of f with the characteristic function $\chi_{I(0,1)}$ of the unit interval in R^n. Similarly, if $\phi_\varepsilon(x) = \varepsilon^{-n}\phi(x/\varepsilon)$, $\varepsilon > 0$, are the dilates of the kernel ϕ, i.e., "bumps" at the origin of width ε and height ε^{-n}, the convolutions $f*\phi_\varepsilon(x)$ are well-defined, and it is natural to consider, in analogy with the Lebesgue Differentiation Theorem, whether they converge, in some sense, to f as $\varepsilon \to 0$.

Before we address this question we need an observation concerning the dilates of an integrable function.

Proposition 2.4. Suppose $\phi \in L(R^n)$ is nonnegative, and $\varepsilon > 0$. Then

$$\int_{R^n} \phi_\varepsilon(x)\,dx = \int_{R^n} \phi(x)\,dx\,, \quad \text{all } \varepsilon > 0\,, \tag{2.9}$$

and

$$\lim_{\varepsilon\to 0} \int_{\{|x|>M\}} \phi_\varepsilon(x)\,dx = 0\,, \quad \text{all } M > 0\,. \tag{2.10}$$

Proof. We begin by checking (2.9) for the simplest function, namely $\phi = \chi_I$, $I = (a_1,b_1) \times \cdots \times (a_n,b_n)$ an open interval in R^n. In this case, by Tonelli's theorem we have

$$\begin{aligned}\int_{R^n} \phi_\varepsilon(x)\,dx &= \frac{1}{\varepsilon^n}\int_{(\varepsilon a_1,\varepsilon b_1)} \cdots \int_{(\varepsilon a_n,\varepsilon b_n)} dx_n \cdots dx_1 \\ &= \frac{1}{\varepsilon^n}\varepsilon(b_1-a_1)\cdots\varepsilon(b_n-a_n) = \int_{R^n}\phi(x)\,dx\,,\end{aligned}$$

and (2.9) holds.

Next we show that this is also true if $\phi = \chi_N$, N a null subset of the boundary of an open interval I. This is not hard: By Tonelli's theorem one of the sections

$$N_i = \{x_i \in R : x = (x_1,\ldots,x_i,\ldots,x_n) \in N\}\,, \quad 1 \le i \le n\,,$$

is also null, and consequently, by 3.9 in Chapter V, we have

$$\int_R \chi_{N_i}(x_i/\varepsilon)\,dx_i = 0 \quad \text{all } \varepsilon > 0\,.$$

Now, by Tonelli's theorem again, we get that (2.9) holds in this case as well, both integrals there being 0. Combining these observations it readily follows that (2.9) is true for $\phi = \chi_E$, E a subset of R^n consisting of an open interval I and any part of its boundary, and by MCT that it is also true for limits of sums of such functions, i.e., $\phi = \chi_{\mathcal{O}}$, $\mathcal{O}$ open in R^n.

It now follows that (2.9) holds for $\phi = \chi_H$, H a G_δ subset of R^n, or a null subset of R^n, and consequently, any measurable subset of R^n. (2.10) is therefore also true for an arbitrary simple function, and by taking limits, for every integrable function.

As for (2.10), fix $M > 0$, and let $y = x/\varepsilon$. Thus, by (2.9), we have

$$\frac{1}{\varepsilon^n}\int_{\{|x|>M\}} \phi(x/\varepsilon)\,dx = \int_{\{|y|>M/\varepsilon\}} \phi(y)\,dy\,,$$

and since $M/\varepsilon \to \infty$ as $\varepsilon \to 0$, it readily follows that the integral on the right-hand side above tends to 0 with ε. ■

What is the meaning of these properties of ϕ_ε? (2.10) indicates the "dilation invariance" of the norm in $L^1(R^n)$. (2.11) means that for small ε, the area under the graph of ϕ_ε is essentially concentrated in a neighbourhood of the origin. Thus, for any integrable function ϕ with integral equal to 1 we expect that, as in the case of the Lebesgue Differentiation Theorem, the effect of letting $\varepsilon \to 0$ in the convolution

$$f * \phi_\varepsilon(x) = \int_{R^n} f(x-y)\phi_\varepsilon(y)\,dy = \int_{R^n} f(x-\varepsilon y)\phi(y)\,dy\,,$$

will be to emphasize the values of $f(x-y)$ corresponding to small y. Our next results show that, in fact, $f * \phi_\varepsilon(x)$ converges as $\varepsilon \to 0$ to $f(x)$ in various senses. For this reason the family ϕ_ε is called an "approximate identity."

We begin by proving the norm convergence.

Theorem 2.5. Suppose ϕ is a nonnegative integrable function with integral 1, and let $f \in L^p(R^n)$, $1 \le p < \infty$. Then

$$\lim_{\varepsilon\to 0} \|f * \phi_\varepsilon - f\|_p = 0\,. \tag{2.11}$$

Proof. Under the conditions of the theorem, by (2.9) we have

$$f(x) = f(x)\int_{R^n} \phi(y)\,dy = \int_{R^n} f(x)\phi_\varepsilon(y)\,dy\,, \quad \text{all } \varepsilon > 0\,.$$

Whence it readily follows that

$$\begin{aligned} |f * \phi_\varepsilon(x) - f(x)| &= \left|\int_{R^n} (f(x-y) - f(x))\phi_\varepsilon(y)\,dy\right| \\ &\le \int_{R^n} |f(x-y) - f(x)|\phi_\varepsilon(y)\,dy\,, \end{aligned} \tag{2.12}$$

and consequently $\|f * \phi_\varepsilon - f\|_p$ is dominated by

$$\sup \int_{R^n}\int_{R^n} |f(x-y) - f(x)|\phi_\varepsilon(y)\,dy\,g(x)\,dx\,, \tag{2.13}$$

where the sup in (2.13) is taken over those g's in $L^q(R^n)$, $1/p + 1/q = 1$, with $\|g\|_q \le 1$. Now, by Tonelli's theorem, each integral in (2.13) equals

$$\begin{aligned} &\int_{R^n}\int_{R^n} |f(x-y) - f(x)||g(x)|\,dx\,\phi_\varepsilon(y)\,dy \\ &\qquad \le \int_{R^n} \|f(\cdot - y) - f(\cdot)\|_p\|g\|_q\phi_\varepsilon(y)\,dy \\ &\qquad \le \left(\int_{\{|y|\le M\}} + \int_{\{|y|>M\}}\right) \|f(\cdot - y) - f(\cdot)\|_p\phi_\varepsilon(y)\,dy \\ &\qquad = A + B\,, \end{aligned}$$

say. Since $\|f(\cdot - y) - f(\cdot)\|_p \to 0$ as $|y| \to 0$, given $\eta > 0$, we can choose M so that $\|f(\cdot - y) - f(\cdot)\|_p \le \eta$ if $|y| \le M$, and consequently

$$A \le \eta \int_{\{|y|\le M\}} \phi_\varepsilon(y)\,dy \le \eta\,, \quad \text{all } M\,.$$

Moreover, since $\|f(\cdot - y) - f(\cdot)\|_p \le 2\|f\|_p$, and having fixed M as above, by (2.10) we get

$$B \le 2\|f\|_p \int_{\{|y|>M\}} \phi_\varepsilon(y)\,dy \to 0\,, \quad \text{as } \varepsilon \to 0\,.$$

Whence combining these estimates we get that the integrals in (2.13) can be made arbitrarily small provided $\varepsilon > 0$ is small enough, and this completes the proof. ■

An interesting consequence of this result is

Corollary 2.6. $C_0^\infty(R^n)$ is dense in $L^p(R^n)$, $1 \le p < \infty$.

Proof. Suppose $f \in L^p(R^n)$, $1 \le p < \infty$. Given $\eta > 0$, choose M so large that $\int_{\{|y|>M\}} |f(y)|^p\,dy \le \eta^p$. Next pick a nonnegative kernel $\phi \in C_0^\infty(R^n)$ with integral 1, and let $f^\varepsilon = (f\chi_{B(0,M)}) * \phi_\varepsilon$. Now, by Theorem 2.3, $f^\varepsilon \in L^p(R^n) \cap C^\infty(R^n)$; but there is something else we can say. Indeed, since both $f\chi_{B(0,M)}$ and ϕ vanish off a compact set K, say, the convolution $f^\varepsilon(x) = \int_{R^n} (f\chi_{B(0,M)})(x-y)\phi_\varepsilon(y)\,dy$ vanishes unless there are points x and y such that $x - y \in K$ and $y/\varepsilon \in K$. Hence, $f^\varepsilon(x) = 0$ unless x is of the form

$$x = x_1 + \varepsilon x_2\,, \quad x_1, x_2 \in K\,,$$

and this is a bounded set of points in R^n. Thus $f^\varepsilon \in C_0^\infty(R^n)$.

Finally,

$$\begin{aligned} \|f - f^\varepsilon\|_p &\le \|f\chi_{B(0,M)} - f^\varepsilon\|_p + \|f(1 - \chi_{B(0,M)})\|_p \\ &\le \|f\chi_{B(0,M)} - f^\varepsilon\|_p + \eta\,, \end{aligned}$$

and by Theorem 2.5 the right-hand side above can be made arbitrarily small with ε. ■

There are substitute results for $f \in L^\infty(R^n)$; one of them is

Theorem 2.7. Suppose ϕ is a nonnegative integrable function with integral 1, and let $f \in L^\infty(R^n)$. Then

$$\lim_{\varepsilon \to 0} f * \phi_\varepsilon(x) = f(x)\,, \tag{2.14}$$

at every point x of continuity of f.

Proof. As before, by (2.12) we have

$$\begin{aligned} |f * \phi_\varepsilon(x) - f(x)| &\le \int_{R^n} |f(x-y) - f(x)|\phi_\varepsilon(y)\,dy \\ &\le \left(\int_{\{|y|\le M\}} + \int_{\{|y|>M\}}\right) |f(x-y) - f(x)|\phi_\varepsilon(y)\,dy \\ &= A + B\,, \end{aligned}$$

say. Now, if f is continuous at x, given $\eta > 0$, there exists $M > 0$ such that $|f(x-y) - f(x)| \le \eta$ if $|y| \le M$. With this choice of M we have

$$A + B \le \eta + 2\|f\|_\infty \int_{\{|y|>M\}} \phi_\varepsilon(y)\,dy\,,$$

where the right-hand side above tends to 0 as $\varepsilon \to 0$. This readily implies that (2.14) is true. ∎

Still we must address the harder question concerning the pointwise convergence to f of the convolutions of f with the approximate identities ϕ_ε for integrable, or more generally, p-integrable functions. We begin by proving

Theorem 2.8. Suppose ϕ is a nonnegative integrable function with integral equal to 1, and let $f \in L^p(R^n)$, $1 \le p < \infty$. If in addition ϕ satisfies

$$\phi(y) \le c/|y|^{n+\eta}\,, \quad \eta > 0\,, \text{all } y \in R^n\,, \tag{2.15}$$

then at each point x of continuity of f we have

$$\lim_{\varepsilon\to 0} f * \phi_\varepsilon(x) = f(x)\,. \tag{2.16}$$

Proof. The proof follows along the lines to that of Theorem 2.7. For, in the notation of that theorem, and with the same choice of M as there, we still have that A can be made arbitrarily small at a point of continuity x of f. As for B, it is majorized by

$$\int_{\{|y|>M\}} |f(x-y)|\phi_\varepsilon(y)\,dy + |f(x)| \int_{\{|y|>M\}} \phi_\varepsilon(y)\,dy = B_1 + B_2\,,$$

say. (2.10) establishes that $\lim_{\varepsilon\to 0} B_2 = 0$. Also, if $p > 1$, by Hölder's inequality with indices p and its conjugate q, we have

$$\begin{aligned} B_2 &\le \left(\int_{R^n} |f(x-y)|^p\,dy\right)^{1/p} \left(\int_{\{|y|>M\}} \phi_\varepsilon(y)^q\,dy\right)^{1/q} \\ &= \|f\|_p \frac{\varepsilon^{n/q}}{\varepsilon^n} \left(\int_{\{|y|>M/\varepsilon\}} \phi(y)^q\,dy\right)^{1/q} = \|f\|_p \Phi(\varepsilon)\,, \end{aligned}$$

say. Consequently, any condition on ϕ that ensures that $\lim_{\varepsilon\to 0}\Phi(\varepsilon) = 0$, will give (2.16); we show next that (2.15) is one such condition. Indeed, if (2.15) holds, then

$$\begin{aligned} \Phi(\varepsilon) &\le c\frac{\varepsilon^{n/q}}{\varepsilon^n} \left(\int_{\{|y|>M/\varepsilon\}} |y|^{-(n+\eta)q}\,dy\right)^{1/q} \\ &= c\frac{\varepsilon^{n/q}}{\varepsilon^n} \left(\frac{M}{\varepsilon}\right)^{n/q-(n+\eta)} = c\varepsilon^\eta\,, \end{aligned}$$

which clearly tends to 0 with ε.

Finally, if $p = 1$, then

$$B_2 \le |f(x)| \int_{\{|y|>M/\varepsilon\}} \phi(y)\, dy\,,$$

which, since ϕ is integrable, also tends to 0 with ε. ∎

To complete the analogy with the Lebesgue Differentiation Theorem, we discuss the a.e. convergence of $f * \phi_\varepsilon$ to f. The results are now more complicated, cf. 4.18 below and Theorem 3.1 in Chapter XVII, but are surprisingly simple in case ϕ vanishes off a compact set.

Theorem 2.10. Suppose ϕ is a nonnegative bounded integrable function with integral equal to 1, which vanishes off $B(0,1)$, and let $f \in L^p(R^n)$, $1 \le p \le \infty$. Then, at each point x of the Lebesgue set of f, and in particular a.e., we have

$$\lim_{\varepsilon \to 0} f * \phi_\varepsilon(x) = f(x)\,.$$

Proof. As before, and in the notation of Theorem 2.7, since ϕ_ε vanishes off the set $\{|y| \le \varepsilon\}$, the choice $M = \varepsilon$ gives that $B = 0$. As for A, since ϕ is bounded, it is dominated by

$$A \le c\frac{1}{\varepsilon^n} \int_{\{|y|\le\varepsilon\}} |f(x-y) - f(x)|\, dy\,,$$

which goes to 0 with ε at precisely those points x in the Lebesgue set of f. ∎

The reader will note that the convergence results presented above may be extended to the following setting: We may assume that ϕ is an integrable function with integral one such that $|\phi(x)| \le \psi(x)$ for all $x \in R^n$, where ψ satisfies the conditions that we required the previously nonnegative function ϕ to verify.

3. ABSTRACT FUBINI'S THEOREM

In this section we present the abstract version of Fubini's theorem. Now, in the case of Euclidean space the problem at hand was facilitated by the fact that the Lebesgue measure is defined on the various spaces involved. Thus, given measures μ, ν defined on $(X,\mathcal{M})$ and $(Y,\mathcal{N})$ respectively, the first order of business is to construct a "product measure" on $\mathcal{M} \times \mathcal{N}$, the σ-algebra introduced in Chapter IV, one that will make statements such as Fubini's theorem true. First some definitions.

A measurable rectangle is any subset of $X \times Y$ of the form $A \times B$, $A \in \mathcal{M}$, $B \in \mathcal{N}$. Finite unions of pairwise disjoint measurable rectangles are called elementary sets, and are often denoted by Q. If $E \subseteq X \times Y$, we define the section E_x of E (at level $x \in X$) as the subset of Y given by

$$E_x = \{y \in Y : (x,y) \in E\}\,, \quad x \in X\,. \tag{3.1}$$

Similarly, the section E^y of E (at level $y \in Y$) is defined as

$$E^y = \{x \in X : (x,y) \in E\}\,, \quad y \in Y\,. \tag{3.2}$$

How do sections behave with respect to measurability?

Proposition 3.1. Every section of a measurable set $E \in \mathcal{M} \times \mathcal{N}$ is measurable. Specifically, $E_x \in \mathcal{N}$ for all $x \in X$, and $E^y \in \mathcal{M}$ for every $y \in Y$.

Proof. Let $\mathcal{F}$ denote the class of those $E \in \mathcal{M} \times \mathcal{N}$ such that $E_x \in \mathcal{N}$ for all $x \in X$; we intend to show that $\mathcal{F}$ is a σ-algebra of subsets of Y that contains all measurable rectangles and which therefore coincides with $\mathcal{M} \times \mathcal{N}$.

First note that if $E = A \times B$ is a measurable rectangle, then $E_x = B$ when $x \in A$ and $E_x = \emptyset$ otherwise, and consequently, every measurable rectangle belongs to $\mathcal{F}$. In particular $X \times Y \in \mathcal{F}$.

Further, since $\mathcal{N}$ is a σ-algebra, it readily follows that if $E \in \mathcal{F}$, then

$$((X \times Y) \setminus E)_x = \{y \in Y : (x,y) \notin E\} = Y \setminus E_x \in \mathcal{N} \quad \text{all } x \in X\,,$$

and $\mathcal{F}$ is closed under complementation.

Finally, if $E_n \in \mathcal{F}$, $n = 1, 2, \ldots$, and $E = \bigcup_n E_n$, since

$$E_x = \bigcup_n (E_n)_x \in \mathcal{N}\,, \quad \text{all } x \in X\,,$$

$\mathcal{F}$ is also closed under countable unions. Thus $\mathcal{F}$ is the σ-algebra $\mathcal{M} \times \mathcal{N}$. The proof for the E^y's is the same. ■

The statement of Proposition 3.1 is one about characteristic functions of measurable sets. For arbitrary measurable functions the situation is as follows: If f is a function defined on $X \times Y$, we call the function

$$f_x(y) = f(x,y)\,, \quad x \in X\,,$$

the X-section of f at level x. Similarly, the Y-section of f at level y is defined by

$$f^y(x) = f(x,y)\,, \quad y \in Y\,.$$

We begin by showing that sections of measurable functions are measurable; the measurability of functions defined on $X \times Y$ is always understood to be with respect to the σ-algebra $\mathcal{M} \times \mathcal{N}$.

Proposition 3.2. The sections of measurable functions are measurable. More precisely, if f is a measurable function defined on $X \times Y$, f_x is a measurable function on $(Y, \mathcal{N})$ for every $x \in X$, and f^y is a measurable function in $(X, \mathcal{M})$ for every $y \in Y$.

Proof. Let f be measurable, and given an open set $\mathcal{O}$ note that for each $x \in X$ we have

$$\begin{aligned} f_x^{-1}(\mathcal{O}) &= \{y \in Y : f_x(y) \in \mathcal{O}\} = \{y \in Y : f(x,y) \in \mathcal{O}\} \\ &= \{y \in Y : (x,y) \in f^{-1}(\mathcal{O})\} = \left(f^{-1}(\mathcal{O})\right)_x \,. \end{aligned}$$

Now, since $f^{-1}(\mathcal{O}) \in \mathcal{M} \times \mathcal{N}$, the measurability of the set on the right-hand side above follows from Proposition 3.1, and that of f_x from Proposition 1.3 in Chapter VI. The proof for f^y is the same. ∎

We are now ready to prove the basic result needed to introduce the product measure.

Theorem 3.3. Let $(X, \mathcal{M}, \mu)$, $(Y, \mathcal{N}, \nu)$ be σ-finite measure spaces, and suppose $E \in \mathcal{M} \times \mathcal{N}$. Then for each $x \in X$, $\nu(E_x)$ is a measurable function on $(X, \mathcal{M})$, and for each $y \in Y$, $\mu(E^y)$ is a measurable function on $(Y, \mathcal{N})$. Furthermore

$$\int_X \nu(E_x)\, d\mu = \int_Y \mu(E^y)\, d\nu\,. \tag{3.3}$$

Proof. The measurability of E_x and E^y has been established in Proposition 3.1, so we begin by computing the integrands of the integrals in (3.3).

Now, as noted above, if $E = \bigcup_n (A_n \times B_n)$ is an elementary set, then $\chi_{E_x}(y) = \sum_n \chi_{A_n}(x)\chi_{B_n}(y)$, and by the additivity of ν it readily follows that

$$\nu(E_x) = \int_X \chi_{E_x}\, d\nu = \sum_n \nu(B_n)\chi_{A_n}(x)\,.$$

Clearly $\nu(E_x)$ is a measurable function on $(X, \mathcal{M})$, and

$$\int_X \nu(E_x)\, d\mu = \sum_n \nu(B_n)\mu(A_n)\,. \tag{3.4}$$

In a similar fashion it follows that $\mu(E^y)$ is a measurable function on $(Y, \mathcal{N})$ and that its integral over Y with respect to ν is equal to the right-hand side of (3.4). Therefore the assertion of the theorem is true for all elementary sets in $\mathcal{M} \times \mathcal{N}$; we now show that the collection $\mathcal{F}$ of subsets of $\mathcal{M} \times \mathcal{N}$ for which (3.3) is true is a monotone class which, on account of 4.19 in Chapter IV, coincides with $\mathcal{M} \times \mathcal{N}$.

Let $\{E_n\} \subseteq \mathcal{F}$ be a nondecreasing sequence, and write $E = \bigcup_n E_n$; we must show that $E \in \mathcal{F}$ as well. First note that $\{(E_n)_x\} \subseteq \mathcal{N}$ is a nondecreasing sequence that converges to $E_x \in \mathcal{N}$, and consequently by (3.3) in Chapter IV,

$$\lim_{n\to\infty} \nu((E_n)_x) = \nu(E_x)\,, \quad \text{all } x \in X\,.$$

Whence $\nu(E_x)$ is a limit of measurable functions on $(X, \mathcal{M})$, and is therefore measurable. Further, by MCT it readily follows that

$$\lim_{n\to\infty} \int_X \nu((E_n)_x)\, d\mu = \int_X \nu(E_x)\, d\mu\,. \tag{3.5}$$

Similarly $\mu(E^y)$ is measurable on $(Y, \mathcal{N})$, and

$$\lim_{n\to\infty} \int_Y \mu((E_n)^y)\, d\nu = \int_Y \mu(E^y)\, d\nu\,. \tag{3.6}$$

Now, since for each n the integrals that appear on the left-hand side of (3.5) and (3.6) are equal, so are their limits. In other words, the integrals on the right-hand side of (3.5) and (3.6) are equal, and (3.3) holds in this case.

Suppose next that $\{E_n\} \subseteq \mathcal{F}$ is a nonincreasing sequence of sets, and put $E = \bigcap_n E_n$; we must show that $E \in \mathcal{F}$. The preceding argument, invoking now (3.4) in Chapter IV instead, certainly goes through if X and Y have finite measure. To see that the same is true in the σ-finite case, let $\{X_k\}$, $\{Y_k\}$ be sequences of sets of finite measure such that $X = \bigcup X_k$ and $Y = \bigcup Y_k$. Then, since for each k the nonincreasing sequence $\{E_n \cap (X_k \times Y_k)\}$ converges to $E \cap (X_k \times Y_k)$ and $\mu(X_k)$, $\nu(Y_k) < \infty$, (3.3) is true for $E \cap (X_k \times Y_k)$, $k = 1, 2, \ldots$ But since the nondecreasing sequence $\{E \cap (X_k \times Y_k)\}$ converges to E, the conclusion of the theorem also holds for E.

We have thus shown that $\mathcal{F}$ is a monotone class, and the proof is complete. ■

The following example shows that the σ-finiteness of the measures was necessary. Let $I = [0,1]$ and consider the measure spaces $(I, \mathcal{L}, |\cdot|)$ and $(I, \mathcal{P}(I), \nu)$, where ν denotes the counting measure on $(I, \mathcal{P}(I))$. Further, let $E = \{(x,y) \in I \times I : x = y\}$ be the "diagonal set" in $I \times I$; it is not difficult to show that $E \in \mathcal{L} \times \mathcal{P}(I)$. Now, since for each real x, y the sets E_x and E^y consist of a single point, we have

$$\int_I \nu(E_x)\, dx = \infty\,, \quad \text{and} \quad \int_I |E^y|\, d\nu = 0\,.$$

If $(X, \mathcal{M}, \mu)$ and $(Y, \mathcal{N}, \nu)$ are as in Theorem 3.3, we define the set function $\mu \times \nu$ on $(X \times Y, \mathcal{M} \times \mathcal{N})$ by

$$(\mu \times \nu)(E) = \int_X \nu(E_x)\, d\mu = \int_Y \mu(E^y)\, d\nu\,, \quad E \in \mathcal{M} \times \mathcal{N}\,. \tag{3.7}$$

The equality of the integrals in (3.7) is assured by Theorem 3.3. We call $\mu \times \nu$ the "product" of the measures μ and ν, and it follows without much difficulty from MCT that $\mu \times \nu$ indeed is a measure. Observe that also $\mu \times \nu$ is σ-finite.

Now, since $\nu(E_x) = \int_Y \chi_E(x,\cdot)\, d\nu$ and $\mu(E^y) = \int_X \chi_E(\cdot, y)\, d\mu$, (3.7) actually states that a Tonelli-like identity is true for the characteristic functions of measurable sets. In fact, the general statement holds as well, to wit,

Theorem 3.4. Let $(X, \mathcal{M}, \mu)$, $(Y, \mathcal{N}, \nu)$ be σ-finite measure spaces, and let f be a nonnegative extended real-valued measurable function defined on $(X \times Y, \mathcal{M} \times \mathcal{N})$. Then $\int_Y f_x(y)\, d\nu$ is a measurable function on $(X, \mathcal{M})$, $\int_X f^y(x)\, d\mu$ is a measurable function on $(Y, \mathcal{N})$, and

$$\int_{X \times Y} f\, d(\mu \times \nu) = \int_X \int_Y f_x(y)\, d\nu\, d\mu = \int_Y \int_X f^y(x)\, d\mu\, d\nu\,. \tag{3.8}$$

Proof. By (3.7) the theorem is true for characteristic functions of measurable sets, and hence (3.8) holds for all nonnegative simple functions. By Theorem 1.12 in Chapter VI we know that f is the limit of nondecreasing sequence of simple functions, and consequently (3.8) follows by MCT as in Theorem 1.8. ■

Corollary 3.5. Under the assumptions of Theorem 3.4, if

$$\int_X \int_Y |f|_x(y)\, d\nu\, d\mu < \infty\,,$$

then $f \in L(X \times Y, \mu \times \nu)$.

Proof. Apply Theorem 3.4 to $|f|$. ■

Theorem 3.6 (Fubini). Under the assumptions of Theorem 3.4, if $f \in L(X \times Y, \mu \times \nu)$, then $f_x \in L(X, \mu)$ μ-a.e. on X, $f^y \in L(Y, \nu)$ ν-a.e. on Y, the functions

$$\int_Y f_x(y)\, d\nu \in L(X, \mu)\,, \qquad \int_X f^y(x)\, d\mu \in L(Y, \nu)$$

and (3.8) holds.

Proof. Write $f = f^+ - f^-$. By Theorem 3.4 we have

$$\int_{X \times Y} f^+\, d(\mu \times \nu) = \int_X \int_Y (f^+)_x(y) d\nu\, d\mu = \int_Y \int_X (f^+)^y(x)\, d\mu\, d\nu\,.$$

Since $f \in L(X \times Y, \mu \times \nu)$, the left-hand side above is finite and so

$$\int_Y (f^+)_x(y)\, d\nu \in L(X, \mu)\,, \quad \text{and} \quad \int_X (f^+)^y(x)\, d\mu \in L(Y, \mu)\,.$$

This implies that $(f^+)_x(\cdot) \in L(Y, \nu)$ μ-a.e. and that $(f^+)^y(\cdot) \in L(X, \mu)$ ν-a.e. The same result holds if we replace f^+ by f^-, and thus the theorem follows. ■

A word about the product measure: If $(X, \mathcal{M}, \mu)$ and $(Y, \mathcal{N}, \nu)$ are complete measure spaces, it does not follow that $(X \times Y, \mathcal{M} \times \mathcal{N}, \mu \times \nu)$ is complete. For instance, if $\mu = \nu = |\cdot|$, the Lebesgue measure on R, let $A = \{x\}$ consist of a single point, and B be any non-Lebesgue measurable subset of R. Then $A \times B \subset A \times R$, $|A \times R| = 0$ but $A \times B \notin \mathcal{L} \times \mathcal{L}$. Thus $|\cdot| \times |\cdot|$ is not a complete measure, and, in particular, it is not the Lebesgue measure on the plane. However, the completion of $|\cdot| \times |\cdot|$ is the Lebesgue measure on the plane, cf. 4.27 below. The statement of Fubini's theorem in this context is left to the reader, cf. 4.28 below.

4. PROBLEMS AND QUESTIONS

4.1 Prove that if E is a Lebesgue measurable subset of the rectangle $[0,1] \times [0,1]$, and if $|E_x| \leq 1/2$ for almost all $x \in [0,1]$, then $|\{y \in [0,1] : |E^y| = 1\}| \leq 1/2$.
Can you think of n-dimensional extensions?

4.2 Show that if an extended real-valued function f defined on R^{n+m} has the property that f_x is Borel measurable for every $x \in R^n$ and f^y is continuous for every $y \in R^m$, then f is Borel measurable.

4.3 Let E be a dense measurable subset of R^n, and f be an extended real-valued function defined on R^{n+m}. Show that if f_x is Lebesgue measurable for all $x \in E$ and f^y is continuous for almost every $y \in R^m$, then f is Lebesgue measurable.

4.4 Calculate

$$\int_{[0,\infty)} \int_{[0,\sqrt{\pi}]} \frac{x^3 y^3 \cos(y^2)}{(x^4 + y^4)^{3/2}} \, dy \, dx \, .$$

4.5 Given that $f \in L(R)$ and that $\int_R \int_R f(4x) f(x+y) \, dx \, dy = 1$, calculate $\int_R f(x) \, dx$.

4.6 Let f be an extended real-valued measurable function defined on $I = [0,1]$. Show that if the function $F(x,y) = f(x) - f(y) \in L(I \times I)$, then $f \in L(I)$.

4.7 For $t > 0$, let

$$\phi(t) = \begin{cases} 1 & \text{if } 2n \leq t < 2n+1 \text{ for some } n \\ -1 & \text{otherwise,} \end{cases}$$

let $I = [0,1]$, and for $(x,y) \in I \times I$ put $f(x,y) = (1/y)\phi(x/y)$ and $g(x) = \int_I f(x,y) \, dx$. Does $f \in L(I \times I)$? Does $g \in L(I)$?

4.8 Given $f \in L(R)$, let

$$\phi_h(x) = \frac{1}{2h} \int_{x-h}^{x+h} f(t) \, dt \, , \quad h > 0 \, .$$

Prove that $\phi \in L(R)$ and $\int_R |\phi_h(x)| \, dx \leq \|f\|_1$.

4.9 Let F be a closed subset of the line, and let $\delta(x) = \delta(x,F) = \inf\{|x-y| : y \in F\}$ denote the distance of x from F. If $\lambda > 0$ and f is a nonnegative function, $f \in L(R \setminus F)$, prove that the function

$$M_\lambda(f,x) = \int_R \frac{\delta(y)^\lambda f(y)}{|x-y|^{1+\lambda}} \, dy \, , \quad x \in R \, ,$$

is integrable over F, and so finite a.e. there, and ∞ if $x \notin F$. M_λ is called the Marcinkiewicz function corresponding to F, and it is an indispensable tool in the theory of Fourier series. The particular case $f = \chi_I$, where I is a bounded interval of the line, is of interest.

4.10 If f is a nonnegative extended real-valued Borel measurable function on R^{n+m}, show that $\int_{R^n} f(x,y)\,dx$ and $\int_{R^m} f(x,y)\,dy$ are Borel measurable, and

$$\int_{R^{n+m}} f = \int_{R^m}\int_{R^n} f(x,y)\,dx\,dy = \int_{R^n}\int_{R^m} f(x,y)\,dy\,dx\,.$$

4.11 Let E be a domain in the plane bounded by the continuous curves $y = \phi(x)$, $y = \psi(x)$ for $x \in I = [a,b]$, where $\phi(x) < \psi(x)$. Prove that if f is a Borel measurable, integrable function defined on E, then

$$\int_E f = \int_I \int_{[\phi(x),\psi(x)]} f(x,y)\,dy\,dx\,.$$

4.12 Let $I = I(0,1)$ denote the unit interval in R^n, $0 < \eta < n$, and suppose $b(x,y)$ is an essentially bounded function defined on $I \times I$. Show that if $f \in L(I)$, the function

$$F(x) = \int_I \frac{b(x,y)}{|x-y|^\eta} f(y)\,dy\,, \quad x \in I\,,$$

is finite a.e. on I. In fact, $F \in L(I)$. What is an estimate of $\|F\|_1$ in terms of $\|f\|_1$?

4.13 Prove that $L^p(R^n) * L^q(R^n) \subseteq L^\infty(R^n) \cap C(R^n)$, $1 < p,q < \infty$, $1/p + 1/q = 1$.

4.14 Prove that $L^p(R^n) * L^q(R^n) \subseteq L^r(R^n)$, $1 < p,q,r < \infty$, $1 + 1/r = 1/p + 1/q$. This result is known as Young's convolution theorem.

4.15 Suppose $\phi \in L^1(R^n) \cap L^\infty(R^n)$ has the property that $\int_{R^n} \phi = 1$ and $\lim_{|x|\to\infty}(\phi(x)/|x|^n) = 0$. Now, if $f \in L^p(R^n)$, $1 \le p < \infty$, show that at each point of continuity x of f, $\lim_{\varepsilon\to 0} f * \phi_\varepsilon(x) = f(x)$.

4.16 Verify that the assumptions of 4.15 are satisfied on the line by the Poisson kernel $P(x) = (1/\pi)(1/1 + x^2)$, the Fejér kernel $K(x) = (1/\pi)(\sin x/x)^2$, and the Gauss-Weierstrass kernel $W(x) = (1/\sqrt{\pi})e^{-x^2}$.

4.17 Suppose that f is integrable over the shell $0 \le r \le |x| \le R < \infty$, and that $\phi \in C([r,R])$. Then, if $F(\rho) = \int_{\{r\le|x|\le\rho\}} f(x)\phi(|x|)\,dx$, $r \le \rho \le R$, show that

$$\int_{\{r\le|x|\le R\}} f(x)\phi(|x|)\,dx = \int_r^R \phi(\rho)\,dF(\rho)\,,$$

the integral on the right-hand side above being a Riemann-Stieltjes integral.

4.18 Suppose $\phi \in L^1(R^n) \cap L^\infty(R^n)$ has the property that $\int_{R^n}\phi = 1$ and that for some $\eta > 0$, $|\phi(x)|/|x|^{n+\eta} \le c$ for all $|x|$ large. Now, if $f \in L^p(R^n)$, $1 \le p < \infty$, show that at each point x in the Lebesgue set of f, we have $\lim_{\varepsilon\to 0} f * \phi_\varepsilon(x) = f(x)$.

4.19 A sequence $\{\phi_k\} \subseteq L(R^n)$ is called an "approximate unit" if: (a) $\phi_k \ge 0$, for all k, (b) $\|\phi_k\|_1 = 1$, for all k, and (c) For each neighbourhood G of 0 we have $\lim_{k\to\infty}\int_{R^n\setminus G}\phi_k = 0$.
If $\{\phi_k\}$ is an approximate unit, and if $1 \le p < \infty$, prove that $\lim_{k\to\infty}\|f * \phi_k - f\|_p = 0$ for all $f \in L^p(R^n)$.

4.20 Show that $E = R^2 \setminus \{(x,y) \in R^2 : x - y \text{ is rational}\}$ contains no measurable rectangle of positive Lebesgue measure.

4.21 Prove that the operation of convolution is associative in $L(R^n)$. Specifically, if f, g, and h are integrable, show that $f*(g*h)(x) = (f*g)*h(x)$ a.e.

4.22 Suppose f and g are nonnegative integrable functions defined on R^n so that both f and g are strictly positive on some set of positive measure (not necessarily the same for f and g). Prove that $f*g > 0$ on a set of positive measure.

4.23 (Minkowski's Integral Inequality). Under all appropriate measurability conditions on f, show that if $1 \le p < \infty$ we have

$$\left(\int_Y\left(\int_X |f(x,y)|\,d\mu\right)^p d\nu\right)^{1/p} \le \int_X\left(\int_Y |f(x,y)|^p d\nu\right)^{1/p} d\mu\,.$$

If we write this inequality in the form

$$\left\|\int_X |f(\cdot,y)|\,d\nu\right\|_p \le \int_X \|f(x,\cdot)\|_p d\mu\,,$$

then it is also true for $p = \infty$.

4.24 Concerning the example following Theorem 3.3, show that the integral $\int_{I\times I}\chi_E\, d(|\cdot|\times\nu)$, where E is the "diagonal" set given there, is different from either of the iterated integrals.

In what follows $(X,\mathcal{M},\mu)$ and $(Y,\mathcal{N},\nu)$ are measure spaces.

4.25 If λ is a measure on $\mathcal{M}\times\mathcal{N}$ such that $\lambda(A\times B)=\mu(A)\nu(B)$ for all measurable rectangles $A\times B$, show that $\lambda=\mu\times\nu$.

4.26 If the measure spaces involved are complete and σ-finite, and if $\mu\times\nu(E)=0$, show that for every $F\subseteq E$ we have

$$\mu(F^y)=0\ \nu\text{-a.e.}, \quad \text{and} \quad \nu(F_x)=0\ \mu\text{-a.e.}$$

4.27 The measure space $(X\times Y,\mathcal{M}\times\mathcal{N},\mu\times\nu)$ is seldom complete, even when the measure spaces involved are both complete. Prove that this is the case if there exists a set $A\subset X$ such that $A\notin\mathcal{M}$, and a nonempty set $B\in\mathcal{N}$ such that $\nu(B)=0$.
In particular, if μ denotes the Lebesgue measure on the line, $(R^2,\mathcal{L}\times\mathcal{L},\mu\times\mu)$ is an incomplete measure space.

4.28 An alternative statement of the Fubini-Tonelli theorem is the following: Suppose $(X,\mathcal{M},\mu)$ and $(Y,\mathcal{N},\nu)$ are complete σ-finite measure spaces, and let $(X\times Y,\mathcal{F},\lambda)$ denote the completion of $(X\times Y,\mathcal{M}\times\mathcal{N},\mu\times\nu)$, cf. Theorem 3.3 in Chapter V. If f is measurable (with respect to $\mathcal{F}$) and either (a) $f\geq 0$ or (b) $f\in L^1(\lambda)$, then f_x is $\mathcal{N}$-measurable μ-a.e., f^y is $\mathcal{M}$-measurable ν-a.e., and, in case (b) holds, also $f_x\in L^1(\nu)$ and $f^y\in L^1(\mu)$, in the a.e. sense. Moreover, the functions $\int_X f_x\,d\nu$ and $\int_X f^y\,d\mu$ are measurable and

$$\int_{X\times Y} f\,d\lambda=\int_X\int_Y f_x\,d\nu\,d\mu=\int_Y\int_X f^y\,d\mu\,d\nu\,.$$

Prove it.

4.29 The requirement that f be measurable cannot be dispensed for the validity of Fubini's theorem. To see this let $X=Y$ be well-ordered sets with ordinal Ω, $\mathcal{M}=\mathcal{N}$ be the σ-algebra consisting of those sets which are either at most countable or so that their complement is at most countable, and let $\mu=\nu$ be the measure defined for $A\in\mathcal{M}$ by $\mu(A)=0$ if A is at most countable and $\mu(A)=1$ otherwise.
Show that if $E=\{(x,y)\in X\times Y : x\prec y\}$, then E_x and E^y are measurable for all x, y, and that both iterated integrals of χ_E exist and are unequal. Hence $E\notin\mathcal{M}\times\mathcal{N}$, and Fubini's thoerem does not hold in this case.

4.30 If one accepts the Continuum Hypothesis the construction in 4.29 leads to the following situation: There is a subset E of $X = [0,1] \times [0,1]$ such that E_x is at most countable for all $x \in [0,1]$, $[0,1] \setminus E^y$ is at most countable for all $y \in [0,1]$, but E is not Lebesgue measurable.

4.31 The following result describes the behaviour of absolute continuity and singularity with respect to product measures. Let μ and μ^* be σ-finite measures on $(Y, \mathcal{N})$. Prove that if $\mu \ll \mu^*$ and $\nu \ll \nu^*$, we have $\mu \times \nu \ll \mu^* \times \nu^*$ and

$$\frac{d(\mu \times \nu)}{d(\mu^* \times \mu^*)}(x,y) = \frac{d\mu}{d\mu^*}(x)\frac{d\nu}{d\nu^*}(y), \quad \text{all } (x,y) \in X \times Y .$$

Also, if $\mu \perp \mu^*$ or $\nu \perp \nu^*$, then $\mu \times \nu \perp \mu^* \times \nu^*$.

4.32 In the notation of 4.28, prove that if $\mu = \mu_a + \mu_s$ is the Lebesgue decomposition of μ with respect to μ^*, and similarly $\nu = \nu_a + \nu_s$ that of ν with respect to ν^*, then the Lebesgue decomposition of $\mu \times \nu$ is given by $(\mu \times \nu)_a = \mu_a \times \nu_a$ and

$$(\mu \times \nu)_s = (\mu_a \times \nu_s) + (\mu_s \times \nu_a) + (\mu_s \times \nu_s) .$$

4.33 Let μ_1 be a finite Borel measure on R^{n_1}, and μ_2 a finite Borel measure on R^{n_2}. If $\mu_1 \times \mu_2$ is absolutely continuous with respect to the Lebesgue measure λ on $R^{n_1+n_2}$, does it necessarily follow that $d(\mu_1 \times \mu_2)/d\lambda = f \cdot g$, where f is a Lebesgue measurable function defined on R^{n_1}, and g is a Lebesgue measurable function defined on R^{n_2}?

CHAPTER XIV

Normed Spaces and Functionals

In this chapter we study the basic properties of linear spaces, and in particular of those spaces which are normed, and of those which are complete in the metric induced by the norm, or Banach spaces. The existence of continuous linear functionals on these spaces is established by the Hahn-Banach Theorem.

1. NORMED SPACES

The time has come to set up a general framework to address some of the important questions we have posed, including the existence of bounded linear functionals on various linear spaces. We begin by introducing the necessary definitions.

Suppose X is a vector space over the field of real, or complex, scalars; since the theory in both cases follows along similar lines we consider them simultaneously. A scalar valued function defined on X is called a functional.

We are first interested in a particular kind of functional, namely a seminorm. A nonnegative functional p defined on X is called a seminorm provided the following two properties are satisfied:

(i) (Triangle Inequality) $p(x+y) \le p(x) + p(y)$, $\quad x, y \in X$.

(ii) (Absolute Homogeneity) $p(\lambda x) = |\lambda| p(x)$, $\quad \lambda$ scalar, $x \in X$.

Of course, in (ii) above, $|\lambda|$ denotes the absolute value of λ when X is a vector space over the reals and the modulus of λ when the scalar field are the complex numbers.

It follows from (ii) that $p(0) = 0$. We say that the seminorm p is a norm provided that

(iii) (Uniqueness) $p(x) = 0$ implies $x = 0$.

Norms are often denoted by $\|\cdot\|$, or variants thereof. To emphasize that X is endowed with a norm we call X a normed linear space.

We have already encountered many instances of normed linear spaces. The finite-dimensional spaces R^n and C^n may, of course, be normed in different ways. For instance, if $z = (z_1, \ldots, z_n) \in C^n$, then the expressions

$$\|z\|_p = \left(\sum_{i=1}^{n} |z_i|^p\right)^{1/p}, \quad 1 \le p < \infty \tag{1.1}$$

and

$$\|z\|_\infty = \max\{|z_1|, \ldots, |z_n|\} \tag{1.2}$$

are norms on C^n. Observe that

$$\|z\|_\infty \le \|z\|_p, \quad \text{and} \quad \|z\|_p \le n^{1/p}\|z\|_\infty, \quad \text{all } z \in C^n.$$

In general, if $\|\cdot\|_1$ and $\|\cdot\|_2$ are norms on X, we say that $\|\cdot\|_1$ is weaker than $\|\cdot\|_2$ if for some constant k we have $\|x\|_1 \le k\|x\|_2$ for all $x \in X$. We also say that the norms are equivalent if we have both

$$\|x\|_1 \le k\,\|x\|_2 \quad \text{and} \quad \|x\|_2 \le k'\,\|x\|_1, \quad \text{all } x \in X.$$

All norms on C^n are equivalent. On the other hand, if I is a bounded interval of R^n, then the L^p-norm on $L^p(I)$ is weaker than the L^q-norm on $L^p(I)$ iff $p \le q$, and consequently, these norms are not equivalent.

In the general context of function spaces defined on subsets of R^n there are other instances of normed linear spaces that are of interest to us. For example, the space $B(I)$ consisting of those real, or complex-valued, bounded functions f defined on I may be normed by

$$\|f\| = \max_{x \in I} |f(x)|. \tag{1.3}$$

Clearly this expression is also a norm on the subspace $C(I)$ of $B(I)$ consisting of those functions which are continuous.

Now, if for a multi-index $\alpha = (\alpha_1, \ldots, \alpha_n)$ of nonnegative integers we let $|\alpha| = \alpha_1 + \cdots + \alpha_n$, and

$$D^\alpha f = \frac{\partial^{|\alpha|} f}{\partial x_1^{\alpha_1} \ldots \partial x_n^{\alpha_n}},$$

then $C^k(I) = \{f \in C(I) : D^\alpha f \in C(I), |\alpha| \le k\}$ may be normed by

$$\|f\|_k = \sum_{|\alpha| \le k} \|D^\alpha f\|. \tag{1.4}$$

Another example is the class of BV functions defined on an interval $I = [a,b]$ of R; it is not hard to see that the expression

$$\|f\| = V(f;a,b) + |f(a)|, \tag{1.5}$$

is a norm.

Note that if X is a normed linear space, the function d on $X \times X$ given by

$$d(x,y) = \|x - y\|, \quad x, y \in X,$$

defines a metric on X. For, $d(x,y) = 0$ implies $\|x-y\| = 0$, which in view of (iii) above is equivalent to $x = y$. The symmetry of d is obvious from the definition. Finally, the triangle inequality is a simple consequence of (i):

$$d(x,y) = \|x-y\| \le \|x-z\| + \|x-y\| = d(x,z) + d(z,y), \quad \text{all } x,y,z \in X.$$

d is called the metric induced by the norm $\|\cdot\|$.

Among the normed linear spaces, a particularly important role is played by those spaces which are complete metric spaces in the metric induced by the norm; these are the so-called Banach spaces. For instance, $C(I)$ normed by (1.3) is a Banach space, but it is not a Banach space when normed by $\|\cdot\|_1$, the Lebesgue L^1 norm. Now, $C(I)$ normed with $\|\cdot\|_1$ is densely embedded in the Banach space $L^1(I)$ and an interesting question to ponder is whether any normed linear space which is not complete may be densely embedded in a Banach space; we will answer this question in the next section.

In the meantime, inspired by the proof of Theorem 1.3 in Chapter XII, we consider a useful criterion to decide when a normed linear space X is complete. Observe that in a normed linear space X it is possible to assign a sum s to the (formal) series

$$\sum_{n=1}^{\infty} x_n = x_1 + \cdots + x_n + \cdots, \quad x_n \in X, \text{all } n.$$

Indeed, the series $\sum x_n$ is said to converge to the sum s if the sequence $\{s_m\}$ of the partial sums $s_m = x_1 + \cdots + x_m$, $m = 1, 2, \ldots$, converges to s in the norm, or metric, of X. Along the same lines we say that the series $\sum x_n$ is "absolutely convergent" if the numerical series

$$\sum_{n=1}^{\infty} \|x_n\| = \|x_1\| + \cdots + \|x_n\| + \cdots \tag{1.6}$$

converges. It is not hard to see that in a Banach space every absolutely convergent series converges. For this it suffices to verify that the sequence of partial sums $\{s_n\}$ is Cauchy in the metric of X. First observe that if $m > n$, then $s_m - s_n = x_{n+1} + \cdots + x_m$, and consequently, by (a simple extension of) (i) above, we also have

$$\|s_m - s_n\| \leq \|x_{n+1}\| + \cdots + \|x_m\| . \tag{1.7}$$

Since the series in (1.6) converges, the right-hand side of (1.7) can be made as small as desired provided n is sufficiently large; thus $\{s_m\}$ is Cauchy in X and consequently, convergent to a limit $s \in X$, say. Furthermore, by the continuity of the norm, cf. 4.7 below, we also have $\|s\| \leq \sum_{n=1}^{\infty} \|x_n\|$.

These remarks lead to the following useful result.

Theorem 1.1. Let X be a normed linear space. Then X is a Banach space iff every absolutely convergent series converges.

Proof. It only remains to show that the condition is sufficient. Let $\{x_n\}$ be a Cauchy sequence of elements of X and choose a sequence $n_{k+1} > n_k$, $k = 1, 2, \ldots$, such that

$$\|x_n - x_m\| \leq 1/2^k , \quad \text{all } m, n \geq n_k , \quad k = 1, 2, \ldots$$

Put $x_{n_0} = 0$, and let $y_k = x_{n_k} - x_{n_{k-1}}$ for $k = 1, 2, \ldots$ By the above estimate it follows that $\sum_{k=1}^{\infty} \|y_k\| \leq \|x_{n_1}\| + \sum_k 2^{-k} < \infty$, and the series $\sum y_k$ is absolutely convergent. By assumption this series converges in X, and since partial sums $\sum_{k=1}^{m} y_k$ of the series equal x_{n_m}, the subsequence $\{x_{n_m}\}$ converges in X. In order to complete the proof it suffices to invoke the well-known fact that if a Cauchy sequence in a metric space has a convergent subsequence, then the sequence itself converges to the same limit. ■

As an illustration of this result we show that normed by (1.5) the space of BV functions on an interval $I = [a,b]$ is a Banach space. Indeed, let $\{f_n\}$ be a sequence of BV functions on I such that $\sum \|f_n\| < \infty$. In particular, for $a \leq x \leq b$ we have

$$\sum V(f_n; a, x) < \infty , \quad \text{and} \quad \sum |f_n(a)| < \infty ,$$

and consequently by (1.5) in Chapter III it follows that

$$\sum |f_n(x)| \leq \sum V(f_n; a, x) + \sum |f_n(a)| < \infty , \quad \text{all } x \in I .$$

Now, since the series with terms $\{f_n(x)\}$ converges absolutely for each $x \in I$, it also converges there. Let $f(x) = \sum f_n(x)$ denote its sum; we must show that f is BV on I, and that $\lim_{m\to\infty} \| \sum_{n=1}^m f_n - f\| = 0$.

First observe that since in summing a series with nonnegative terms we may interchange freely the order of summation, given a partition $\mathcal{P}$ of I we have

$$\sum_{\text{over } \mathcal{P}} |f(x_{k+1}) - f(x_k)| \le \sum_n \sum_{\text{over } \mathcal{P}} |f_n(x_{k+1}) - f_n(x_k)|$$
$$\le \sum_n V(f_n; a, b) \le \sum \|f_n\| < \infty\,,$$

and consequently f is BV on I, and $V(f; a, b) \le \sum V(f_n; a, b)$.

Moreover, since $|f(a) - \sum_{n-1}^m f_n(a)| \to 0$ as $m \to \infty$, and

$$V\left(f - \sum\nolimits_{n=1}^m f_n; a, b\right) \le \sum\nolimits_{n=m+1}^\infty V(f_n; a, b) \to 0\,, \quad \text{as } m \to \infty\,,$$

it readily follows that $\|f - \sum_{n=1}^m f_n\| \to 0$ as $m \to \infty$, and we are done.

The advantage of the above argument is apparent: We showed that a normed linear space was complete without dealing with Cauchy sequences, which are often difficult to handle.

2. THE HAHN–BANACH THEOREM

A functional L defined on a linear space X is said to be linear if for every $x, y \in X$ and scalar λ, we have

$$L(x + \lambda y) = Lx + \lambda Ly\,. \tag{2.1}$$

When X is finite dimensional, the conjugate, or dual, space X^* consisting of all the linear functionals on X, plays an important role in the development of the general theory of linear spaces. One of the main results is that the natural embedding of X into X^{**} is an isomorphism. In general, it is not possible to extend this result to infinite dimensional linear spaces; in fact, the result itself is not quite as relevant because most of the functionals that arise in concrete analytic situations and examples are also bounded. This requires, of course, that X be normed and we make this clear in our next definition.

We say that a linear functional L defined on a normed linear space X is bounded if there is a constant c, independent of $x \in X$, so that

$$|Lx| \le c\|x\|\,, \quad \text{all } x \in X\,. \tag{2.2}$$

The study of bounded linear functionals originated in two closely related areas: The solution of linear systems of infinitely many equations with infinitely many unknowns, and the so-called summability methods of divergent series, in particular, the Fourier series of integrable functions.

Now, even when the boundedness condition is imposed, it is found that the duality theory for infinite dimensional linear spaces is more complex than the finite dimensional case. We have already encountered some of these results as we examined the theory of the Lebesgue L^p spaces, $0 < p \leq \infty$, in Chapter XII.

The Hahn-Banach Theorem, or theorems actually, are an indispensable tool in the theory of duality. In the case of arbitrary linear spaces, where no topology is apparent, the Hahn-Banach Theorem assures a plentiful supply of linear functionals, and in the case of normed linear spaces, under some general "domination" assumptions, a supply of bounded linear functionals. To elucidate this point we discuss first a simple question of geometric nature.

We say that a subset C of a normed linear space X is convex if for every $x, y \in C$, the set $\{\eta x + (1-\eta)y : 0 \leq \eta \leq 1\}$ is contained in C. The question of interest to us may be loosely stated as follows: If $C \neq R^2$ is a closed convex subset of the plane which contains the origin as an interior point, can we draw a line so that C lies entirely on one side of the line? The answer to this question is not intuitively obvious. First some definitions.

Given a convex subset C of a normed linear space X which contains 0 as an interior point, there is a natural functional, called the Minkowski functional of C, which is denoted by p_C, and which satisfies the following properties:

(i) p_C is nonnegative and finite everywhere.

(ii) (Positive homogeneity) For each $x \in X$ and $\eta \geq 0$ we have

$$p_C(\eta x) = \eta p_C(x)\,.$$

(iii) (Triangle Inequality) For any $x, y \in X$ we have

$$p_C(x+y) \leq p_C(x) + p_C(y)\,.$$

(iv) For every $x \in X \setminus C$ we have $p_C(x) \geq 1$.

This is how we go about defining p_C: Given $x \in X$, put

$$p_C(x) = \inf\{1/\lambda : \lambda > 0 \text{ and } \lambda x \in C\}\,. \tag{2.3}$$

We claim that p_C satisfies (i)–(iv) above. First note that since by the continuity of the norm $\|\lambda x\| \to 0$ as $\lambda \to 0$, and since 0 is an interior point of C, $\lambda x \in C$ for sufficiently small λ. Thus the inf in (2.3) is finite and (i) holds. As for (ii), since $\lambda 0 \in C$ for every λ, it follows that $p_C(0) = 0$, and consequently, we may assume that $\eta \neq 0$ and $x \neq 0$. Now,

$$\begin{aligned} p_C(\eta x) &= \inf\{1/\lambda : \lambda > 0,\ \lambda\eta x \in C\} = \inf\{\eta/\lambda : \lambda > 0,\ \lambda x \in C\} \\ &= \eta \inf\{1/\lambda : \lambda > 0,\ \lambda x \in C\} = \eta p_C(x)\,, \end{aligned}$$

which is precisely (ii). To prove (iii), given $\varepsilon > 0$, let $\mu, \nu > 0$, be such that

$$\mu x \in C\,, \quad 1/\mu \leq p_C(x) + \varepsilon/2\,, \tag{2.4}$$

and

$$\nu y \in C\,, \quad 1/\nu \leq p_C(y) + \varepsilon/2\,, \tag{2.5}$$

and put

$$1/\lambda = 1/\mu + 1/\nu\,. \tag{2.6}$$

Now, since $0 < \eta = \lambda/\mu < 1$ and C is convex, it readily follows that $\lambda(x+y) = \eta(\mu x) + (1-\eta)(\nu y) \in C$, and by (2.3), (2.6), (2.4) and (2.5) we have $p_C(x+y) \leq 1/\lambda \leq p_C(x) + p_C(y) + \varepsilon$. But $\varepsilon > 0$ is arbitrary, and consequently, (iii) holds. Finally, if $x \in X \setminus C$, since C is convex we cannot have $\lambda x \in C$ for some $\lambda \geq 1$. Whence, if $\lambda x \in C$, it follows that $\lambda < 1$ and, as asserted, $p_C(x) \geq 1$, thus proving (iv).

By means of the Minkowski functional we may answer the question posed above concerning convex sets of the plan. The idea, after some simple arguments, is to consider the linear functional L defined on the one-dimensional subspace of X consisting of all elements of the form λx_0, $x_0 \notin C$, by the formula $L(\lambda x_0) = \lambda p_{C_1}(x_0)$, where C_1 is a convex set related to C, and observing that in this case $L(\lambda x_0) \leq p_{C_1}(\lambda x_0)$ for all real λ. This expression exhibits the domination alluded to above, and the question is whether L can be extended to the plane satisfying the same inequality.

We make these remarks precise with the aid of the following result.

Theorem 2.1 (Hahn-Banach Theorem). Suppose X is a real linear space and p is a functional on X which satisfies the triangular inequality, and so that $p(\lambda x) = \lambda p(x)$ for all $x \in X$ and $\lambda > 0$. Further, let X_0 be a linear subspace of X and L_0 a linear functional on X_0 such that

$$L_0 x \leq p(x)\,, \quad \text{all } x \in X_0\,. \tag{2.7}$$

Then there is a linear functional L defined on X that extends L_0, i.e., $Lx = L_0 x$ for $x \in X_0$, and so that

$$Lx \leq p(x)\,, \quad \text{all } x \in X\,. \tag{2.8}$$

Proof. The idea of the proof is to invoke Zorn's Lemma to construct a maximal extension of L_0, and then to show that this extension satisfies (2.8). Let $\mathcal{X}$ be the collection of all pairs of the form (Y, L) where

(i) Y is a linear subspace of X and $X_0 \subseteq Y \subseteq X$.

(ii) L is a linear functional on Y, $L|X_0 = L_0$, and $Lx \le p(x)$ for all $x \in Y$.

Note that $(X_0, L_0) \in \mathcal{X}$. On $\mathcal{X}$ we introduce a partial ordering as follows: We say that (Y, L) precedes (Y', L'), and we write $(Y, L) \prec (Y', L')$, if $Y \subseteq Y'$ and $L'|Y = L$.

In order to apply Zorn's Lemma we must first check that any linearly ordered family $\{(Y_s, L_s)\}$, say, of elements in $\mathcal{X}$ has an upper bound. But this is not hard: Indeed, put $Y = \bigcup_s Y_s$, and consider the functional L on Y so that $L|Y_s = L_s$. Since the family is ordered it readily follows that (Y, L) is an upper bound, and we are in a position to invoke the conclusion of Zorn's Lemma, to wit, $\mathcal{X}$ has a maximal element (X_1, L_1), say.

There are two possibilities: Either $X_1 = X$, and in this case we are done, or else X_1 is a proper linear subspace of X. Next we show that the latter possibility does not occur, for otherwise we would reach a contradiction. Indeed, if the latter possibility occurs, let $x_0 \in X \setminus X_1$ and consider the linear subspace X_2 of X spanned by X_1 and $\{x_0\}$, i.e., X_2 consists of all linear combinations of the form $x_1 + \lambda x_0$, where $x_1 \in X_1$ and λ is a real number. We claim that L_1 may be extended to a linear functional L_2 on X_2 which satisfies $L_2 x \le p(x)$ for all $x \in X_2$, thus contradicting the maximality of (Y, L).

Denote by L_2 a candidate for such an extension of L to X_2, and observe that if $L_2 x_0 = \eta$, an arbitrary real scalar, we have

$$L_2(x_1 + \lambda x_0) = L_2 x_1 + \lambda\eta = L x_1 + \lambda\eta \le p(x_1) + \lambda\eta\,.$$

If we can produce a scalar η so that for all x_1 in X_1 and scalars λ the inequality

$$p(x_1) + \lambda\eta \le p(x_1 + \lambda x_0) \tag{2.9}$$

is true, it then follows that $(X_2, L_2) \in \mathcal{X}$, and that (Y, L) strictly precedes it, thus contradicting its assumed maximality.

Observe that if (2.9) holds, we also have

$$\eta \le (p(x_1 + \lambda x_0) - p(x_1))/\lambda\,, \quad \lambda > 0 \tag{2.10}$$

and

$$\eta \ge (p(x_1 + \lambda x_0) - p(x_1))/\lambda\,, \quad \lambda < 0\,. \tag{2.11}$$

By setting $\lambda = -\mu, \mu > 0$, in (2.11), (2.10) and (2.11) may be combined into the single expression

$$\frac{p(x_1) - p(x_1 - \mu x_0)}{\mu} \leq \eta \leq \frac{p(x_1 + \lambda x_0) - p(x_1)}{\lambda}, \tag{2.12}$$

which should now hold for all $\lambda, \mu > 0$.

Thus, the existence of η is equivalent to the validity of the inequality

$$\frac{p(x_1) - p(x_1 - \mu x_0)}{\mu} \leq \frac{p(x_1 - \lambda x_0) - p(x_1)}{\lambda}, \quad \text{all } \lambda, \mu > 0, \tag{2.13}$$

for then η may be chosen to be any real number lying between the sup of the left-hand side of (2.13) and the inf of the right-hand side there.

To show that (2.13) holds is not hard. First observe that it is equivalent to

$$p(x_1) \leq \frac{\mu p(x_1 + \lambda x_0) + \lambda p(x_1 - \mu x_0)}{(\lambda + \mu)}. \tag{2.14}$$

Next note that since

$$x_1 = \frac{\mu x_1 + \lambda \mu x_0}{(\lambda + \mu)} + \frac{\lambda x_1 - \lambda \mu x_0}{(\lambda + \mu)},$$

and since p is subadditive and positively homogeneous, it follows that

$$p(x_1) \leq \frac{\mu}{(\lambda + \mu)} p(x_1 + \lambda x_0) + \frac{\lambda}{(\lambda + \mu)} p(x_1 - \mu x_0),$$

which is precisely (2.14). Thus, reversing the steps, also (2.13) holds and L can be extended to a subspace of X containing Y and satisfying (2.8), thus contradicting the maximality of (Y, L). Whence Y is actually X, and the proof is complete. ∎

Next we consider the Hahn-Banach Theorem for complex linear spaces, the proof presented here is due to Bohnenblust and Sobczyk. We begin by exploring the relationship between real and complex functionals.

Lemma 2.2. Suppose X is a complex linear space, and let L be a (complex) linear functional defined on X. Then $L_1 x = \Re(Lx)$ is a real linear functional defined on X and

$$Lx = L_1 x - iL_1(ix), \quad \text{all } x \in X. \tag{2.15}$$

Conversely, if L_1 is a real linear functional defined on X, then the functional L defined by (2.15) is a complex linear functional on X.

Proof. That L_1 is a real functional on X if L is a complex functional on X is a simple verification left to the reader. Now, since for any complex number z we have that $\Re(iz) = -\Im(z)$, it readily follows that

$$\begin{aligned} Lx &= \Re(Lx) + i\Im(Lx) = L_1x + i(-\Re(iLx)) \\ &= L_1x - i\Re(L(ix)) = L_1x - iL_1(ix), \end{aligned}$$

and (2.15) follows.

Finally, if L_1 is a real functional defined on X and L is given by (2.15), in order to show that L is a complex linear functional on X it suffices to check that $L(ix) = iLx$ for all $x \in X$. But

$$\begin{aligned} L(ix) &= L_1(ix) - iL_1(i(ix)) = L_1(ix) - iL_1(-x) \\ &= i(L_1x - iL_1(ix)) = iLx . \quad \blacksquare \end{aligned}$$

We are now ready to present the complex version of the Hahn-Banach Theorem.

Theorem 2.3. Let X be a complex linear space, p a seminorm on X, X_0 a linear subspace of X and L_0 a complex linear functional defined on X_0 such that

$$|L_0x| \le p(x), \quad \text{all } x \in X_0 . \tag{2.16}$$

Then there is a linear functional L defined on X which extends L_0, i.e., $Lx = L_0x$ for $x \in X_0$, and so that

$$|Lx| \le p(x), \quad \text{all } x \in X . \tag{2.17}$$

Proof. Let $L_1 = \Re L_0$; by Proposition 2.2, L_1 is a real linear functional defined on X_0, and by (2.16) we have $L_1x \le |L_0x| \le p(x)$ for $x \in X_0$. We are now in a position to invoke Theorem 2.1 and extend L_1 to a real linear functional L_2 defined on X with the property that $L_2x = L_1x$, $x \in X_0$, and

$$L_2x \le p(x), \quad \text{all } x \in X . \tag{2.18}$$

Since p is a seminorm, replacing x by $-x$ if necessary in (2.18), we note that we have $|L_2x| \le p(x)$ as well. Inspired by Lemma 2.2, let

$$Lx = L_2x - iL_2(ix) .$$

L is a complex linear functional defined on X, and since L_2 extends L_1, the restriction of L to X_0 coincides with L_0. It only remains to check (2.17):

Since for each $x \in X$ so that $Lx \neq 0$ we have $|Lx| = \lambda Lx = L(\lambda x)$, where $\lambda = \overline{Lx}/|Lx|$ is complex number of modulus 1, it follows that $L(\lambda x)$ is real, and consequently we also have

$$|Lx| = L(\lambda x) = L_2(\lambda x) \leq p(\lambda x) = p(x),$$

and (2.17) holds. ■

We focus our discussion next in the normed linear spaces; we begin with some definitions.

A functional L defined on a normed linear space X is said to be continuous if

$$\|x_n - x\| \to 0 \quad \text{implies} \quad |Lx_n - Lx| \to 0.$$

For linear functionals L, which are the functionals of interest to us, the notion of continuity is equivalent to that of continuity at a single point of X. For, suppose that L is continuous at a point x_0 of X and let $x_n \to x \in X$. Then we have $x_n - x + x_0 \to x_0$, and consequently, $|L(x_n - x + x_0) - Lx_0| \to 0$. But, since L is linear, it is obvious that $L(x_n - x + x_0) - Lx_0 = Lx_n - Lx$, and our assertion follows.

Also, for linear functionals, the concepts of boundedness and continuity are interchangeable.

Proposition 2.4. Suppose L is a linear functional on a normed linear space X. Then L is bounded iff L is continuous.

Proof. Suppose first that L is bounded; since L is also linear we have

$$|Lx_n - Lx| = |L(x_n - x)| \leq c\|x_n - x\|, \tag{2.19}$$

and the right-hand side, and consequently also the left-hand side, of (2.19) goes to 0 with $\|x_n - x\|$. Whence, L is continuous.

Conversely, suppose that L is a continuous linear functional on X which is not bounded. Then, by (2.2), for each positive integer n there is $y_n \neq 0$ in X, so that $|Ly_n| > n\|y_n\|$. Put $x_n = (1/n\|y_n\|)y_n$, and observe that the sequence $\{x_n\} \subseteq X$ satisfies

$$\|x_n\| \to 0 \quad \text{and yet} \quad |Lx_n| > 1 \quad \text{for all } n.$$

But this is not possible if L is continuous. ■

Although not intuitively apparent, there are linear functionals that are not bounded. To see this consider an infinite dimensional linear space

X and, referring to Section 3 in Chapter II, let H be a Hamel basis for X over the ambient scalar field. It is a straightforward application of Zorn's Lemma to prove that any linearly independent subset of X is contained in a Hamel basis for X. In particular, any linear space has a Hamel basis. Now, for each x in X we can find a unique elements $h_1, \ldots, h_n$ in H and scalars $\lambda_1, \ldots, \lambda_n$, say, such that $x = \sum_{i=1}^n \lambda_i x_i$. Define $\|x\|_\infty$ to be the maximum of the the numbers λ_i, $i = 1, \ldots, n$. Clearly $\|x\|_\infty$ is a norm in X, and consequently, any linear space over the real or complex scalar field can be given a norm.

There is an interesting case for which a Hamel basis can be exhibited. Let $I = [0,1]$ and let $\mathcal{P}(I) \subset C(I)$ denote the class of all polynomial functions on I. Then $H = \{1, x, x^2, \ldots\}$ is a Hamel basis for $\mathcal{P}(I)$. Let now H_1 be a Hamel basis for $C(I)$ that contains H and choose any element $h_1 \in H_1 \setminus H$. Put $Lh_1 = 1$ and $Lh = 0$ for all h in H_1, $h \neq h_1$, and extend L to all of $C(I)$ by requiring that it be linear. It is clear that L cannot be continuous with respect to the uniform norm on $C(I)$. Indeed, if this were the case, then by 4.15 below the set $\{f \in C(I) : Lf = 0\}$ would be a closed subspace of $C(I)$. But this set contains $\mathcal{P}(I)$, which, by the Weierstrass theorem, cf. Corollary 2.3 in Chapter XVII, is dense in $C(I)$. Hence if L were continuous, it would have to be identically 0, contrary to the fact that $Lh_1 = 1$.

We are now ready to introduce the conjugate, or dual, space X^* of a normed linear space X, i.e., the space consisting of all continuous linear functionals defined on X. More precisely, given a normed linear space X, let

$$X^* = \{L : L \text{ is a continuous linear functional on } X\}.$$

It is readily seen that X^* is itself a linear space over the scalar field of X; $L_1 + \lambda L_2$ is defined as the continuous linear functional on X given by

$$(L_1 + \lambda L_2)(x) = L_1 x + \lambda L_2 x\,, \quad \text{all } x \in X\,.$$

We also have

Proposition 2.5. Suppose X is a normed space. Then X^*, normed by

$$\|L\| = \sup_{x \neq 0} \frac{|Lx|}{\|x\|}\,, \tag{2.20}$$

is a Banach space.

Proof. It is clear that the expression in (2.20) is a seminorm on X^*. Now, if $\|L\| = 0$, it follows that $|Lx| = 0$ for each $x \in X$, L is the 0 functional and consequently, $\|L\|$ is a norm on X^*.

To show that normed by (2.20) X^* becomes a Banach space, by Theorem 1.1 it suffices to prove that if $\sum \|L_n\| < \infty$, then $\sum L_n$ converges in X^*. First observe that for each $x \in X$ we have

$$\sum |L_n x| \le \left(\sum \|L_n\|\right) \|x\| < \infty . \tag{2.21}$$

Thus the numerical series with terms $(L_n x)$ converges absolutely for each x in X, and since the scalar field is complete, also $\sum L_n x$ converges, even unconditionally, to a sum Lx, say. First we show that L is a bounded linear functional on X. Indeed, given x, y in X and a scalar λ, we have

$$\begin{aligned} L(x + \lambda y) &= \sum L_n(x + \lambda y) = \lim_{m\to\infty} \sum_{n=1}^{m} L_n(x + \lambda y) \\ &= \lim_{m\to\infty} \sum_{n=1}^{m} (L_n x + \lambda L_n y) = Lx + \lambda Ly , \end{aligned}$$

and the linearity of L follows. Moreover, by (2.21) it also readily follows that $|Lx|/\|x\| \le \sum \|L_n\|$, $x \ne 0$, and consequently, L is bounded. Finally, since for $x \in X$ we have

$$\left| Lx - \sum_{n=1}^{m} L_n x \right| \le \left(\sum_{n=m+1}^{\infty} \|L_n\|\right) \|x\| ,$$

we get that

$$\left\| L - \sum_{n=1}^{m} L_n \right\| \le \sum_{n=m+1}^{\infty} \|L_n\| \to 0 , \quad \text{as } m \to \infty ,$$

and $L = \lim_{m\to\infty} \sum_{n=1}^{m} L_n$ (in X^*). Since all the assumptions of Theorem 1.1 are now satisfied, we get that X^* is complete. ■

It is interesting to point out that the conclusion of Proposition 2.5 holds whether X itself is complete or not, as the proof only makes use of the completeness of the field of scalars.

After this brief digression we turn to prove a version of the Hahn-Banach Theorem that deals with continuous linear functionals.

Theorem 2.6 (Hahn-Banach Theorem). Suppose X is a normed linear space, and let L_0 be a bounded linear functional defined on a subspace X_0 of X. Then there exists a bounded linear functional L defined on X such that

$$L|X_0 = L_0 \quad \text{and} \quad \|L\| = \|L_0\| . \tag{2.22}$$

Proof. We consider first the case when X is a real linear space. Since L_0 is a bounded linear functional on X_0, it follows that

$$L_0 x \leq |L_0 x| \leq \|L_0\| \, \|x\| , \quad \text{all } x \in X_0 . \tag{2.23}$$

Note that the expression on the right-hand side of (2.22) may be thought of as a seminorm on X. More precisely, if for x in X we put $p(x) = \|L_0\| \, \|x\|$, then p is a seminorm on X, and (2.23) actually states that the assumptions of Theorem 2.1 are satisfied. By the conclusion of that theorem there is a linear functional L defined on X such that

$$Lx = L_0 x , \quad x \in X_0, \quad \text{and} \quad Lx \leq p(x), \quad \text{all } x \in X . \tag{2.24}$$

The estimate in (2.24) may be rewritten as

$$Lx \leq \|L_0\| \, \|x\| , \tag{2.25}$$

and since L is linear we also have

$$-Lx = L(-x) \leq \|L_0\| \, \| - x\| = \|L_0\| \, \|x\| . \tag{2.26}$$

Thus combining (2.25) and (2.26) it follows that $|Lx| \leq \|L_0\| \, \|x\|$ for all $x \in X$, and consequently, $\|L\| \leq \|L_0\|$. Furthermore, since the restriction of L to X_0 is L_0, we also have

$$\|L\| \geq \sup_{x \neq 0, x \in X_0} \frac{|L_0 x|}{\|x\|} = \|L_0\| ,$$

and $\|L\| = \|L_0\|$. This completes the proof in the real case. As for the complex case, the proof follows along similar lines once we invoke Theorem 2.3. ■

Many important topics in the theory of linear spaces rely on the notion of convexity; as a first application of the Hahn-Banach Theorem we formalize the discussion preceeding Theorem 2.1; first a definition.

Given subsets X_0 and X_1 of a linear space X, a linear functional L defined on X is said to separate X_0 and X_1 if

$$\sup_{x \in X_1} Lx \leq \inf_{x \in X_0} Lx .$$

The lack of symmetry in this definition is only apparent as the roles of X_0 and X_1 are interchanged when L is replaced by $-L$. It follows at once from this definition that L separates X_0 and X_1 iff L separates $X_0 - X_1 = \{z : z = x_0 - x_1, x_0 \in X_0, x_1 \in X_1\}$ and $\{0\}$ iff L separates $X_0 - x = \{z : z = x_0 - x, x_0 \in X_0\}$ and $X_1 - x$ for every $x \in X$.

We then have

Theorem 2.7. Let C_0, C_1 be two disjoint, nonempty convex subsets of a real normed linear space X, and suppose that at least one of the sets, C_0 say, has a nonempty interior. Then there exists a nontrivial linear functional L on X that separates C_0 and C_1.

Proof. Let x_0 be an interior point to C_0; by considering if necessary $C_0 - x_0$ and $C_1 - x_0$, which are also convex, we may assume that 0 is an interior point to C_0. Let x_1 be a point of C_1, then $-x_1$ is an interior point to the convex set $C_0 - C_1 = \{z : x = x - y, x \in C_0, y \in C_1\}$ and 0 is an interior point to the convex set $C = C_0 - C_1 + x_1 = \{x : z = x + x_1, x \in C_0 - C_1\}$. Moreover, since C_0 and C_1 are disjoint we also have

$$0 \notin C_0 - C_1\,, \quad x_1 \notin C = C_0 - C_1 + x_1\,. \tag{2.27}$$

Let p_C be the Minkowski functional corresponding to C; from (2.27) it follows that $p_C(x_1) \geq 1$.

Let $X_1 = \{x_1\}$ be the one-dimensional subspace of X spanned by x_1; X_1 consists of all elements of the form λx_1, λ real, and consider the linear functional L_1 defined on X_1 by

$$L_1(\lambda x_1) = \lambda p_C(x_1)\,, \quad \lambda \text{ real}\,.$$

Since $p_C(\lambda x_1) = \lambda p_C(x_1)$ if $\lambda \geq 0$, while

$$L_1(\lambda x_1) = \lambda L_1 x_1 < 0 \leq p_C(\lambda x_1)\,, \quad \text{if } \lambda < 0\,,$$

we also have $L_1(\lambda x_1) \leq p_C(\lambda x_1)$, for all real λ.

We are now in a position to invoke the Hahn-Banach Theorem, and extend L_1 to a linear functional L defined on the whole space X satisfying the condition

$$Lx \leq p_C(x)\,, \quad \text{all } x \in X\,. \tag{2.28}$$

Since $p_C(x) \leq 1$ on C, while $Lx_1 = L_1 x_1 \geq 1$, by (2.28) it follows that L separates C and $\{x_1\}$. But as observed above this is equivalent to the statement that L separates $C_0 - C_1$ and $\{0\}$, which is in turn equivalent to the fact that L separates C_0 and C_1. ■

We discuss next further applications of the Hahn-Banach Theorem to different settings.

3. APPLICATIONS

We begin by discussing three interesting applications of the Hahn-Banach Theorem: The determination of when a linear subspace is dense in a linear space, the general form of the converse to Hölder's inequality, and the construction of a natural embedding of a normed space into a Banach space.

First we prove

Proposition 3.1. Let Y be a linear subspace of a normed linear space X, and suppose $x \in X$ is such that $d(x,Y) = \inf_{y\in Y} \|x-y\| = \eta > 0$. Then there is a bounded linear functional L on X with norm $\|L\| = 1/\eta$ which separates x from Y. More precisely, we have

$$Lx = 1, \quad \text{and} \quad Ly = 0 \quad \text{for all } y \in Y.$$

Proof. Let Y_1 be the subspace of X spanned by Y and $\{x\}$; each element of y_1 of Y_1 can be written uniquely as $y_1 = y + \lambda x$, with $y \in Y$ and a scalar λ. Now, if $y_1 = y + \lambda x$, note that

$$|\lambda| \le \|y_1\|/\eta. \tag{3.1}$$

Indeed, if $\lambda = 0$ there is nothing to prove. Otherwise, if $\lambda \neq 0$, since $(-1/\lambda)y \in Y$, it readily follows that $\eta \le \|y_1\|/|\lambda|$, and (3.1) holds.

We define now the linear functional L_1 on Y_1 as follows: If $y_1 = y + \lambda x$, then put $L_1 y_1 = \lambda$. By (3.1) it follows that $|L_1 y_1| \le \|y_1\|/\eta$, and consequently, we have $\|L_1\| \le 1/\eta$. To show that equality actually holds here let $\{y_n\} \subseteq Y$ be such that $\lim_{n\to\infty} \|x - y_n\| = \eta$. It is clear that

$$1 = L_1(x - y_n) \le \|L_1\| \, \|x - y_n\|$$

where the right-hand side above tends to $\|L_1\|\eta$ as $n \to \infty$. Thus we also have $1/\eta \le \|L_1\|$, and consequently, $\|L_1\| = 1/\eta$.

We are now in a position to invoke Theorem 2.6. By that result there exists a linear functional L defined on X with $\|L\| = 1/\eta$ that extends L_1. Since it is also clear that $Lx = L_1 x = 1$ and $Ly = L_1 y = 0$ for $y \in Y$, the functional L does the job. ∎

Corollary 3.2. Let X be a normed linear space. For any $0 \neq x \in X$ there exists a linear functional L defined on X with $\|L\| = 1$ and such that $Lx = \|x\|$. In particular, if x and y are distinct points of X, there exists $L \in X^*$ such that $Lx \neq Ly$.

Proof. Suppose $x \neq 0$. Then by Proposition 3.1, with $Y = \{0\}$ there, there exists a functional $L' \in X^*$ such that $\|L'\| = 1/\|x\|$ and $L'x = 1$. The first part of the conclusion follows now upon setting $L = \|x\|L'$. As for the second part, it follows from the first with x replaced by $x - y \neq 0$. ■

Next we show a "density" result, it roughly states that if Y is a dense subspace of X, then the only bounded linear functional that vanishes on Y is the trivial, or zero, functional.

Proposition 3.3. Suppose X is a normed linear space, and let Y be a subspace of X which is not dense in X. Then there exists a nontrivial linear functional L defined on X which vanishes on Y.

Proof. Since Y is not dense in X, there is $x \in X$ which satisfies $\inf_{y \in Y} \|x - y\| > 0$. To obtain L apply now Proposition 3.1. ■

The next result we discuss is an extension to the converse to Hölder's inequality in the spirit of Proposition 2.3 in Chapter XII.

Proposition 3.4. Suppose X is a normed linear space, and let $x \in X$. Then we have

$$\|x\| = \sup_{L \neq 0} \frac{|Lx|}{\|L\|} = \sup_{\|L\|=1} |Lx| . \tag{3.2}$$

Proof. Since for each $L \in X^*$ we have $|Lx| \leq \|L\| \, \|x\|$, it readily follows that either sup in (3.2) above is less than or equal to $\|x\|$. As for the opposite inequality, note that by Corollary 3.2 there is a bounded linear functional L of norm 1 defined on X so that $Lx = \|x\|$. For this functional we have $\|x\| = |Lx|/\|L\|$, and we have finished. ■

Next we discuss the embedding of a normed space into a Banach space, but first a definition.

The natural map, denoted by J_X, of a normed linear space X into its second conjugate space X^{**} (the Banach space of bounded linear functionals on X^*) is defined by

$$(J_X x)L = Lx , \quad \text{all } L \in X^* . \tag{3.3}$$

It is not hard to check that for each $x \in X$, $J_X x$ is a bounded linear functional on X^*. To show that $J_X x$ is a linear functional on X^*, let

$L_1, L_2 \in X^*$, λ a scalar, and note that by (3.3) we have

$$\begin{aligned}(J_X x)(L_1 + \lambda L_2) &= (L_1 + \lambda L_2)(x) = L_1 x + \lambda L_2 x \\ &= (J_X x) L_1 + \lambda (J_X x) L_2 .\end{aligned}$$

To show that $J_X x$ is actually bounded we make use of (3.2): If $\|J_X x\|$ denotes the norm of $J_X x$ as an element of X^{**}, then by (3.3) and (3.2) it follows that

$$\|J_X x\| = \sup_{L \neq 0} \frac{|(J_X x) L|}{\|L\|} = \sup_{L \neq 0} \frac{|Lx|}{\|L\|} = \|x\| .$$

In fact, we have shown that J_X is also norm preserving, and consequently, one-to-one. In other words, the natural map establishes a linear isometric embedding from X into X^{**}. These properties of the natural map lead to a simple proof of the following result.

Theorem 3.5. Every normed linear space is a dense subspace of a Banach space.

Proof. Given a normed linear space X, let $X_1 = J_X(X) \subseteq X^{**}$ denote the image of X into X^{**} under the natural map. Since, as established above, X and X_1 are isometrically isomorphic, we may think of X as X_1, and prove the conclusion for X_1 instead. Let X_2 denote the closure of X_1 in X^{**}; X_2 is a closed subspace of a complete space, and consequently it is also complete. Moreover, since by construction X_1 is dense in X_2, we are done. ■

If the range of the natural map J_X is all of X^{**}, then X is said to be reflexive. For instance, from the definition of the natural map and the representation of the dual space to the Lebesgue L^p spaces given in Theorem 2.5 in Chapter XII, it follows that $L^p(\mu)$ is reflexive when $1 < p < \infty$. The reader should be warned that, in general, the equivalence of a normed linear space with its second conjugate does not guarantee the reflexivity of the space.

On the other hand, $L^1(\mu)$ is not in general reflexive, and to see this we make use of the following observation. A normed linear space X is said to be separable, if there exists a countable dense subset of X. For instance, $L^p(R^n)$ is separable if $1 \leq p < \infty$, and is not separable if $p = \infty$.

We then have

Proposition 3.6. If the conjugate X^* of a normed linear space X is separable, then X is also separable.

Proof. Let $\{L_n\}$ be an at most countable dense subset of X^*, and $\{x_n\}$ a sequence of elements in X such that $|L_n x_n| \geq \|L_n\|/2$, $\|x_n\| = 1$, for all n. We claim that the linear subspace Y of X spanned by the x_n's is dense in X. Suppose this is not the case. Then, by Proposition 3.3, there is a nontrivial linear functional $L \in X^*$ such that $Lx = 0$ for every $x \in Y$. Since by assumption $\{L_n\}$ is dense in X^*, there is a sequence $\{L_{n_m}\}$ that converges to L. Now, since for each m we have

$$\|L_{n_m} - L\| \geq |(L_{n_m} - L)(x_{n_m})| = |L_{n_m} x_{n_m}| \geq \|L_{n_m}\|/2,$$

it readily follows that $\lim_{m\to\infty} L_{n_m} = 0$. But this is impossible since $\lim_{m\to\infty} L_{n_m} = L \neq 0$, and consequently, Y is dense in X. Finally, since the set consisting of all finite linear combinations of the x_n's with rational coefficients is countable and dense in X, X is separable. ∎

Since $\ell^\infty = (\ell^1)^*$ is not separable but ℓ^1 is, the converse to the above proposition is not true. Nevertheless, we have

Corollary 3.7. The conjugate space of a reflexive separable space is also separable.

Proof. Suppose X is a normed linear space which is reflexive and separable. Then $X^{**} = J_X(X)$ is also separable and, by Proposition 3.6, X^* is separable. ∎

Since as pointed out above $(\ell^1)^* = \ell^\infty$ and ℓ^1 is separable but ℓ^∞ is not, ℓ^1 is not reflexive. It is therefore of interest to describe the dual space to $L^\infty(\mu)$, a task we left open in Chapter XII. We begin by discussing a related result of independent interest, namely the dual space to $C(I)$.

Let $I = [0,1]$, and L be a continuous linear functional defined on $C(I)$. Since $C(I)$ is a (closed) linear subspace of $L^\infty(I)$, by Theorem 2.6 there is a bounded linear functional L_1 defined on $L^\infty(I)$ that satisfies

$$L_1 f = Lf \quad \text{if } f \in C(I), \quad \text{and} \quad \|L_1\| = \|L\|. \tag{3.4}$$

Now, for each $x \in I$ we define a bounded function ϕ_x by

$$\phi_x(t) = \begin{cases} 1 & \text{if } 0 \leq t \leq x \\ 0 & \text{if } x < t \leq 1, \end{cases}$$

and put $g(x) = L_1\phi_x$, $x \in I$. Since $\psi_0(t) = 0$ for all $t \in I$, it follows that $g(0) = L_1 0 = 0$. What else can we say about g?

Let $0 = x_0 < x_1 < \ldots < x_n = 1$ be a partition of I, and put $\varepsilon_i = \operatorname{sgn}(g(x_i) - g(x_{i-1}))$. We then have

$$A = \sum_{i=1}^{n} |g(x_i) - g(x_{i-1})| = \sum_{i=1}^{n} \varepsilon_i (g(x_i) - g(x_{i-1}))$$
$$= \sum_{i=1}^{n} \varepsilon_i (L_1 \phi_{x_i} - L_1 \phi_{x_{i-1}}) = L_1 \phi ,$$

where $\phi = \sum_{i=1}^{n} \varepsilon_i (\phi_{x_i} - \phi_{x_{i-1}})$ is a bounded function with $\|\phi\|_\infty = 1$. Now, by (3.4) we get $A \leq \|L_1\| \, \|\phi\|_\infty = \|L\|$, and consequently, g is BV on I, and $V(g; 0, 1) \leq \|L\|$.

Given $f \in C(I)$, define the bounded functions

$$f_n = \sum_{k=1}^{n} f(k/n) \left(\phi_{k/n} - \phi_{(k-1)/n)}\right) , \quad n = 1, 2, \ldots$$

and note that since f is uniformly continuous it follows that $\|f_n - f\|_\infty \to 0$ as $n \to \infty$. Whence by the continuity of L_1 and (3.4) we have

$$\lim_{n \to \infty} L_1 f_n = L_1 f = Lf . \tag{3.5}$$

On the other hand, since $L_1 f_n$ may be rewritten as

$$L_1 f_n = \sum_{k=1}^{n} f(k/n)(g(k/n) - g((k-1)/n)) ,$$

by Theorem 2.6 in Chapter III we obtain

$$\lim_{n \to \infty} L_1 f_n = \int_0^1 f \, dg . \tag{3.6}$$

Thus combining (3.5) and (3.6) we conclude that

$$Lf = \int_0^1 f \, dg . \tag{3.7}$$

Furthermore, by (3.2) in Chapter III, we also have

$$|Lf| \leq \max_{x \in I} |f(x)| V(g; 0, 1) , \tag{3.8}$$

and consequently $\|L\| = V(g; 0, 1)$.

Two observations: First, since $g(0) = 0$, it follows that $V(g;0,1) = \|g\|$, the norm on BV introduced in (1.5). Also, by (3.8), for BV functions g with $g(0) = 0$, the integral in (3.7) determines a bounded linear functional L on $C(I)$ with $\|L\| \le \|g\|$. The only difficulty here is that the expression in (3.7) does not uniquely determine the functional L, cf. 4.24-4.26 in Chapter III, and, as in the case of the Lebesgue L^p spaces, some kind of normalization is needed. The details are left to the reader, cf. 4.36 below.

We close this section with the description of the conjugate space to $L^\infty(\mu)$. It is not intuitively clear how the bounded linear functionals on $L^\infty(\mu)$ look. On the one hand, it is obvious that functions $g \in L^1(\mu)$ induce such functionals by means of

$$L_g f = \int_X fg\, d\mu\,, \quad f \in L^\infty(\mu)\,,$$

but it is not hard to see that not all functionals are of this form. Indeed, let $I = [-1,1]$, and

$$Y = \left\{ f \in L^\infty(I):\ \lim_{r\to 0+} \frac{1}{r} \int_{(0,r)} f\, dy \text{ exists} \right\}.$$

Then Y is a nonempty subspace of $L^\infty(I)$, and

$$Lf = \lim_{r\to 0+} \frac{1}{r} \int_{(0,r)} f\, dy$$

is a bounded linear functional on Y with $\|L\| = 1$. Now, by the Hahn-Banach Theorem, L can be extended to a bounded linear functional on $L^\infty(I)$, also of norm 1. For simplicity denote this extension also by L and observe that L cannot be of the form

$$Lf = \int_I fg\, dy\,, \quad \text{all } f \in L^\infty(I)\,, \tag{3.9}$$

for any $g \in L^1(I)$. Indeed, if (3.9) is true for an integrable function g, let $f_\eta = \chi_{R\setminus(0,\eta)} \operatorname{sgn} g$, where $0 < \eta \le 1$; it is clear that $f_\eta \in Y$ and $Lf_\eta = 0$. It then readily follows that

$$Lf_\eta = \int_{[\eta,1]} |g|\, dy = 0\,, \quad \text{all } \eta > 0\,.$$

But this implies that $g = 0$ a.e. on $(0,1]$, and a similar argument gives that $g = 0$ a.e. on $[-1,0]$. In other words, $g = 0$ a.e., and L is then the zero functional, contrary to the fact that $\|L\| = 1$.

The analytic representation of the conjugate space to $L^\infty(\mu)$ requires that we extend the notion of integral to include integration with respect to a signed additive set function. Because it suffices for this application, we restrict our attention to the case when both the function to be integrated and the set function with respect to which the integration is carried out, are bounded. First some definitions.

Let $\mathcal{A}$ be an algebra of subsets of X and ψ a bounded nonnegative set function defined on $\mathcal{A}$. Given a bounded function $g: X \to R$ and a partition $\mathcal{P}$ of X consisting of pairwise disjoint measurable sets $E_1, \dots, E_n$, put $m_k = \inf_{E_k} g$, $M_k = \sup_{E_k} g$, and consider the lower and upper sums of g corresponding to $\mathcal{P}$ with respect to ψ, defined by the expressions

$$s(g,\psi,\mathcal{P}) = \sum_{k=1}^{n} m_k \psi(E_k) \quad \text{and} \quad S(g,\psi,\mathcal{P}) = \sum_{k=1}^{n} M_k \psi(E_k)$$

respectively. The usual properties of lower and upper sums hold in this case as well. They are:

(i) If a partition $\mathcal{P}'$ refines a partition $\mathcal{P}$, then we have

$$s(g,\psi,\mathcal{P}) \le s(g,\psi,\mathcal{P}') \quad \text{and} \quad S(g,\psi,\mathcal{P}') \le S(g,\psi,\mathcal{P}).$$

(ii) No lower sum exceeds an upper sum, even when they are formed with two different partitions.

In case the quantities

$$\sup_{\mathcal{P}} s(g,\psi,\mathcal{P}) \quad \text{and} \quad \inf_{\mathcal{P}} S(g,\psi,\mathcal{P})$$

coincide, we define the integral $\int_X g\,d\psi$ of g over X with respect to ψ as that common value.

The class of functions for which the integral exists is rather wide and, as we now show, it includes the bounded measurable functions. By the way, since $\mathcal{A}$ is not necessarily a σ-algebra, we say that a function is measurable provided all four conditions in Proposition 1.1 in Chapter VI are satisfied.

Proposition 3.8. If $\psi(X) < \infty$ and g is bounded and measurable, then $\int_X g\,d\psi$ exists.

Proof. By (i) and (ii) above it suffices to show that there are partitions $\mathcal{P}$ of X for which the lower and upper sums are arbitrarily close to each other. Let $m < M$ be real numbers such that $m < g(x) < M$ for all $x \in X$, suppose η is an arbitrary constant, $0 < \eta < M - m$, and divide the interval (m, M) by means of the points $m = t_0 < t_1 < \ldots < t_n = M$ into a finite number of subintervals, each of length less than or equal to η. Form now the sets

$$E_k = \{t_{k-1} \leq g < t_k\}, \quad k = 1, \ldots, n,$$

and observe that they are pairwise disjoint and measurable. Let $\mathcal{P}$ denote the partition of X into the E_k's; if any E_k is empty, simply drop it. Further note that since this family is finite we have $\psi(X) = \sum \psi(E_k)$. Moreover, since for each k we have

$$t_{k-1} \leq m_k \leq M_k \leq t_k, \quad M_k - m_k \leq \eta,$$

it readily follows that

$$\begin{aligned} S(g, \psi, \mathcal{P}) - s(g, \psi, \mathcal{P}) &= \sum (M_k - m_k)\psi(E_k) \\ &\leq \eta \sum \psi(E_k) = \eta\psi(X). \end{aligned}$$

Thus, by means of an appropriate choice of η, the difference between the upper and lower sums above can be made arbitrarily small, which is what we set out to prove. ■

It is interesting to point out that, in general, the class of functions for which the integral exists includes functions that are not measurable. Indeed, if $X = N$ and $\mathcal{A}$ is the algebra of those subsets E of N which are either finite or so that $N \setminus E$ is finite, then

$$\psi(E) = \begin{cases} 0 & \text{if } E \text{ is finite} \\ \infty & \text{if } N \setminus E \text{ is finite,} \end{cases}$$

is an additive set function defined on $\mathcal{A}$, the function g = characteristic function of the odd integers is not measurable, and yet $\int_N g \, d\psi = \infty$ exists.

It is possible to define the integral of g with respect to a signed additive set function ψ over $\mathcal{A}$ as follows: If ψ_+ and ψ_- denote the positive and negative variations of ψ respectively, cf. 4.8 in Chapter IV, let

$$\int_X g \, d\psi = \int_X g \, d\psi_+ - \int_X g \, d\psi_-, \tag{3.10}$$

provided the expression on the right-hand side of (3.10) is well-defined. Now, from (3.10) it follows that the basic properties of the Riemann-Stieltjes and Lebesgue integrals hold in this context, with slight or no change. We need two specific properties of the integral, to wit, linearity and boundedness; we state them next, their proof is left to the reader.

If the integral of g_1 and that of g_2 with respect to ψ exist and λ is a scalar, then the integral of $g_1 + \lambda g_2$ with respect to ψ exists, and we have

$$\int_X (g_1 + \lambda g_2)\, d\psi = \int_X g_1\, d\psi + \lambda \int_X g_2\, d\psi\,.$$

Also, if $|g(x)| \leq M$ for all $x \in X$, then

$$\left| \int_X g\, d\psi \right| \leq M|\psi|(X)\,, \tag{3.11}$$

We are now ready to give a description of the dual to $L^\infty(\mu)$. Suppose $(X, \mathcal{M}, \mu)$ is a measure space, let L be a bounded linear functional defined on $L^\infty(\mu)$, and for $E \in \mathcal{M}$ put

$$\psi(E) = L\chi_E\,. \tag{3.12}$$

From the linearity of L it readily follows that ψ is an additive set function defined on $\mathcal{M}$. Moreover, since L is bounded we also have

$$|\psi(E)| \leq \|L\|\, \|\chi_E\|_\infty\,. \tag{3.13}$$

Now, if $\mu(E) = 0$ we have $\|\chi_E\|_\infty = 0$, and consequently, by (3.13) we obtain $\psi(E) = 0$. Moreover, since we also have that $\mu(A) = 0$ for any $A \subseteq E$, $A \in \mathcal{M}$, it follows that $\psi(A) = 0$ for those sets, and, by 4.8 in Chapter IV, we get that $\psi_+(E) = \psi_-(E) = 0$, and $|\psi|(E) = 0$.

Suppose now that $f \in L^\infty(\mu)$ and consider a representative in the equivalence class of f, which we call f again, which is bounded everywhere by $\|f\|_\infty$. Let $M > \|f\|_\infty$, and divide the interval $[-M, M]$ by means of the points $-M = t_0 < t_1 < \ldots < t_n = M$ into a finite number of subintervals, each of length less than or equal to an arbitrary real number η. Form now the partition of X consisting of the measurable sets

$$E_k = \{t_{k-1} \leq f < t_k\}, \quad k = 1, \ldots, n,$$

and let h be the measurable function $h = \sum_{k=1}^n t_{k-1}\chi_{E_k}$. Now, if $x \in E_k$, then $|f(x) - h(x)| = |f(x) - t_{k-1}| \leq \eta$, and consequently, we have

$$\|f - h\|_\infty = \sup_{x \in X} |f(x) - h(x)| \leq \eta\,. \tag{3.14}$$

Further, since L is linear and bounded, by (3.14) it follows that

$$\begin{aligned} |Lf - Lh| &= |L(f - h)| \\ &\leq \|L\|\, \|f - h\|_\infty \leq \|L\|\, \eta\,. \end{aligned} \tag{3.15}$$

On the other hand, both f and h have an integral with respect to ψ on X, and since $\int_X h\, d\psi = \sum t_{k-1}\psi(E_k) = Lh$, we get

$$\int_X (f-h)\, d\psi = \int_X f\, d\psi - \int_X h\, d\psi = \int_X f\, d\psi - Lh\,.$$

Whence it follows that

$$\begin{aligned}\left|\int_X f\, d\psi - Lh\right| &= \left|\int_X (f-h)\, d\psi\right| \\ &\le \|f-h\|_\infty |\psi|(X) \le \eta\, |\psi|(X)\,,\end{aligned}$$

which combined with (3.15) allows us to estimate $\left|Lf - \int_X f\, d\psi\right|$ by

$$|Lf - Lh| + \left|Lh - \int_X f\, d\psi\right| \le (\|L\| + |\psi|(X))\,\eta\,.$$

Since η is arbitrary this can only happen if

$$Lf = \int_X f\, d\psi\,, \quad \text{all bounded } f\,. \tag{3.16}$$

This is the first step in obtaining the representation of L.

There is, of course, the question of the uniqueness of the representation: We must be sure that for each $f \in L^\infty(\mu)$ the right-hand side in (3.16) above is independent of the bounded representative we choose in the equivalence class of f. This amounts to proving that if f is equivalent to 0, then we have $\int_X f\, d\psi = 0$. Now, in this case, X can be partitioned into two disjoint measurable sets E_1 and E_2, say, so that $f = 0$ on E_1 and $\mu(E_2) = 0$. By the definition of the integral it is clear that $\int_{E_1} f\, d\psi = 0$, and since as observed above we also have $|\psi|(E_2) = 0$, if c is a bound for f, by (3.11) we get that

$$\left|\int_{E_2} f\, d\psi\right| \le c|\psi|(E_2) = 0\,.$$

By the linearity of the integral it now follows that $\int_X f\, d\psi = 0$, which insures that the right-hand side of (3.16) is well-defined at the level of bounded equivalent functions of $L^\infty(\mu)$ functions.

The stage is now set for

Theorem 3.9. Let $(X, \mathcal{M}, \mu)$ be a measure space. The dual to $L^\infty(\mu)$ can be described as follows: Each bounded linear functional L on $L^\infty(\mu)$ is of the form

$$Lf = \int_X f\, d\psi\,, \quad \text{all bounded } f \in L^\infty(\mu)\,, \tag{3.17}$$

where ψ is a bounded additive set function defined on $\mathcal{M}$ satisfying the condition $|\psi|(E) = 0$ whenever $\mu(E) = 0$. Furthermore, the norm $\|L\|$ is

$$\|L\| = |\psi|(X)\,. \tag{3.18}$$

Proof. As discussed above, if L is a bounded linear functional defined on $L^\infty(\mu)$, then there is a bounded additive function ψ defined on $\mathcal{M}$ such that (3.17) holds. Moreover, this representation is independent of the bounded representative of each $f \in L^\infty(\mu)$, and (3.11) implies that $\|L\| \leq |\psi|(X)$. Thus to verify (3.18) it suffices to prove that we also have $|\psi|(X) \leq \|L\|$. Now, by 4.9 in Chapter IV, given $\varepsilon > 0$, there exist measurable subsets E_1, E_2 of X such that

$$|\psi(X)| \leq \psi(E_1) - \psi(E_2) + \varepsilon\,.$$

Put $f = \chi_{E_1} - \chi_{E_2}$ and observe that since f only takes the values 0 and ± 1, we have $\|f\|_\infty = 1$ and consequently,

$$\begin{aligned}\|L\| \geq |Lf| \geq Lf &= L\chi_{E_1} - L\chi_{E_2} \\ &= \psi(E_1) - \psi(E_2) \geq |\psi|(X) - \varepsilon\,.\end{aligned}$$

Since $\varepsilon > 0$ is arbitrary, the above estimate implies that $\|L\| \geq |\psi|(X)$, (3.18) holds, and the integral representation of L has been established.

On the other hand, if ψ is a bounded additive set function defined on $\mathcal{M}$ with the property that $|\psi|(E) = 0$ whenever $\mu(E) = 0$, then it is not hard to see that (3.17) defines a bounded linear functional L on $L^\infty(\mu)$ with norm $\|L\| = |\psi|(X)$. Indeed, if f is a bounded representative of a function $f \in L^\infty(\mu)$, consider the measurable partition of X consisting of the sets $B = \{|f| > \|f\|_\infty\}$, and $X \setminus B$. Since $\mu(B) = 0$, it follows that $|\psi|(B) = 0$ and by (3.11) we get that $\int_B f\, d\psi = 0$. Whence we have $\int_X f\, d\psi = \int_{X \setminus B} f\, d\psi$ and, by (3.11) again, it follows that

$$|Lf| \leq |\psi|(X \setminus B)\|f\|_\infty \leq |\psi|(X)\|f\|_\infty\,.$$

These observations imply at once that L is a bounded linear functional defined on $L^\infty(\mu)$ with $\|L\| \leq |\psi|(X)$. The opposite inequality holds as before, L also satisfies (3.18), and we have finished. ■

4. PROBLEMS AND QUESTIONS

4.1 Is every metric on a linear space induced by a norm?

4.2 Let X be a normed linear space, and $B = \{x \in X : \|x\| < 1\}$. Show that the closure of B is the set $\{x \in X : \|x\| \leq 1\}$.

4.3 Suppose M is a closed subspace of a normed linear space X and define an equivalence relation R on $X \times X$ by xRy iff $x - y$ belongs to M. If X/M denotes the set of equivalence classes and $x + M$ the equivalence class corresponding to $x \in X$, show that X/M is a linear space over the scalar field of X with the operations

$$(x + M) + (\lambda y + M) = (x + \lambda y) + M\,, \quad x, y \in M\,, \lambda \text{ scalar}\,.$$

For future reference, the dimension of X/M is called the codimension of M.
Further, X/M is also a normed space with norm $\|x+M\| = d(x, M)$. Are these conclusions true if M is not necessarily closed?

4.4 Let X be a normed linear space, and M be a closed subspace of X. Prove that X is complete iff M and X/M are complete. Also, show that X is separable iff M and X/M are separable.

4.5 If M is a finite-dimensional proper subspace of a normed linear space X, prove there exists an element $x \in X$, $\|x\| = 1$, such that $d(x, M) = 1$.

4.6 (F. Riesz) Let X be a normed linear space and M be a proper closed linear subspace of X. Show that given $\varepsilon > 0$, there exists an element $x \in X$, $\|x\| = 1$, such that $d(x, M) > 1 - \varepsilon$.

4.7 Let X be a normed linear space and suppose $\lim_{n\to\infty} \|x_n - x\| = 0$. Show that $\lim_{n\to\infty} \|x_n\| = \|x\|$.

4.8 Suppose $x_1, \ldots, x_n$ are linearly independent elements of a normed linear space X. Show that there exists a constant $c > 0$ with the property that for every choice of scalars $\lambda_1, \ldots, \lambda_n$ we have

$$\|\lambda_1 x_1 + \cdots + \lambda_n x_n\| \geq c(|\lambda_1| + \cdots + |\lambda_n|)\,.$$

4.9 Referring to the construction of the norm on a linear space following Proposition 2.4, suppose we put $\|x\|_p = \left(\sum_{i=1}^n |\lambda_i|^p\right)^{1/p}$, $1 \leq p < \infty$ there. Is $\|\cdot\|_p$ a norm?

4.10 Let $(X, \mathcal{M}, \mu)$ be a measure space and $1 \le p, q \le \infty$. Show that $L^p(\mu) + L^q(\mu) = \{f : f$ can be written as $f = g + h$, $g \in L^p(\mu)$, $h \in L^q(\mu)\}$, is a linear space. Further, normed by

$$\|f\| = \inf\{\|g\|_p + \|h\|_q : f = g + h\},$$

$L^p(\mu) + L^q(\mu)$ is a Banach space. Along similar lines, $L^p(\mu) \bigcap L^q(\mu)$ normed by $\|f\| = \max\{\|f\|_p, \|f\|_q\}$, is also a Banach space. Can you characterize the conjugate space to $L^p(\mu) + L^q(\mu)$? to $L^p(\mu) \bigcap L^q(\mu)$?

4.11 A sequence (x_n) of elements of a Banach space X is said to be a Schauder basis for X if for each $x \in X$ there is a unique sequence of scalars (λ_n) such that $\lim_{m \to \infty} \|x - \sum_{n=1}^{m} \lambda_n x_n\| = 0$. Show that ℓ^p has a Schauder basis if $1 \le p < \infty$, but ℓ^∞ does not.

4.12 Prove that if a Banach space has a Schauder basis, then it is separable.

4.13 Let $c_0 = \{(x_n) \in \ell^\infty : \lim_{n \to \infty} x_n = 0\}$. Show that c_0 is a closed linear subspace of ℓ^∞, and that it has a Schauder basis. Is c_0 reflexive?

4.14 For each positive integer n let e_n be the sequence with 1 in the nth place and zeros elsewhere. Prove that $\{e_n\}$ is a Schauder basis for ℓ^1, but it is not a Hamel basis for ℓ^1.

4.15 Let X be a linear normed space, and L a nontrivial linear functional on X. Prove the following three conditions are equivalent: (a) L is continuous, (b) The null space of L is a proper, closed linear subspace of X, and, (c) The null space of L is not dense in X.

4.16 Let X be a linear normed space over C. If a linear functional L on X is not continuous, prove that $\{Lx : \|x\| \le 1\}$ is all of C.

4.17 Let $L \ne 0$ be a linear functional on a linear space X and x_0 any fixed element of X/N, where N is the null space of L. Show that any $x \in X$ has a unique representation $x = \lambda x_0 + y$, where λ is a scalar and $y \in N$.

4.18 Referring to 4.17, show that any two elements $x_1, x_2 \in X$ belong to the same element of X/N iff $Lx_1 = Lx_2$. Further, the codimension of N is equal to 1.

4.19 Show that two linear functionals L_1, L_2 which are defined on the same linear space and have the same null space are proportional.

4.20 If Y is a subspace of a linear space X and the codimension of Y is equal to 1, then every element of X/Y is called a hyperplane

parallel to Y. Show that for any linear functional $L \neq 0$ on X the set $Y_1 = \{x \in X : Lx = 1\}$ is a hyperplane parallel to the null space N of L.
Further, show that the norm $\|L\|$ of L can be interpreted geometrically as the reciprocal of the distance of the hyperplane Y_1 to the origin.

4.21 Let X be a normed linear space, and suppose L is a bounded linear functional on X with norm 1. Given $\varepsilon > 0$, show there exists $x_\varepsilon \in X$ such that $\|x_\varepsilon\| = 1$ and $Lx_\varepsilon > 1 - \varepsilon$. Give an example to show that there need not exist $x \in X$ such that $\|x\| = 1$ and $Lx = 1$.

4.22 Let X be a normed linear space and let $\{x_n\} \subseteq X$. Prove that $x \in X$ is the limit of finite linear combinations of the x_n's iff $Lx = 0$ for all continuous linear functionals L on X such that $Lx_n = 0$ for all n.

4.23 Let Y be a subset of a (real) normed linear space X, and L_0 a functional defined on Y. Show that a necessary and sufficient condition for L_0 to have a bounded linear extension to X is that there exists a constant k with the property that $\|\sum_{n=1}^{m} \lambda_n L_0 x_n\| \leq k\,\|\sum_{n=1}^{m} \lambda_n x_n\|$ for any $x_1, \ldots, x_n$ in Y and scalars $\lambda_1, \ldots, \lambda_n$.

4.24 Let $1 < p, q < \infty$, be conjugate indices, i.e., $1/p + 1/q = 1$. Suppose $g \in L^q(R^n)$ has the property that $\int_{R^n} fg\,dx = 0$ for each f in $D = \{f \in L^p(R^n) \cap L(R^n) : \int_{R^n} f\,dx = 0\}$. Prove that $g = 0$ a.e. As a consequence, show that D is dense in $L^p(R^n)$.
Is a similar statement true if we consider a bounded interval I instead of R^n? Also, what can we say about the case $p = 1$?

4.25 Suppose $1 < p, q < \infty$ are conjugate indices, and $f \notin L^p(X, \mu)$. Show that the set $\{g \in L^q(X, \mu) : fg \in L(X, \mu)$ and $\int_X fg\,d\mu = 0\}$, is dense in $L^q(X, \mu)$.

4.26 Let $X = L^2(\mu) \times L^2(\mu) = \{(f, g) : f, g \in L^2(\mu)\}$ be the linear space normed by

$$\|(f, g)\| = \left(\|f\|_2^3 + \|g\|_2^3\right)^{1/3}.$$

Show that X is a Banach space and describe X^*.

4.27 Let $X \neq \{0\}$ be a normed linear space. Show that $X^* \neq \{0\}$. Moreover, prove that if X has n linearly independent elements, so does X^*.

4.28 Show that if $Lx = Ly$ for every $L \in X^*$, then $x = y$.

4.29 Prove that if a normed linear space X is reflexive, so is X^*.

4.30 Prove that the "completion" of the normed linear space described in Theorem 3.5 is unique up to isomorphisms.

4.31 Let Y be a closed subspace of a normed linear space X. Show that if every $L \in X^*$ which vanishes on Y vanishes also on X, then $Y = X$.

4.32 Let X be a normed linear space. A sequence $\{x_n\} \subseteq X$ is said to converge weakly to an element $x \in X$ if $\lim_{n\to\infty} Lx_n = Lx$ for all $L \in X^*$. Prove that no sequence can have two distinct weak limits. Further, a sequence $\{x_n\} \subseteq X$ is said to be weakly Cauchy if $\{Lx_n\}$ is a Cauchy (scalar) sequence for every $L \in X^*$, and X is said to be weakly sequentially complete if every weakly Cauchy sequence converges weakly. Prove that if X is weakly sequentially complete, then it is complete.

4.33 Prove that a reflexive Banach space is weakly sequentially complete.

4.34 Show that any closed subspace of a weakly sequentially complete Banach space is itself weakly sequentially complete.

4.35 Show that ℓ^1 is weakly sequentially complete.

4.36 Describe a normalization of BV functions that allows for the identification of the dual of $C(I)$.

4.37 If I is a compact interval of R^n and μ is a finite Borel measure on I, then $Lf = \int_I f\,d\mu$ is a positive bounded linear functional on $C(I)$. Positive here means that $Lf \geq 0$ whenever $f \geq 0$. Prove now the following result, a particular case of the so called Riesz Representation Theorem: Suppose I is a compact interval of R^n and L is a positive bounded linear functional on $C(I)$. Then there is a unique Borel measure μ such that $Lf = \int_I f\,d\mu$ for every $f \in C(I)$.

4.38 Fix an integer N. Does there exist a function f which is BV in $I = [0,1]$ such that $\int_0^1 p(x)\,df(x) = \sum_{n=1}^{N} p^{(n)}(n/N)$ for all polynomials p of degree less than or equal to N?

4.39 Let $I = [0,1]$ and consider the sequence $\{f_n\} \subset C(I)$ defined by $f_n(x) = nx$ if $0 \leq x \leq 1/n$, $= 2 - nx$ if $1/n \leq x \leq 2/n$, and $= 0$ otherwise. Show that $\{f_n\}$ converges weakly to 0 in $C(I)$, but that $\lim_{n\to\infty} \|f_n\| \neq 0$.

CHAPTER **XV**

The Basic Principles

In this chapter we consider the three basic principles concerning continuous linear transformations that provide the foundation for many results in linear analysis. These principles are: The Uniform Boundedness Principle, The Open Mapping Theorem, and the Closed Graph Theorem.

1. THE BAIRE CATEGORY THEOREM

Baire's theorem concerning the structure of complete metric spaces is an essential ingredient in proving the validity of the basic principles alluded to above. To state it we need some definitions.

Let (X, d) be a metric space. A set $E \subseteq X$ is said to be nowhere dense if its closure $\overline{E}$ has empty interior. The sets of first category in X are those that are countable unions of nowhere dense sets; these sets are also called meager. All other sets are said to be of second category in X, or nonmeager.

For instance, the rational numbers Q are of first category in R, and the irrational numbers I are of second category in R.

We begin by proving

Theorem 1.1 (Baire's Category Theorem). A complete metric space X is of second category in itself.

Proof. Suppose, to the contrary, that X is of first category in itself and let

$$X = \bigcup_{n=1}^{\infty} X_n, \quad X_n \text{ nowhere dense, all } n.$$

Take a point x_0 in X and consider the (nonempty) open ball $B(x_0,1)$ centered at x_0 of radius 1. Since the interior of $\overline{X}_1$ is empty, $\overline{X}_1$ does not contain $B(x_0,1)$; let then x_1 be a point in $B(x_0,1)\setminus\overline{X}_1$ and $0 < r_1 < 1/2$ be such that

$$\overline{B(x_1,r_1)} \subset B(x_0,1)\,, \quad \text{and} \quad \overline{B(x_1,r_1)} \cap \overline{X}_1 = \emptyset\,.$$

Similarly, since X_2 is nowhere dense, $\overline{X}_2$ does not contain $B(x_1,r_1)$ and, as before, there are a point $x_2 \in B(x_1,r_1)\setminus\overline{X}_2$ and $0 < r_2 < 1/4$ such that

$$\overline{B(x_2,r_2)} \subset B(x_1,r_1)\,, \quad \text{and} \quad \overline{B(x_2,r_2)} \cap \overline{X}_2 = \emptyset\,.$$

Continuing in this fashion step by step we get a decreasing sequence of closed balls $\{\overline{B(x_n,r_n)}\}$ with the property that

$$\overline{B(x_n,r_n)} \cap \overline{X}_m = \emptyset\,, \quad 1 \le m \le n\,, \quad \text{and} \quad 0 < r_n \le 1/2^n\,.$$

Now, by a well-known result in the theory of metric spaces, actually an extension of the Nested sequence theorem on the real line, since the $\overline{B(x_n,r_n)}$'s form a monotone decreasing sequence of nonempty closed sets whose diameters tend to 0, and since (X,d) is complete, there exists one, and only one, point $x \in X$ so that

$$\bigcap_{n=1}^{\infty} \overline{B(x_n,r_n)} = \{x\}\,.$$

By construction $x \notin \overline{X}_n$ for all n; thus $x \notin \bigcup_n X_n = X$, and this is a contradiction. ■

Theorem 1.1 is often cast in the following form: If $\mathcal{O}_n = X \setminus \overline{X}_n$ denotes the complement of $\overline{X}_n$, then each $\mathcal{O}_n$ is open and dense in X, and the conclusion of Baire's Category Theorem is that $\bigcap_n \mathcal{O}_n \neq \emptyset$. More precisely, the intersection of every countable family of dense open subsets of X is dense in X.

The Baire Category Theorem is useful in proving that a set is nonempty. In fact, the category method furnishes a whole class of examples, and it often makes it possible to construct an explicit example by successive approximations. We exemplify this by showing that, in the sense of category, almost all continuous functions are nowhere differentiable. In fact, as we prove below, it is exceptional for a continuous function to have a finite one-sided derivative anywhere in an interval.

To make this precise let $I = [0,1]$, consider $C(I)$ with the uniform metric, and let E_n denote the class of functions $f \in C(I)$ such that for some x in $[0,1-1/n]$ we have

$$|f(x+h)-f(x)| \leq nh\,, \quad \text{all } 0<h<1-x\,.$$

We claim that for each n, E_n is closed and nowhere dense in $C(I)$. To see that E_n is closed let $f \in \overline{E}_n$, and let $\{f_k\}$ be a sequence in E_n that converges to f. Then, there is a sequence (x_k) such that $0 \leq x_k \leq 1-1/n$, and

$$|f_k(x_k+h)-f_k(x_k)| \leq nh\,, \quad \text{all } 0<h<1-x_k\,.$$

By the Bolzano-Weierstrass theorem we may also assume that $x_k \to x$ for some $0 \leq x \leq 1-1/n$ since this condition is satisfied if we replace (x_k) by a suitable subsequence. Now, if $0<h<1-x$, note that the inequality $0<h<1-x_k$ holds for sufficiently large k, and that

$$\begin{aligned}|f(x+h)-f(x)| &\leq |f(x+h)-f(x_k+h)|+|f(x_k+h)-f_k(x_k+h)|\\ &\quad+|f_k(x_k+h)-f_k(x_k)|+|f_k(x_k)-f(x_k)|+|f(x_k)-f(x)|\\ &= A_1+A_2+A_3+A_4+A_5\,,\end{aligned}$$

say. Clearly, $A_2, A_4 \leq \|f-f_k\|$. Also by the choice of x_k, we get $A_3 \leq nh$. Moreover, since f is uniformly continuous in I, we have $\lim_{k\to\infty} A_1, A_5 = 0$. Thus, letting $k \to \infty$, it follows that

$$|f(x+h)-f(x)| \leq nh\,, \quad \text{all } 0<h<1-x\,,$$

f belongs to E_n, and E_n is closed.

Next, since any continuous function on I can be approximated arbitrarily closely by a piecewise linear continuous function g, to show that E_n is nowhere dense in $C(I)$ it suffices to prove that given any such function g and $\varepsilon>0$, there is a function h in $C(I)\setminus E_n$ such that $\|g-h\| \leq \varepsilon$. This is not hard, in fact, a "proof by pictures" using "saw-tooth" functions works, cf. 6.1 below.

Hence the set $E = \bigcup E_n$ is of first category in $C(I)$. This is the set of all continuous functions that have bounded right difference quotients at some point of $[0,1)$ and it contains the set of all functions in $C(I)$ that have a finite right-sided derivative somewhere there.

The use of the Baire Category Theorem in this, and other contexts, amounts to the verification of the fact that a set is not empty by showing that an element of the set can be found as the limit of a suitable sequence. In fact, the above proof hints that a nowhere differentiable function can be exhibited as the sum of a (uniformly convergent) series of "saw-tooth" functions, cf. 6.2 below.

2. THE SPACE $\mathcal{B}(X,Y)$

The time has come to consider the theory of linear mappings from a linear space into another linear space over the same field of scalars. More precisely, let X, Y be normed linear spaces over the same field of scalars, usually but not necessarily the real or complex numbers, and let T be a mapping, or operator, with domain $D(T)$ in X and range $R(T)$ contained in Y. We say that T is linear if for all x_1, x_2 in $D(T)$ and scalars λ we have

$$T(x_1 + \lambda x_2) = Tx_1 + \lambda Tx_2 . \tag{2.1}$$

A word about (2.1) above: The sign "+" denotes the addition in X on the left-hand side of (2.1) and the addition in Y on the right-hand side there; a similar remark applies to other notations throughout this chapter.

A mapping $T: X \to Y$ is said to be continuous at a point x_0 in X if given $\varepsilon > 0$, there exists $\delta = \delta(x_0, \varepsilon)$ such that

$$\|Tx - Tx_0\| \le \varepsilon \quad \text{whenever} \quad \|x - x_0\| \le \delta .$$

Closely related to this concept is that of boundedness: A mapping $T: X \to Y$ is said to be bounded if

$$\|T\| = \sup_{\|x\| \neq 0} \frac{\|Tx\|}{\|x\|} < \infty . \tag{2.2}$$

As in the case of linear functionals we have

Proposition 2.1. Let X and Y be normed linear spaces and T be a linear operator defined on X and range in Y. The following statements are equivalent:

(i) T is continuous at a point $x_0 \in X$.
(ii) T is uniformly continuous on X.
(iii) T is bounded.

The proof of the proposition follows along the lines to that of Proposition 2.4 in Chapter XIV and is therefore left to the reader.

We denote by $\mathcal{B}(X,Y)$ the collection of all the bounded linear mappings defined on X with range in Y; when $X = Y$ we simply write $\mathcal{B}(X)$.

We begin by giving an example of a mapping in $\mathcal{B}(X,Y)$, and computing its norm. Let $X = Y = C(I)$, where I is a compact interval of the line, and let (the kernel) k be a continuous real-valued function defined on $I \times I$. We consider the mapping T on $f \in C(I)$ given by

$$Tf(x) = \int_I k(x,y) f(y)\, dy .$$

From the continuity of k and LDCT it readily follows that also $Tf \in C(I)$. The question is whether $T \in \mathcal{B}(C(I))$, and if so to compute $\|T\|$. First observe that

$$|Tf(x)| \le \|f\| \sup_{x \in I} \int_I |k(x,y)|\, dy\,,$$

and consequently $T \in \mathcal{B}(C(I))$, and

$$\|T\| \le \sup_{x \in I} \int_I |k(x,y)|\, dy\,. \tag{2.3}$$

Next we show that equality holds in (2.3). Since the function

$$g(x) = \int_I |k(x,y)|\, dy$$

is continuous on I, it attains its maximum value at some point x_0 in I. Now, the function h defined by $h(y) = \operatorname{sgn}(k(x_0, y))$ is bounded and measurable, and consequently, integrable on I. It follows from Theorem 1.3 in Chapter VIII that there is a sequence $\{\phi_n\} \subseteq C(I)$ such that

$$\|\phi_n\|_\infty \le 1\,, \quad \text{and} \quad \lim_{n \to \infty} \|\phi_n - h\|_1 = 0\,.$$

On the one hand we have

$$\max_{x \in I} \int_I |k(x,y)|\, dy = \int_I |k(x_0,y)|\, dy = \int_I k(x_0,y) h(y)\, dy\,, \tag{2.4}$$

and, on the other hand, by LDCT it follows that

$$\lim_{n \to \infty} T\phi_n(x_0) = \int_I k(x_0,y) h(y)\, dy\,. \tag{2.5}$$

Moreover, since for each n we have $T\phi_n(x_0) \le \|T\phi_n\| \le \|T\|$, combining this estimate with (2.4) and (2.5) we get that the inequality opposite to (2.3) also holds, and consequently equality holds there.

It is not hard to see that $\mathcal{B}(X, Y)$ is a linear space over the ambient scalar field. For, given $T_1, T_2 \in \mathcal{B}(X, Y)$ and a scalar λ, since $T_1 + \lambda T_2$ is the linear mapping given by

$$(T_1 + \lambda T_2)x = T_1 x + \lambda T_2 x\,, \quad \text{all } x \in X\,, \tag{2.6}$$

it readily follows that $\|T_1 + \lambda T_2\| \le \|T_1\| + |\lambda|\, \|T_2\|$, and $T_1 + \lambda T_2$ also belongs to $\mathcal{B}(X, Y)$.

Moreover, we also have

Proposition 2.2. Let X, Y be normed spaces over the same field of scalars. Then, normed by (2.2), $\mathcal{B}(X,Y)$ is a normed space. Furthermore, if $X \neq \{0\}$, $\mathcal{B}(X,Y)$ is a Banach space iff Y is complete.

Proof. That $\mathcal{B}(X,Y)$ is a normed space, and that it is complete if Y is complete, follows along the lines of the particular case of functionals, cf. Proposition 2.5 in Chapter XIV, so we say no more.

Now, suppose that $\mathcal{B}(X,Y)$ is a Banach space and let $\{y_n\}$ be a sequence of elements in Y such that $\sum_n \|y_n\| < \infty$; we must show that $\sum_n y_n$ converges in Y. Let now $x_0 \in X$, $\|x_0\| = 1$, invoke the Hahn-Banach theorem to construct a bounded linear functional L on X so that $Lx_0 = \|x_0\| = 1$, and define the sequence $\{T_n\} \subseteq \mathcal{B}(X,Y)$ by

$$T_n x = (Lx) y_n\,, \quad x \in X\,.$$

Since

$$\|T_n\| = \sup_{x \neq 0} \frac{\|T_n x\|}{\|x\|} \leq \|L\|\,\|y_n\|\,, \quad \text{all } n\,,$$

it follows at once that $\sum_n \|T_n\| < \infty$, and since $\mathcal{B}(X,Y)$ is complete $\sum_n T_n$ converges to a sum $T \in \mathcal{B}(X,Y)$, say, in the norm of $\mathcal{B}(X,Y)$. In particular, we have

$$\Big\|\sum_{n=1}^{m} T_n x_0 - T x_0\Big\| \leq \Big\|\sum_{n=1}^{m} T_n - T\Big\|\,\|x_0\|\,. \tag{2.7}$$

But the right-hand side of (2.7) tends to 0 with m, and so does the left-hand side there. By the definition of the T_n's this means that

$$\sum_{n=1}^{m} T_n x_0 = L x_0 \sum_{n=1}^{m} y_n = \sum_{n=1}^{m} y_n \to T x_0\,, \quad \text{as } m \to \infty\,,$$

and $\sum_n y_n$ converges to $T x_0$ in Y. ∎

We operate with elements in $\mathcal{B}(X,Y)$ pretty much like with numerical functions, including the taking of inverses. Indeed, suppose T is a one-to-one linear operator in $\mathcal{B}(X,Y)$. The inverse T^{-1} of T is the map from $R(T)$ into X given by $T^{-1}(Tx) = x$ for all $x \in X$. It is clear that T^{-1} is linear, and we also have

Proposition 2.3. Let $T \in \mathcal{B}(X,Y)$. Then T^{-1} exists and is continuous iff there exists a constant $c > 0$ such that

$$\|Tx\| \geq c\|x\|\,, \quad \text{all } x \in X\,. \tag{2.8}$$

Proof. Suppose (2.8) is true and observe that if $x \neq 0$, then also $Tx \neq 0$, and T is one-to-one. Moreover, given $y \in R(T)$, let $y = Tx$, $x \in X$, and note that

$$\|T^{-1}y\| = \|T^{-1}(Tx)\| = \|x\| \leq c\|Tx\| = c\|y\|,$$

and $T^{-1} \in \mathcal{B}(R(T), X)$. On the other hand, if T^{-1} exists and is continuous, we have

$$\|x\| = \|T^{-1}(Tx)\| \leq \|T^{-1}\| \, \|Tx\|, \quad x \in X,$$

and (2.8) follows upon taking $c = 1/\|T^{-1}\|$. ∎

3. THE UNIFORM BOUNDEDNESS PRINCIPLE

A family $\mathcal{F} \subseteq \mathcal{B}(X,Y)$ can be "bounded" in at least two different senses. First, since $\mathcal{B}(X,Y)$ is normed, $\mathcal{F}$ can be bounded in the norm, i.e., $\sup\{\|T\| : T \in \mathcal{F}\}$ is finite. If this is the case we say that $\mathcal{F}$ is a norm bounded set. On the other hand, it can also happen that $\sup\{\|Tx\| : T \in \mathcal{F}\}$ is finite for each x in X. When this is the case we say that $\mathcal{F}$ is pointwise bounded on X. A norm bounded set is certainly pointwise bounded on X. The remarkable fact is that, when X is complete, the converse is true. This is the content of the Uniform Boundedness Principle, and of the Resonance Theorem, which we present next.

Theorem 3.1 (The Uniform Boundedness Principle). Let X be a Banach space, Y a normed linear space, and $\mathcal{F} = \{T_i : i \in I\} \subseteq \mathcal{B}(X,Y)$ a pointwise bounded family on X. Then

$$\lim_{x \to 0} \|T_i x\| = 0, \quad \text{uniformly in } I.$$

Proof. We show that given $\varepsilon > 0$, there exists $\delta > 0$ such that

$$\|T_i x\| \leq \varepsilon, \quad \text{whenever } \|x\| \leq \delta, \text{ all } i \in I. \tag{3.1}$$

Let

$$X_k = \{x \in X : \sup_{i \in I} \|T_i x\| \leq k\}, \quad k = 1, 2 \ldots$$

Since the T_i's are continuous, each X_k is closed. Moreover, since by assumption $X = \bigcup_k X_k$, by the Baire Category Theorem some X_k, X_{k_0} say, contains an open ball $B(x_0, r) = \{x \in X : \|x - x_0\| < r\}$. This implies, in particular, that

$$\sup_{i \in I} \|T_i x\| \leq k_0 , \quad \text{all } x \in B(x_0, r) .$$

Now it is all a matter of translations: Observe that if $x \in B(0, r)$, then $x - x_0 \in B(x_0, r)$ and consequently, since $x_0 \in X_{k_0}$, we have

$$\begin{aligned} \|T_i x\| &= \|T_i(x - x_0) + T_i x_0\| \\ &\leq k_0 + \|T_i x_0\| \leq 2k_0 , \quad \text{all } i \in I . \end{aligned}$$

Next observe that given $0 \neq x \in X, (r/2\|x\|)x \in B(0, r)$, and consequently, by the above estimate we also have

$$(r/2\|x\|)\|T_i x\| = \|T_i((r/2\|x\|)x)\| \leq 2k_0 , \quad \text{all } i \in I ,$$

or equivalently,

$$\|T_i x\| \leq (4k_0/r)\|x\| , \quad \text{all } i \in I . \tag{3.2}$$

Whence letting $\delta = \varepsilon r/4k_0$, from (3.2) it follows that (3.1) is true, and the proof is complete. ■

As for the Resonance theorem, it states

Theorem 3.2 (Banach-Steinhaus). Under the assumptions of Theorem 3.1, $\mathcal{F}$ is a norm bounded set.

Proof. In the notation of Theorem 3.1, by (3.1) it readily follows that for all $i \in I$ we have

$$\sup_{x \neq 0} \frac{\|T_i x\|}{\|x\|} = \frac{1}{\delta} \sup_{x \neq 0} \|T_i((\delta/\|x\|)x)\| \leq \varepsilon/\delta ,$$

$\sup_{i \in I} \|T_i\| \leq \varepsilon/\delta$, and $\mathcal{F}$ is a norm bounded set. ■

It is interesting to note that the assumption concerning the completeness of X is necessary for the above results to hold. To see this let X be the linear subspace of ℓ^1 consisting of all sequences that have only a finite number of nonzero terms; clearly X is dense, but not closed, in ℓ^1. For each positive integer n, let T_n denote the linear functional defined on sequences $x = (x_1, \ldots x_m, \ldots)$ of X by

$$T_n x = n x_n , \quad x \in X .$$

Now, since $T_n x = 0$ for each $x \in X$ and all sufficiently large n, it is clear that the family $\{T_n\}$ is pointwise bounded on X. On the other hand, if $e_n \in \ell^1$ is the sequence which has a 1 in the nth position and zeros elsewhere, we have $e_n \in X$, $\|e_n\| = 1$, and

$$\|T_n\| \geq T_n e_n = n\,, \quad \text{all } n\,.$$

Thus $\{T_n\}$ is not a norm bounded set.

Important applications of these results, including the existence of a continuous function whose Fourier series diverges at a point, cf. the discussion following Proposition 1.2 in Chapter XVII, follow from the following restatement of Theorem 3.2 involving sequences of linear operators: Let X be a Banach space, Y a normed linear space, and $\{T_n\}$ a sequence in $\mathcal{B}(X,Y)$. Then the (good) set

$$G = \{x \in X : \limsup \|T_n\| < \infty\} \tag{3.3}$$

either coincides with X or is a set of first category in X.

This formulation enables us to generalize this result somewhat.

Theorem 3.3 (Principle of the Condensation of Singularities). Let $\{T_{m,n}\}, n = 1,2,\ldots$, be a sequence of bounded linear operators from a Banach space space X into a normed linear space $Y_m, m = 1,2,\ldots$ Suppose that for each m there exists $x_m \in X$ such that

$$\limsup_{n\to\infty} \|T_{m,n} x_m\| = \infty\,, \quad m = 1,2\ldots \tag{3.4}$$

Then the (bad) set

$$B = \{x \in X : \limsup_{n\to\infty} \|T_{m,n} x\| = \infty \text{ all } m = 1,2,\ldots\}$$

is of second category in X.

Proof. Consider the sequence $\{G_m\}$ of (good) subsets of X given by

$$G_m = \{x \in X : \limsup_{n\to\infty} \|T_{m,n} x\| < \infty\}\,, \quad m = 1,2,\ldots$$

By the preceding remark and (3.4) it readily follows that each G_m is of first category in X. Since X is complete, by the Baire Category Theorem, we get that $B = X \setminus (\bigcup_m G_m)$ is of second category in X. ∎

4. THE OPEN MAPPING THEOREM

Suppose that X, Y are normed linear spaces over the same field of scalars, and let T be a mapping from X into Y. We say that T is open at $x \in X$ if $T(V)$ contains a neighbourhood of Tx whenever V is a neighbourhood of x. We say that T is open if $T(U)$ is open in Y whenever U is open in X.

It is clear that T is open iff T is open at every $x \in X$. Because of the translation invariance of the neighbourhoods in linear spaces, it is also clear that T is open iff T is open at a single point of X.

An interesting question to consider is whether a one-to-one continuous linear mapping from a Banach space onto another has a continuous inverse. The answer is affirmative and, as we shall see below, it is a special case of our next result.

Theorem 4.1 (Open Mapping Theorem). Let X, Y be Banach spaces, and suppose $T \in \mathcal{B}(X, Y)$. If T maps X onto Y, then T is open.

Proof. The proof amounts to showing that T is open at the origin. To simplify the notation, in what follows we put $B(0, r) = B_r$, the open ball of X of radius r centered at the origin, and similarly, B'_r, the open ball of Y of radius r centered at the origin.

Now, suppose that V is a neighbourhood of 0 in X, we will be done once we show that $T(V)$ contains a neighbourhood of $T0 = 0$ in Y. Since every neighbourhood V of 0 contains a ball B_r for sufficiently small r, it suffices to prove that given an arbitrary ball B_r, $T(B_r)$ contains a ball $B'_{r'}$. We do this in two steps: First we show that given $\varepsilon > 0$, there exists $\eta > 0$ such that the closure $\overline{T(B_\varepsilon)}$ of $T(B_\varepsilon)$ contains B'_η, and then we show that the same is true of $T(B_\varepsilon)$.

First note that since for all $\varepsilon > 0$, $\bigcup_{n=1}^{\infty} B_{n\varepsilon} = X$, and since T is onto, we have

$$Y = T(X) = \bigcup_{n=1}^{\infty} T(B_{n\varepsilon}), \quad \text{all } \varepsilon > 0. \tag{4.1}$$

We are now in a position to invoke the Baire Category Theorem and conclude that at least one of the sets in the union on the right-hand side of (4.1) is not nowhere dense in Y, and consequently, its closure contains a ball. Specifically, there exist an integer n and $r > 0$ such that

$$\overline{T(B_{n\varepsilon})} \supset y_0 + B'_r = B'(y_0, r). \tag{4.2}$$

Moreover, since

$$B'_r \subseteq B'(y_0, r) - B'(y_0, r) = \{y \in Y : y = y_1 - y_2, y_1, y_2 \in B'(y_0, r)\},$$

from (4.2) we get

$$\overline{T(B_{n\varepsilon})} - \overline{T(B_{n\varepsilon})} \supseteq B'_r\,. \tag{4.3}$$

If y belongs to the set on the left-hand side of (4.3), there are sequences $\{x_k\}$ and $\{x'_k\}$ of points in $B_{n\varepsilon}$ such that

$$y = \lim_{k\to\infty} Tx_k - \lim_{k\to\infty} Tx'_k = \lim_{k\to\infty} T(x_k - x'_k)\,.$$

Since the points $x_k - x'_k \in B_{2n\varepsilon}$ for all k, by (4.3) we get that also

$$\overline{T(B_{2n\varepsilon})} \supseteq B'_r\,. \tag{4.4}$$

Finally, since

$$B_{2n\varepsilon} = 2nB_\varepsilon\{x \in X : x = 2nx', x' \in B_\varepsilon\}$$

and $T(2nB_\varepsilon) = 2nT(B_\varepsilon)$, (4.4) gives

$$\overline{T(B_\varepsilon)} \supseteq B'_\eta\,, \quad \eta = r/2n\,, \tag{4.5}$$

and our first assertion is true.

Next, given $\varepsilon > 0$, put $\varepsilon_n = \varepsilon/2^n$, $n = 1, 2, \ldots$, and for each n let η_n be the choice of η corresponding to ε_n in (4.5) above. Whence

$$\overline{T(B_{\varepsilon_n})} \supseteq B'_{\eta_n}\,, \quad n = 1, 2, \ldots \tag{4.6}$$

Clearly we may, and do, assume that $\lim_{n\to\infty} \eta_n = 0$.

Put now $\eta_1 = \eta$, and suppose $y \in B'_\eta$. By (4.6) also $y \in \overline{T(B_{\varepsilon_1})}$, and consequently, y can be approximated as close as we want by points in $T(B_{\varepsilon_1})$. In particular, there exists $x_1 \in B_{\varepsilon_1}$ such that $\|y - Tx_1\| < \eta_1$. In this case, since $y - Tx_1 \in B'_{\eta_2}$, we can find $x_2 \in B_{\varepsilon_2}$ such that $\|y - Tx_1 - Tx_2\| < \eta_3$.

In general, having chosen $x_i \in B_{\varepsilon_i}, 1 \le i \le k$, pick $x_{k+1} \in B_{\varepsilon_{k+1}}$ with the property that

$$\|y - Tx_1 - \cdots - Tx_k - Tx_{k+1}\| < \eta_{k+1}\,. \tag{4.7}$$

We claim that $\sum x_k$ converges to a point $x \in B_\varepsilon$, and that $y = Tx$. If this is the case, then we have $T(B_\varepsilon) \supseteq B'_\eta$, and the second assertion is also true.

To see that $\sum x_k$ converges, since X is complete it suffices to check that $\sum \|x_k\| < \infty$. Now, since $x_k \in B_{\varepsilon_k}$ for all k, it readily follows that $\sum_{k=1}^{\infty} \|x_k\| < \sum_{k=1}^{\infty} \varepsilon/2^{k+1} = \varepsilon$. Whence $\sum_{k=1}^{\infty} x_k$ converges to an element $x \in X$ with $\|x\| < \varepsilon$. Also, by the continuity of T we get that $T(\sum_k x_k) = Tx$, and, since $\lim_{k\to\infty} \eta_k = 0$, from (4.7) it is clear that

$$\lim_{n\to\infty} \left\| y - T\left(\sum_{k=1}^{n} x_k\right)\right\| = 0\,.$$

Thus $y = Tx$, and we have finished. ∎

A word about the hypothesis of Theorem 4.1. The assumption that T is continuous is not essential, cf. 6.29 below, and the assumption that T is onto may be replaced by the assumption that the range $R(T)$ of T is of second category in Y. On the other hand, this last assumption, as well as the completeness of X, are necessary for the map T to be open. We postpone the presentation of specific examples until after the proof of the Closed Graph Theorem.

Now, concerning the inverses, we have

Corollary 4.2. Let X, Y be Banach spaces and assume $T \in \mathcal{B}(X,Y)$ is a one-to-one mapping from X onto Y. Then T^{-1} is a well-defined bounded linear mapping from Y onto X.

Proof. By Theorem 4.1 there exists $\eta > 0$ such that

$$T(B_1) \supseteq B'_\eta\,. \tag{4.8}$$

Now, since T^{-1} is well-defined, (4.8) is equivalent to

$$B_1 \supseteq T^{-1}(B'_\eta)\,. \tag{4.9}$$

Thus, if $0 \neq y \in Y$, we have $(\eta/2\|y\|)y \in B'_\eta$, and by (4.9) it follows that

$$\|T^{-1}((\eta/2\|y\|)y)\| \le 1\,.$$

By the linearity of T^{-1} we get $\|T^{-1}y\| \le (2/\eta)\|y\|$, and consequently, $T^{-1} \in \mathcal{B}(Y,X)$, and $\|T^{-1}\| \le (2/\eta)$. ∎

Corollary 4.3. Suppose that the linear space X is normed by $\|\cdot\|$ and by $\|\cdot\|_1$, and that, endowed with both norms, X is complete. Then if for some constant c we have

$$\|x\| \le c\|x\|_1\,, \quad \text{all } x \in X\,, \tag{4.10}$$

there is a constant k such that

$$\|x\|_1 \le k\|x\|\,, \quad \text{all } x \in X\,. \tag{4.11}$$

More precisely, the norms are equivalent.

Proof. Let T denote the identity map from X, normed by $\|\cdot\|_1$, into X, normed by $\|\cdot\|$; clearly T is linear, one-to-one and onto, and, by (4.10), it is also continuous. By Corollary 4.2, $T^{-1} = T$ is also continuous, and (4.11) holds. ■

The completeness of X under both norms is essential for (4.11) to be true. Referring to the construction following Proposition 2.4 in Chapter XIV, if X is a Banach space, and if we introduce the norm $\|\cdot\|_\infty$ in X, then (4.10) holds. Now, if (4.11) were to hold, then X normed by $\|\cdot\|_\infty$ would become a complete normed space, which it is not.

5. THE CLOSED GRAPH THEOREM

Many important operators in Analysis enjoy the following property: They are well-defined on a dense subspace of a normed linear space X, and yet fail to be continuous. For instance, put $I = [0,1]$, and let $X = Y = C(I)$, and $X_1 = C^1(I) \subset X$; X_1 is dense in the uniform norm of X, a proof of this will be given in Corollary 2.3 in Chapter XVII. Consider now the linear operator $T: X_1 \to Y$ given by

$$Tf = f', \quad \text{or} \quad Tf(x) = f'(x), \quad \text{all } x \in I.$$

Thus T is a densely defined operator, and it is clear that it is not bounded since the sequence $\{f_n\}$ consisting of the functions

$$f_n(x) = x^n, \quad n = 1,2\ldots$$

satisfies $\|Tf_n\| = n$ and $\|f_n\| = 1$ for all n.

The challenge is to incorporate operators such as T into the theory we are developing, and to discover what properties they satisfy. Referring to the differentiation operator T, we are interested in considering sequences $\{f_n\} \subseteq C^1(I)$, $f_n \to f \in C(I)$, and the corresponding sequences $\{Tf_n\} \subseteq C(I)$. As observed above, $\{Tf_n\}$ need not converge, but when it does, i.e., if $\lim_{n\to\infty} Tf_n = g \in C(I)$, then the following is true: Since the sequence $\{f_n'\}$ converges uniformly to g on I, the sequence $\{f_n\}$ converges uniformly to an antiderivative of g on I. But since by assumption also $\{f_n\}$ converges uniformly to f on I, it follows that $f \in C^1(I)$ and that $Tf = g$.

Because of the importance of this example we formalize these considerations into a definition.

Let X, Y be normed spaces, and let $T: D(T) \to Y$ be a linear mapping. We say that T is closed in X, if for any sequence $\{x_n\} \subseteq D(T)$,

$$\lim_{n\to\infty} x_n = x\,, \quad \text{and} \quad \lim_{n\to\infty} Tx_n = y \tag{5.1}$$

imply

$$x \in D(T), \quad \text{and} \quad Tx = y. \tag{5.2}$$

As the differentiation mapping shows, not all closed operators are continuous. The opposite is also true; namely, not all continuous operators are closed. For instance, if X_1 is a proper dense subspace of a normed space $X = Y$, then the identity map $T: X_1 \to Y$ is obviously bounded, but not closed.

The Closed Graph Theorem, a close relation to the Open Mapping Theorem, establishes when a closed mapping is bounded. In order to prove it we find it convenient to consider a more "geometric" setting. First a definition.

Let X, Y be normed linear spaces and let $X \times Y$ be the linear space normed by

$$\|(x, y)\| = \|x\| + \|y\|.$$

Given a linear mapping $T: D(T) \to Y$, the graph $G(T)$ of T is the set

$$G(T) = \{(x, Tx): x \in D(T)\} \subseteq X \times Y.$$

Since T is linear, $G(T)$ is a linear subspace of $X \times Y$. Now, when T is closed, (5.1) and (5.2) imply that $G(T)$ is a closed subspace of $X \times Y$. The converse is also true: If $G(T)$ is closed in $X \times Y$, then T is closed in X. Thus, the concepts T closed and $G(T)$ closed are interchangeable.

It is also clear that if $D(T)$ is a closed subspace of X and T is continuous, then T is closed in X. The remarkable fact is that for Banach spaces the converse to this statement is also true.

More precisely, we have

Theorem 5.1 (Closed Graph Theorem). Let X, Y be Banach spaces, and suppose $T: X \to Y$ is linear. If T is closed in X, then T is continuous in X.

Proof. Since $X \times Y$ is a complete normed space, and since by assumption $G(T)$ is a closed subspace of $X \times Y$, $G(T)$ is also a Banach space in the norm induced by that of $X \times Y$. Consider now the (projection) linear mapping $P: G(T) \to X$ given by

$$P((x, Tx)) = x, \quad x \in X.$$

Note that in addition to being linear, P is one-to-one and onto X. Moreover, since also

$$\|P((x, Tx))\| = \|x\| \leq \|x\| + \|Tx\| = \|(x, Tx)\|,$$

P is also bounded. Thus by Corollary 4.2, the inverse $P^{-1}: X \to G(T)$ of P given by

$$P^{-1}x = (x, Tx), \quad x \in X,$$

is also bounded. Specifically, there exists a constant c such that

$$\|(x, Tx)\| = \|x\| + \|Tx\| \leq c\|x\|, \quad x \in X.$$

This clearly implies that $\|Tx\| \leq c\|x\|$, i.e., that T is bounded. ■

For the validity of the Closed Graph Theorem it is essential that both the domain X and the target space Y be complete, as may be seen by the following examples.

As pointed out above, the differential operator $T: C^1(I) \to C(I)$ is closed but not bounded; in this case the domain $X = C^1(I)$ is not complete. An example along similar lines, roughly speaking it corresponds to differentiation of Fourier series, is the following: Let $\ell^1 = \{(a_n): \|(a_n)\|_1 = \sum_{n=1}^{\infty} |a_n| < \infty\}$, and

$$X = \left\{x = (a_n) \in \ell^1 : \sum_{n=1}^{\infty} n|a_n| < \infty\right\}.$$

Since X is a proper dense subspace of ℓ^1, it is not complete. Let now $T: X \to \ell^1$ be the mapping defined by

$$Tx = (na_n), \quad x = (a_n) \in X.$$

It is easy to check that T is well-defined and closed, and since for the sequence $e_n \in \ell^1$ which has a 1 in the nth position and zeros elsewhere we have

$$\|Te_n\|_1 = n, \quad \text{and} \quad \|e_n\|_1 = 1,$$

it follows that T is not bounded. Now, since T is also one-to-one and onto ℓ^1, it has a well-defined inverse $T^{-1}: \ell^1 \to X$; in fact, we have

$$T^{-1} = (a_n/n), \quad x = (a_n) \in \ell^1.$$

Observe that indeed T^{-1} is defined on the whole of ℓ^1 since $x \in \ell^1$ implies $T^{-1}x \in X$, and

$$\|T^{-1}x\|_1 = \sum |a_n|/n \leq \sum |a_n| = \|x\|_1.$$

In fact, the above remark shows that T^{-1} is bounded, and putting $x = e_1$ there we also get that $\|T^{-1}\| = 1$. Moreover, since $x = T^{-1}(Tx)$, it follows that T^{-1} is onto X. Now, T^{-1} is not open, for if it were open, then $(T^{-1})^{-1} = T$ would be continuous, and, as we saw above, this is not the case. Thus, for the validity of the Open Mapping Theorem it is essential that the target space be complete.

Consider next an infinite dimensional Banach space X, let H be a (necessarily infinite) Hamel basis for X such that $\|h\| = 1$ for all $h \in H$, and let $\|\cdot\|_1$ be the norm on X given by

$$\|x\|_1 = \sum_{i=1}^{n} |a_i|\,, \quad x = \sum_{i=1}^{n} a_i h_i\,, \quad h_i \in H\,,\, 1 \leq i \leq n\,.$$

Let Y denote the linear space X endowed with the metric $\|\cdot\|_1$, and let $T: X \to Y$ be the identity map. T is one-to-one, onto and, as it is readily verified, closed. However T is not bounded, for otherwise the fact that X is complete would imply that Y is also complete, and this is not the case. Thus, for the Closed Graph Theorem to be true it is essential that the target space Y be complete.

Now, $T^{-1}: Y \to X$ is also the identity, and as such it is one-to-one and onto. Moreover, since

$$\|x\| \leq \|x\|_1\,, \quad \text{all } x \in X\,,$$

T^{-1} is also bounded. However, T^{-1} is not open, for if it were open, then $(T^{-1})^{-1} = T$ would be continuous, and as pointed above, this is not the case. Thus, for the Open Mapping Theorem to be true, the domain must be complete.

6. PROBLEMS AND QUESTIONS

6.1 Show that any continuous function on [0,1] can be approximated uniformly and arbitrarily closely by a piecewise linear continuous function.

6.2 Let $\phi(x)$ denote the function that assigns to each real x the distance from x to the nearest integer. Prove that for appropriately chosen sequences (ε_n) and (k_n) the function given by the uniformly convergent series $\sum_{n=1}^{\infty} \varepsilon_n \phi(k_n x)$ is nowhere differentiable.

6.3 Let X, Y be Banach spaces and $L(x, y)$ be a functional on $X \times Y$, continuous and linear in each variable separately. Prove that L is continuous at (0,0), and consequently everywhere.

6.4 If X is a finite dimensional normed linear space, prove that every linear operator $T: X \to X$ is bounded.

6.5 Let $X, Y \neq \{0\}$ be normed linear spaces and suppose the dimension of X is infinite. Show that there is at least one unbounded linear operator $T: X \to Y$.

6.6 If $T \in \mathcal{B}(X,Y)$, $T \neq 0$, and $\|x\| < 1$, then $\|Tx\| < \|T\|$. Is it also true that $\|Tx\| < \|x\|$?

6.7 Suppose $T, T_n \in \mathcal{B}(X,Y)$, $n = 1,2,\ldots$, $\sup \|T_n\| < \infty$, and $\lim_{n\to\infty} \|T_n x - Tx\| = 0$ for every x in a dense subset of X, does it follow that $\lim_{n\to\infty} \|T_n - T\| = 0$?

6.8 Suppose $0 \neq T \in \mathcal{B}(X)$ and $\{x_n\} \subseteq X$ has the property that $\lim_{n\to\infty} \|x_n\| = \infty$. Does it follow that $\lim_{n\to\infty} \|Tx_n\| = \infty$?

6.9 Suppose $T_1, T_2 \in \mathcal{B}(X)$. Show that $T_1 T_2 \in \mathcal{B}(X)$, and

$$\|T_1 T_2\| \leq \|T_1\| \, \|T_2\|$$

6.10 Prove that if X is a Banach space and $T \in \mathcal{B}(X)$ and $\|T\| < 1$, then the "geometric series" $I + T + \cdots + T^n + \cdots$ converges in $\mathcal{B}(X)$. What does it converge to?

6.11 Referring to 6.7, the condition $\|T\| < 1$ is not necessary for $I + T + \cdots + T^n + \cdots$ to converge in $\mathcal{B}(X)$. For, suppose

$$\lim_{n\to\infty} \sqrt[n]{\|T^n\|} = L\,,$$

exists. Show that if $L < 1$ the above series converges and if $L > 1$ it does not.
Further, prove that a necessary and sufficient condition for the series to converge is that for some k we have $\|T^k\| < 1$.

6.12 (Banach) Let X be a Banach space and suppose $T \in \mathcal{B}(X)$ is such that $\|T\| \leq \eta < 1$. Prove that the operator $I - T$ has a continuous inverse $(I - T)^{-1}$ and $\|(I - T)^{-1}\| \leq 1/(1 - \eta)$.

6.13 Let $T_0 \in \mathcal{B}(X,Y)$, where X and Y are Banach spaces, and suppose T_0 has a bounded inverse $T_0^{-1} \in \mathcal{B}(Y,X)$. Show that if an operator $T \in \mathcal{B}(Y,X)$ satisfies $\|T\| < 1/\|T_0^{-1}\|$, then the operator $U = T_0 + T: Y \to X$ has a continuous inverse and

$$\|U^{-1}\| \leq \|T_0^{-1}\|/(1 - \|T_0^{-1}T\|) \leq \|T_0^{-1}\|/(1 - \|T_0^{-1}\| \, \|T\|)\,.$$

6.14 Let X, Y be normed linear spaces and let T_0 be a mapping from a subset M of X into Y. Show that a necessary and sufficient condition for T_0 to have a bounded linear extension to the span of M is that there exists a constant k such that

$$\left\|\sum_{n=1}^{m} \lambda_n T_0 x_n\right\| \le k \left\|\sum_{n=1}^{m} \lambda_n x_n\right\| ,$$

for any $x_1, \ldots, x_m$ in M and scalars $\lambda_1, \ldots, \lambda_m$.

6.15 Let X, Y be normed linear spaces and suppose Y is complete. Show that every continuous linear operator T_0 from a subset M of X into Y has a unique continuous linear extension T to the closure of M into Y, and $\|T\| = \|T_0\|$.
In particular, prove that if a continuous linear operator T from a normed linear space X into a Banach space Y maps a dense subset of X into 0, then $Tx = 0$ for all $x \in X$.

6.16 Let X, Y be Banach spaces and $T: X \to Y$, T linear. Prove that if $L \circ T \in X^*$ for every $L \in Y^*$, then $T \in \mathcal{B}(X, Y)$.

6.17 Suppose X, Y are Banach spaces and $\{T_n\} \subseteq \mathcal{B}(X, Y)$. Prove that if for each $L \in Y^*$ we have

$$\sup |LT_n x| < \infty , \quad \text{all } x \in X ,$$

then $\sup \|T_n\| < \infty$.

6.18 Assume $\{T_n\} \subseteq \mathcal{B}(X)$ and $\lim_{n\to\infty} T_n x = Tx$ exists for each x in X. Show that T is a bounded linear operator on X and that $\|T\| \le \limsup_{n\to\infty} \|T_n\|$.

6.19 Let X be a Banach space, and $T, T_1, T_2, \ldots$ be bounded linear operators defined on X with the property that $\lim_{n\to\infty} T_n x = Tx$ for all $x \in X$. Prove that there exists a constant $c > 0$ such that $\sup \|T_n\| \le c$.

6.20 As an application of the Uniform Boundedness Principle show that the space X of polynomials $p(x) = \sum_{n=0}^{\infty} a_n x^n$, where $a_n = 0$ for all but finitely many n's, normed by $\|p\| = \max |a_n|$, is not complete.

6.21 Let X be a normed linear space, Y a Banach space, and suppose $T \in \mathcal{B}(X, Y)$. If N denotes the null space of T we may define a map $T^*: X/N \to Y$ as follows: For each class $x + N$, let $T^*(x + N) = Tx$. Prove that $T^* \in \mathcal{B}(X/N, Y)$, and that if X is a Banach space, then T^* is an isomorphism.

6.22 Let X be an infinite-dimensional Banach space, and $\{x_n\}$ a linearly independent set in X. Show that for each $n = 1, 2, \ldots$, the linear span $\{x_1, \ldots, x_n\}$ is a nowhere dense subset of X. As a consequence of this result prove that the dimension of every infinite-dimensional Banach space is at least $\aleph_1$.

6.23 Prove that if X is an infinite-dimensional Banach space, there is an embedding of ℓ^∞ into X.

6.24 Prove that every separable Banach space is isomorphic to some quotient space of ℓ^1.

6.25 Let $1 < p < \infty$. If $\sum_n a_n b_n$ converges for every sequence (b_n) such that $\sum_n |b_n|^p < \infty$, prove that $\sum_n |a_n|^q < \infty$, where q is the index conjugate to p.

6.26 Prove that there is no sequence of positive real numbers (a_n) such that $\sum_n a_n |b_n|$ converges iff the sequence (b_n) is bounded.

6.27 Let X be a Banach space, and assume Y and Z are closed subspaces of X. If each $x \in X$ has a unique representation of the form $x = y + z$ with $y \in Y$ and $z \in Z$, show that there exists a constant c such that for all $x = y + z$ we have $\|y\|, \|z\| \le c\|x\|$.

6.28 Let $I = [0,1]$. Show that $L^p(I)$ is properly contained in $L(I)$ for each $p > 1$, and conclude that in the metric of $L(I)$, $L^p(I)$ is a set of first category in $L(I)$.

6.29 Let X be a Banach space and Y a normed linear space of second category. Prove that if the linear mapping $T: X \to Y$ is closed and onto, then T takes open sets into open sets.

6.30 Let I be a compact interval of the line, and K a closed subinterval of I with the following property: For every function $g \in C(K)$ there exists a function $f \in C(I)$ such that $f|K = g$. Show that there exists a constant $c > 0$ with the property that for every $g \in C(K)$ there exists $f \in C(I)$ such that $f|K = g$, and $\max_I |f| \le \max_K |g|$.

6.31 Let $I = [0,1]$ and suppose A is a closed subspace of $C(I)$. Suppose that for each $f \in A$, $Tf = \phi f \in A$, where ϕ is a real-valued function defined on I. Show that T is continuous in the norm of A. Is ϕ necessarily continuous?

6.32 Let X be a Banach space, and suppose that T_1, T_2 are linear operators from X into itself, and that $T_2 \in \mathcal{B}(X)$ is also one-to-one. Show that $T_1 \in \mathcal{B}(X)$ iff $T_2 T_1 \in \mathcal{B}(X)$.

6.33 Let $I = [0,1]$, $1 < p < \infty$, and suppose Tf is the linear operator defined on $L^p(I)$ by $Tf(x) = \int_I k(x,y)f(y)\,dy$, $x \in I$. Specifically, assume that for almost every $x \in I$, the function $k(x,y)f(y)$ is integrable as a function of $y \in I$, and $Tf \in L^p(I)$. Prove that T is bounded.

6.34 Let X,Y be normed linear spaces and suppose $T: X \to Y$. Prove that if T is closed and one-to-one, then T^{-1} is also closed.
Also, if $T \in \mathcal{B}(X,Y)$ and the domain $D(T)$ of definition of T is closed, then T is closed.

6.35 Suppose X,Y are Banach spaces, and let $T \in \mathcal{B}(X,Y)$. Show that $R(T)$ is closed in Y iff there exists a constant $c > 0$ such that $\inf\{\|x - y\| : Ty = 0\} \leq c\|Tx\|$ for all $x \in X$.

6.36 Let Y be a subspace of a normed linear space X. A mapping $P \in \mathcal{B}(X,Y)$ is said to be a projection of X onto Y if P maps X onto Y and $P^2 = P$.
Suppose now Y is a closed subspace of a Banach space X. Show that there exists a projection P of X onto Y iff there exists a closed subspace Z of X such that $X = Y \oplus Z$, i.e., $Y \cap Z = \{0\}$ and every $x \in X$ can be written uniquely as $x = y + z$ with y in Y and z in Z. If this is the case, there also exists a constant $k > 0$ such that

$$\|y + z\| \geq k\|y\|, \quad y \in Y, z \in Z.$$

6.37 Let X,Y be normed linear spaces and suppose $T: D(T) \subseteq X \to Y$. T is said to be closable if there exists a linear extension of T to all of X which is closed in X; the domain $D(T)$ of T is not required to be dense in X.
Prove that the following are equivalent:
(i) T is closable.
(ii) For any $y \neq 0$ in Y, $(0,y)$ is not in the closure of the graph of T.
(iii) T has a minimal closed linear extension, i.e., there exists a closed linear extension T^* of T such that any closed linear extension of T is a closed linear extension of T^*.

6.38 Let X,Y be Banach spaces and $T: X \to Y$ be closed. Prove that T has a bounded inverse iff T is one-to-one and has a closed inverse.

CHAPTER **XVI**

Hilbert Spaces

In this chapter we consider those normed spaces where, as in the case of the Euclidean space R^n, there is an inner product which is connected with the norm by a simple relation: The square of the norm of an element is the inner product of that element with itself. Some examples to keep in mind are, of course, R^n and C^n, and what we will have the occasion to verify is a universal model of a Hilbert space, $L^2(\mu)$.

1. THE GEOMETRY OF INNER PRODUCT SPACES

A complex vector space X is said to be an inner product space provided there is a complex-valued map defined on $X \times X$ denoted by $\langle \cdot, \cdot \rangle$, and called an inner product on X, or plainly an inner product, which satisfies the following properties:

(i) $\langle x_1 + \lambda x_2, y \rangle = \langle x_1, y \rangle + \lambda \langle x_2, y \rangle$, all x_1, x_2 in X, and $\lambda \in C$.

(ii) $\langle x, y \rangle = \overline{\langle y, x \rangle}$, all x, y in X.

(iii) $\langle x, x \rangle \geq 0$, and $\langle x, x \rangle = 0$ iff $x = 0$.

Of course we may also consider real inner product spaces, and in this case we restrict our attention to $\lambda \in R$, and require that $\langle \cdot, \cdot \rangle$ be real valued.

Property (i) is known as the linearity of the inner product, and together with (ii) it implies "conjugate linearity" in the second variable, to wit,

(iv) $\langle x, \lambda y \rangle = \overline{\lambda} \langle x, y \rangle$, all $x, y \in X$ and $\lambda \in C$.

An immediate consequence of these properties is that $\langle x, 0 \rangle = \langle 0, x \rangle = 0$ for all $x \in X$.

Inner product spaces satisfy an important inequality, which we prove next.

Proposition 1.1 (The Cauchy-Schwarz Inequality). For any x, y in X we have

$$|\langle x, y\rangle|^2 \leq \langle x, x\rangle \langle y, y\rangle . \tag{1.1}$$

Proof. Note that for any x, y in X and scalars λ, by (iii), (iv) and (ii) above it follows that

$$\begin{aligned} 0 &\leq \langle x+\lambda y, x+\lambda y\rangle = \langle x,x\rangle + \langle x, \lambda y\rangle + \langle \lambda y, x\rangle + \langle \lambda y, \lambda y\rangle \\ &= \langle x,x\rangle + \overline{\lambda}\langle x,y\rangle + \lambda\langle y,x\rangle + \lambda\overline{\lambda}\langle y,y\rangle \\ &= \langle x,x\rangle + \overline{\lambda}\langle x,y\rangle + \lambda\overline{\langle x,y\rangle} + |\lambda|^2\langle y,y\rangle \\ &= \langle x,x\rangle + 2\Re\left(\overline{\lambda}\langle x,y\rangle\right) + |\lambda|^2\langle y,y\rangle . \end{aligned}$$

Now, if $y = 0$, (1.1) is trivially true since both sides there are 0. On the other hand, if $y \neq 0$, then $\langle y, y\rangle > 0$, and putting $\lambda = -\langle x, y\rangle / \langle y, y\rangle$ the above estimate becomes

$$\langle x,x\rangle - \frac{2|\langle x,y\rangle|^2}{\langle y,y\rangle} + \frac{|\langle x,y\rangle|^2}{\langle y,y\rangle^2}\langle y,y\rangle = \langle x,x\rangle - \frac{|\langle x,y\rangle|^2}{\langle y,y\rangle} \geq 0 ,$$

which is clearly equivalent to (1.1). ■

If in an inner product space X we set

$$\|x\| = \sqrt{\langle x,x\rangle}, \quad x \in X , \tag{1.2}$$

then X is turned into a normed space. Indeed, with the exception of the triangle inequality, all other properties of the norm follow at once from (1.2) and the properties of the inner product. As for the triangle inequality, given $x, y \in X$, observe that

$$\|x+y\|^2 = \langle x+y, x+y\rangle = \|x\|^2 + 2\Re\langle x,y\rangle + \|y\|^2 , \tag{1.3}$$

which, by the Cauchy-Schwarz inequality, is dominated by

$$\|x\|^2 + 2\|x\|\,\|y\| + \|y\|^2 = (\|x\| + \|y\|)^2 .$$

In other words, we get $\|x+y\|^2 \leq (\|x\| + \|y\|)^2$, which is equivalent to the triangle inequality.

An easy consequence of the Cauchy-Schwarz inequality is the continuity of the inner product. More precisely, if $x_n \to x$ and $y_n \to y$, then $\langle x_n, y_n\rangle \to \langle x, y\rangle$. This observation follows from the estimate

$$\begin{aligned} |\langle x_n, y_n\rangle - \langle x,y\rangle| &= |\langle x_n, y_n\rangle \pm \langle x_n, y\rangle - \langle x,y\rangle| \\ &\leq |\langle x_n, y_n - y\rangle| + |\langle x_n - x, y\rangle| \\ &\leq \|x_n\|\,\|y_n - y\| + \|x_n - x\|\,\|y\| , \end{aligned}$$

and the fact that convergent sequences are bounded.

An inner product space X endowed with the norm introduced in (1.2) is said to be a pre-Hilbert space. A complete pre-Hilbert space is called a Hilbert space. For instance, $L^2(\mu)$ is a Hilbert space with inner product given by

$$\langle f, g\rangle = \int_X f\overline{g}\, d\mu\,, \quad f, g \in L^2(\mu)\,.$$

ℓ^2 is the prototype of a Hilbert space. It was introduced by Hilbert in the early 1900's in his work on integral equations. The axiomatic definition of a Hilbert space was not given until much later by J. von Neumann(1903-1957) in the mid 1920's, in a paper dealing with the mathematical foundations of Quantum Mechanics.

In what follows we restrict our attention to Hilbert spaces since, as we show next, every inner product space can be "completed", and the completion is a Hilbert space, unique up to isomorphisms.

In the present context, a linear mapping $T: X \to Y$ from an inner product space X onto another inner product space Y over the same field of scalars is said to be an isomorphism if it preserves inner products, i.e.,

$$\langle Tx, Ty\rangle = \langle x, y\rangle\,, \quad \text{all } x, y \in X\,.$$

Thus, isomorphisms of inner product spaces preserve their whole structure, including inner products and norms.

Proposition 1.2. Suppose X is an inner product space. Then there exist a Hilbert space Y and an isomorphism T of X onto a dense subspace of Y. The space Y is unique up to isomorphisms.

Proof. Because the proof follows along familiar lines we only sketch it. By Theorem 3.5 in Chapter XIV there exist a unique, up to isometries, Banach space Y and an isometric (in the linear sense) isomorphism T of X onto Y.

Consider now the complex-valued map $\langle\cdot,\cdot\rangle_1$ defined on $Y \times Y$ as follows: If $x, y \in Y$, and $\{x_n\}$ and $\{y_n\}$ are sequences of elements in X that converge in Y to x and y respectively, let

$$\langle x, y\rangle_1 = \lim_{n\to\infty} \langle x_n, y_n\rangle\,.$$

By the continuity of the inner product it is not hard to see that $\langle\cdot,\cdot\rangle_1$ is well-defined, i.e., it is independent of the approximating sequences chosen, and that it is an inner product on Y. Also, by (1.2), it follows that T is an isomorphism of X onto Y. ■

The notion of Hilbert space is an immediate generalization of Euclidean space, so its "geometry" approaches Euclidean geometry more closely than that of other Banach spaces. For instance, the "parallelogram law" holds: For any $x, y \in X$ we have

$$\|x+y\|^2 + \|x-y\|^2 = 2\left(\|x\|^2 + \|y\|^2\right) . \tag{1.4}$$

Indeed, (1.4) follows at once by adding (1.3) and the expression we obtain replacing y by $-y$ there.

In fact, even more is true.

Proposition 1.3. A normed linear space X is an inner product space iff the "parallelogram law" holds.

Proof. It only remains to check that if (1.4) holds, then X is an inner product space. The proof of this result is entirely computational since we can exhibit explicitly the inner product $\langle\cdot,\cdot\rangle$ associated to the norm $\|\cdot\|$ in X: It is given by the expression

$$4\langle x, y\rangle = \left(\|x+y\|^2 - \|x-y\|^2\right) + i\left(\|x+iy\|^2 - \|x-iy\|^2\right) . \tag{1.5}$$

When X is a real normed space the second summand above is ommited. The details of the straightforward, and tedious, computation needed to verify the properties of the inner product are left to the reader. ■

An interesting application of Proposition 1.3 is that the Lebesgue L^p spaces are inner product spaces only if $p = 2$. For instance, in the case of ℓ^p, consider $x = (1, 1, 0, 0, \ldots)$, and $y = (1, -1, 0, 0, \ldots)$. We then have

$$\|x\|_p = \|y\|_p = 2^{1/p} , \quad \text{and} \quad \|x+y\|_p = \|x-y\|_p = 2 ,$$

and for these elements (1.4) holds iff $p = 2$.

Another important notion is that of orthogonality. Elements x, y in an inner product space X are said to be orthogonal, and we write $x \perp y$, when $\langle x, y\rangle = 0$.

If $x \in X$ is orthogonal to each element of a subset A of X, then x is said to be orthogonal to A, and we write $x \perp A$.

In this context we have

Proposition 1.4 (Pythagorean Theorem). Suppose $\{x_i\}_{i=1}^n$ is a collection of pairwise orthogonal elements of X. Then

$$\left\|\sum_{i=1}^n x_i\right\|^2 = \sum_{i=1}^n \|x_i\|^2 . \tag{1.6}$$

Proof. By definition, the left-hand side of (1.6) equals

$$\left\langle \sum_{i=1}^{n} x_i, \sum_{j=1}^{n} x_j \right\rangle = \sum_{i,j=1}^{n} \langle x_i, x_j \rangle .$$

Also, since the x_i's are pairwise orthogonal, the sum on the right-hand side above equals $\sum_{i=1}^{n} \|x_i\|^2$. ■

Our next goal is to explore whether given a subspace M of a Hilbert space X and $x \in X \setminus M$, we can find $y \in M$ such that

$$d(x, M) = \inf \{ \|x' - x\| : x' \in M \} = \|x - y\| . \tag{1.7}$$

This question is related to that of dropping a perpendicular from x to M, or "projecting" x onto M. Simple examples in R^2 when M is an open segment or an arc show that there may exist no points $y \in M$ which satisfy (1.7), or that there may exist infinitely many such y's.

The following result handles the difficulties raised by these examples.

Proposition 1.5 (Existence of the Minimizing Element). Let X be an inner product space and M a nonempty, complete, convex subset of X. Then for every $x \in X$ there exists a unique $y \in M$ such that

$$d(x, M) = \|x - y\| . \tag{1.8}$$

Proof. If $x \in M$, then (1.8) holds with $y = x$. Otherwise, if $x \notin M$, let $\{y_n\} \subseteq M$ be a minimizing sequence, i.e.,

$$\lim_{n\to\infty} \|x - y_n\| = d(x, M) . \tag{1.9}$$

We claim that the sequence $\{y_n\}$ is Cauchy. Indeed, by the parallelogram law we have

$$\begin{aligned}\|y_n - y_m\|^2 &= \|(y_n - x) - (y_m - x)\|^2 \\ &= 2\|y_n - x\|^2 + 2\|y_m - x\|^2 - \|(y_n - x) + (y_m - x)\|^2 \\ &= 2\|y_n - x\|^2 + 2\|y_m - x\|^2 - 4\|(y_n + y_m)/2 - x\|^2 .\end{aligned}$$

Now, since M is convex it follows that $(y_n + y_m)/2 \in M$, and consequently,

$$d(x, M) \le \|(y_n + y_m)/2 - x\| .$$

Thus the above estimate becomes

$$\|y_n - y_m\|^2 \le 2\|y_n - x\|^2 + 2\|y_m - x\|^2 - 4d(x, M) , \tag{1.10}$$

and, in view of the choice of the y_n's, the right-hand side of (1.10) goes to 0 as $n, m \to \infty$. But this implies that the sequence $\{y_n\} \subseteq M$ is Cauchy and, since M is complete, that it converges to a limit $y \in M$, say. Furthermore, passing to the limit in (1.9) it follows at once that (1.8) holds.

It only remains to check the uniqueness of y. Suppose y and y' are elements of M that satisfy (1.8). Since (1.10) is actually true for arbitrary elements of M, by setting $y_n = y$ and $y_m = y'$ there, we get that $\|y - y'\| \leq 0$. Thus $y = y'$. ∎

Turning from arbitrary convex sets to subspaces we obtain the result alluded to above concerning projections. But first a definition.

Given a subset A of an inner product space X, let the orthogonal complement $A^\perp$ of A be the set of all elements of X orthogonal to A, to wit,

$$A^\perp = \{x \in X : x \perp y \text{ for all } y \in A\} . \tag{1.11}$$

An elementary and important property of the orthogonal complement is

Proposition 1.6. $A^\perp$ is a closed subspace of X.

Proof. First observe that if $x_1, x_2 \in A^\perp$, λ is a scalar and $y \in A$, then we have

$$\langle x_1 + \lambda x_2, y\rangle = \langle x_1, y\rangle + \lambda\langle x_2, y\rangle = 0 ,$$

and consequently, $x_1 + \lambda x_2 \in A^\perp$.

Next suppose that $\{x_n\} \subseteq A$ and $\lim_{n\to\infty} x_n = x$. Now, if $y \in A$, we get that

$$|\langle x, y\rangle| = |\langle x - x_n, y\rangle| \leq \|x - x_n\| \, \|y\| \to 0 \quad \text{as } n \to \infty .$$

Thus, $x \in A^\perp$, and $A^\perp$ is closed. ∎

We are now ready to prove a result of fundamental importance in the theory of Hilbert spaces, the projection theorem.

Theorem 1.7. Let X be a Hilbert space and M a complete subspace of X. Then every element $x \in X$ can be expressed in the form

$$x = x_1 + x_2 , \quad x_1 \in M, x_2 \in M^\perp . \tag{1.12}$$

Furthermore, the representation is unique, and

$$\|x\| = \|x_1\| + \|x_2\| . \tag{1.13}$$

Proof. If $x \in M$ we put $x_1 = x$ and $x_2 = 0$. Otherwise, let x_1 be the unique element of M which satisfies (1.8), i.e.,

$$d(x, M) = \|x - x_1\| . \tag{1.14}$$

Next we verify that $x_2 = x - x_1$ is orthogonal to M, and so it belongs to $M^\perp$. Let $0 \neq y \in M$ and observe that since for each scalar λ also $x_1 + \lambda y \in M$, it readily follows that

$$\|x_2 - \lambda y\|^2 = \|x - (x_1 + \lambda y)\|^2 \geq d(x, M)^2 . \tag{1.15}$$

Now, by a familiar argument using (1.14), (1.15) may be rewritten

$$-\overline{\lambda}\langle x_2, y\rangle - \lambda\langle y, x_2\rangle + |\lambda|^2\langle y, y\rangle \geq 0 .$$

In particular, when $\lambda = \langle x_2, y\rangle / \langle y, y\rangle$ we conclude that

$$-\frac{|\langle x_2, y\rangle|^2}{\langle y, y\rangle} - \frac{|\langle x_2, y\rangle|^2}{\langle y, y\rangle} + \frac{|\langle x_2, y\rangle|^2}{\langle y, y\rangle} \geq 0 ,$$

that is, $|\langle x_2, y\rangle|^2 \leq 0$. But, this can only happen when $x_2 \perp y$, and consequently $x_2 \in M^\perp$. By the Pythagorean theorem, (1.13) is then true.

Finally we prove that the representation in (1.12) is unique. For, if we also have

$$x = x_1' + x_2' , \quad x_1' \in M, x_2' \in M^\perp ,$$

then, comparing this with (1.12), we get

$$x_1 - x_1' = x_2' - x_2 .$$

Now, the element on the left-hand side above belongs to M, while that on the right-hand side belongs to $M^\perp$. Thus,

$$\langle x_1 - x_1', x_2' - x_2\rangle = \langle x_1 - x_1', x_1 - x_1'\rangle = 0 ,$$

and $x_1 - x_1' = 0$. This implies that also $x_2' - x_2 = 0$, and we are done. ∎

The elements $x_1 \in M$ and $x_2 \in M^\perp$ uniquely determined by x are called the projections of x onto M and $M^\perp$ respectively. The operator $P_M: X \to M$ given by $P_M x = x_1$ is called the projection on M; it is not hard to see that P_M is a bounded linear operator onto M, and that $\|P_M\| \leq 1$. As we describe in Section 2, projection operators play an important role in the description of basic properties of the subspaces of X.

Theorem 1.7 may be applied to characterize the bounded linear functionals on a Hilbert space.

Theorem 1.8 (F. Riesz). Let X be a Hilbert space, and suppose L is a bounded linear functional defined on X. Then there exists a unique $y \in X$ such that

$$Lx = \langle x, y \rangle , \quad \text{all } x \in X . \tag{1.16}$$

Furthermore, $\|L\| = \|y\|$.

Proof. Let M be the null space of L, i.e., $M = \{x \in X : Lx = 0\}$; since L is continuous, M is a closed subspace of X, cf. 4.15 in Chapter XIV. If $M = X$, then we choose $y = 0$, and we are done. Otherwise, let $0 \neq x \notin M$, and note that by Theorem 1.7 there is an element $z = (1/\|x - P_M x\|)(x - P_M x)$ in $M^\perp$ with $\|z\| = 1$. Now, given $x \in X$, put $x_1 = (Lz)x - (Lx)z$, and observe that

$$Lx_1 = L((Lz)x) - L((Lx)z) = LzLx - LxLz = 0 ,$$

i.e., $x_1 \in M$. Furthermore, since $z \in M^\perp$ and $\|z\| = 1$, we get

$$\langle (Lz)x - (Lx)z, z \rangle = Lz\langle x, z \rangle - Lx\langle z, z \rangle = Lz\langle x, z \rangle - Lx = 0 .$$

In other words, we have

$$Lx = Lz\langle x, z \rangle = \langle x, (\overline{Lz})z \rangle ,$$

and (1.16) holds with $y = (\overline{Lz})z$.

Suppose now there is another point y', say, such that $Lx = \langle x, y' \rangle$ for all $x \in X$. Then $\langle x, y - y' \rangle = 0$ for all $x \in X$, and by taking $x = y - y'$ we find that $\|y - y'\| = 0$, and so $y = y'$.

Finally, about the norm. Since $\|z\| = 1$ we have $\|y\| = |\overline{Lz}|\,\|z\| \leq \|L\|$. On the other hand,

$$\|Lx\| \leq |\langle x, y \rangle| \leq \|x\|\,\|y\| , \quad \text{all } x \in X ,$$

and consequently, $\|L\| \leq \|y\|$. We have thus proved that $\|L\| = \|y\|$, and we have finished. ∎

A well-known property of finite-dimensional linear spaces is that of being algebraically reflexive. It is, therefore, natural, to consider whether arbitrary Hilbert spaces are reflexive.

In order to answer this question we begin by showing that X^*, which we already know to be complete, is an inner product space as well.

Proposition 1.9. If X is a Hilbert space, then X^* is a Hilbert space.

Proof. Consider the mapping $T: X \to X^*$ defined by $Tx = \langle \cdot, x \rangle$, i.e., Tx is the bounded linear functional on X given by

$$(Tx)y = \langle y, x \rangle, \quad \text{all } y \in X. \tag{1.17}$$

It is apparent that T is one-to-one and that, by Theorem 1.8, it is also norm preserving and onto. Observe that (1.17) may be rewritten

$$Ly = \langle y, T^{-1}L \rangle, \quad \text{for each } L \in X^*, y \in X. \tag{1.18}$$

Now, T establishes an equivalence between X and X^* at the level of sets, but not as linear spaces, since T is not linear but rather conjugate linear. More precisely, we have

$$T(x_1 + \lambda x_2) = Tx_1 + \overline{\lambda} Tx_2, \quad \text{all } x_1, x_2 \text{ in } X, \text{ scalars } \lambda.$$

Nevertheless, this property of T enables us to introduce an inner product $\langle \cdot, \cdot \rangle_*$ on X^* as follows: Given $L_1, L_2 \in X^*$, let

$$\langle L_1, L_2 \rangle_* = \langle T^{-1}L_2, T^{-1}L_1 \rangle. \tag{1.19}$$

A straightforward computation gives that $\langle \cdot, \cdot \rangle_*$ is an inner product on X^*, and that the norm and inner product on X^* are related by (1.2). Thus X^* is a Hilbert space. ■

We are now ready to show

Proposition 1.10. Suppose X is a Hilbert space, then X is reflexive.

Proof. Along the lines of the proof of Proposition 1.9 above, let $\tau: X^* \to X^{**}$ be the mapping on X^* given by

$$\tau L = \langle \cdot, L \rangle_*, \quad L \in X^*. \tag{1.20}$$

It then readily follows that τ is one-to-one, onto and norm-preserving, and consequently, (1.20) may be rewritten

$$x^{**}L = \langle L, \tau^{-1}x^{**} \rangle_*, \quad \text{all } x^{**} \in X^{**}, L \in X^*.$$

But then, by (1.19) and (1.18), for any $x^{**} \in X^{**}$ and $L \in X^*$ we have

$$x^{**}L = \langle T^{-1}\tau^{-1}x^{**}, T^{-1}L \rangle = L(T^{-1}\tau^{-1}x^{**}) = J_X(T^{-1}\tau^{-1}x^{**})L.$$

Since L is arbitrary this can only mean that $x^{**} = J_X(T^{-1}\tau^{-1}x^{**})$, and consequently $T^{-1}\tau^{-1}x^{**} \in X$. Thus, the natural map J_X is onto, and X is reflexive. ■

2. PROJECTIONS

We consider now some of the connections between the class of projections and the geometry of subspaces of X. We begin by noting an obvious property of projections: If $P_M: X \to M$ is the projection of X onto the subspace M, then P_M is a bounded linear operator and, if $M \neq \{0\}$, we have $\|P_M\| = 1$. Indeed, for $x, y \in X$ and scalars λ, by Theorem 1.7 it follows that

$$x = x_1 + x_2\,,\; y = y_1 + y_2\,, \quad x_1\,, y_1 \in M\,, x_2\,, y_2 \in M^{\perp}\,,$$

and

$$x + \lambda y = (x_1 + \lambda y_1) + (x_2 + \lambda y_2)\,, \quad x_1 + \lambda y_1 \in M\,, x_2 + \lambda y_2 \in M^{\perp}\,.$$

Whence, we see that

$$P_M(x + \lambda y) = x_1 + \lambda y_1 = P_M x + \lambda P_M y\,,$$

and P_M is linear.

Further, by (1.13), $\|P_M x\|^2 \leq \|x\|^2$, so that $\|P_M\| \leq 1$. Provided that $M \neq \{0\}$, we can choose $x \in M$ with $\|x\| = 1$. Then $\|P_M\| \geq \|P_M x\| = \|x\| = 1$, and $\|P_M\| = 1$.

It is also possible to characterize the projections.

Proposition 2.1. Let $P: X \to X$ be a linear map from a Hilbert space X into itself. Then P is a projection iff

(i) $\langle Px, y\rangle = \langle x, Py\rangle$, all $x, y \in X$. (2.1)

(ii) $P^2 x = P(Px) = Px$, all $x \in X$. (2.2)

Proof. We do the necessity first. Suppose $P = P_M$ is a projection onto a closed subspace M, and, given $x, y \in M$, write $x = x_1 + x_2$, $y = y_1 + y_2$, with $x_1, y_1 \in M$ and $x_2, y_2 \in M^{\perp}$. We then have

$$\langle Px, y\rangle = \langle x_1, y_1 + y_2\rangle = \langle x_1, y_1\rangle = \langle x_1 + x_2, y_1\rangle = \langle x, Py\rangle\,,$$

which is precisely (2.1). Moreover, since for $y \in M$ we have $Py = y$, it readily follows that

$$P(Px) = Px\,, \quad x \in X\,,$$

and (2.2) is true.

Conversely, suppose P is a linear mapping which satisfies both the conditions in the theorem, and first note that P is bounded. Indeed, by (2.1), (2.2) and the Cauchy-Schwarz inequality we have

$$\|Px\|^2 = \langle Px, Px \rangle = \langle x, P^2x \rangle = \langle x, Px \rangle \leq \|x\| \, \|Px\| \,,$$

and consequently, P is bounded, and $\|P\| \leq 1$.

Next let $M = P(X)$ be the image of X under P. It is clear that M is a linear subspace of X; it is also closed. Indeed, if $\{y_n\} \subseteq M$ and $y_n \to y$, let $\{x_n\} \subseteq X$ be such that $Px_n = y_n$, and note that by (2.2) it follows that

$$Py_n = P(Px_n) = Px_n = y_n \,, \quad \text{all } n \,.$$

Hence by the continuity of P we get

$$y = \lim_{n \to \infty} y_n = \lim_{n \to \infty} Py_n = Py \in M \,,$$

and M is the closed subspace of X, $M = \{x \in X : Px = x\}$.

Finally we show that P is the projection P_M of X onto M. Take any $x \in X$ and write it as $x = Px + (x - Px)$; we want to show that $Px \in M$ and $(x - Px) \in M^{\perp}$. The first assertion obtains since $Px = P(Px) \in M$. Also, if $y \in M$, then $Py = y$ and hence, by (2.1) and (2.2),

$$\begin{aligned} \langle x - Px, y \rangle &= \langle x, y \rangle - \langle Px, Py \rangle = \langle x, y \rangle - \langle x, P^2y \rangle \\ &= \langle x, y \rangle - \langle x, Py \rangle = \langle x, y \rangle - \langle x, y \rangle = 0 \,, \end{aligned}$$

and $x - Px \in M^{\perp}$. ■

Two closed subspaces M, N of a Hilbert space X are said to be orthogonal, and we write $M \perp N$, if

$$\langle x, y \rangle = 0 \,, \quad \text{all } x \in M, y \in N \,.$$

It is not hard to characterize orthogonal subspaces in terms of the projections they determine.

Proposition 2.2. Suppose X is a Hilbert space, and M, N are closed subspaces of X. Then, $M \perp N$ iff $P_M P_N = P_N P_M = 0$.

Proof. First suppose M and N are orthogonal and let $x, y \in X$. Then $P_M x \in M$, $P_N y \in N$, and

$$\langle P_N y, P_M x \rangle = \langle P_M P_N y, x \rangle = 0 \,.$$

Since x, y are arbitrary this can only happen if $P_M P_N = 0$; similarly $P_N P_M = 0$.

Conversely, suppose that $P_M P_N = 0$, and let $x \in M$, $y \in N$. Then,

$$\langle x, y\rangle = \langle P_M x, P_N y\rangle = \langle x, P_M P_N y\rangle = 0\,,$$

and $M \perp N$.

Note that the above assumptions are somewhat redundant in that $P_M P_N = 0$ iff $P_N P_M = 0$. ■

How about the sum of projections? First a definition.

Let M, N be closed linear subspaces of a Hilbert space X, and assume that every element in the vector sum $M + N$ has a unique representation of the form $x + y$, where $x \in M$, $y \in N$. Then we call $M + N$ the direct sum of M and N. If $M \perp N$, we denote this direct sum by $M \oplus N$. We leave it to the reader to verify that $M \oplus N$ is also a closed subspace of X.

If $Y = M \oplus N$, then we say that N is the orthogonal complement of M in Y, and we write $N = Y \ominus M$; symmetrically, $M = Y \ominus N$ denotes the fact that M is the orthogonal complement of N in Y.

For instance, in the projection theorem we have $X = M \oplus M^{\perp}$, $M = X \ominus M^{\perp}$, and $M^{\perp} = X \ominus M$.

Proposition 2.3. Suppose X is a Hilbert space, and M, N are closed subspaces of X. Then the sum $P_M + P_N$ of the projections P_M and P_N is a projection iff $P_M P_N = P_N P_M = 0$. In this case, $P_M + P_N = P_{M \oplus N}$.

Proof. First assume that $P = P_M + P_N$ is a projection. By Proposition 2.1 we have

$$\|Px\|^2 = \langle Px, Px\rangle = \langle P^2 x, x\rangle = \langle Px, x\rangle\,, \quad \text{all } x \in X\,,$$

and similarly,

$$\|P_M x\|^2 = \langle P_M x, x\rangle\,, \quad \|P_N x\|^2 = \langle P_N x, x\rangle\,, \quad \text{all } x \in X\,.$$

Whence, we get

$$\begin{aligned} \|P_M x\|^2 + \|P_N x\|^2 &= \langle P_M x, x\rangle + \langle P_N x, x\rangle \\ &= \langle Px, x\rangle = \|Px\|^2 \le \|x\|^2\,. \end{aligned} \tag{2.3}$$

Consider now an arbitrary element y in X and put $x = P_N y$ in (2.3). Since $P_N x = P_N^2 y = P_N y$, this gives

$$\|P_M P_N y\|^2 + \|P_N y\|^2 \le \|P_N y\|^2\,,$$

which can only be true if $P_M P_N = 0$. By the way, this is equivalent to $P_N P_M = 0$.

Conversely, we verify that $P = P_M + P_N$ satisfies the conditions of Proposition 2.1, and so it is a projection. Since P is a sum of operators that satisfy (2.1), it also satisfies (2.1). As for (2.2), since

$$\begin{aligned} P^2 &= (P_M + P_N)^2 = (P_M + P_N)(P_M + P_N) \\ &= P_M^2 + P_M P_N + P_N P_M + P_N^2 \\ &= P_M^2 + P_N^2 = P_M + P_N = P, \end{aligned}$$

we have that $P^2 = P$, and P is a projection.

Finally, it is clear that $Px = P_M x + P_N x$ varies over $M \oplus N$ as x varies over X. Conversely, if $x = x_1 + x_2 \in M \oplus N, x_1 \in M, x_2 \in N$, since $PP_M = P_M$, $PP_N = P_N$, and since $x_1 = P_M x_1 = PP_M x_1$, $x_2 = P_N x_2 = PP_N x_2$, we have

$$Px = P(x_1 + x_2) = PP_M x_1 + PP_N x_2 = x_1 + x_2 = x\,.$$

Hence $P = P_{M\oplus N}$. ■

How about the product, or composition, and the difference of projections?

Proposition 2.4. Suppose X is a Hilbert space, and M and N are closed subspaces of X. Then the composition $P = P_M P_N$ of the projections P_M and P_N is a projection iff they commute, i.e., $P_M P_N = P_N P_M$. In that case $P = P_{M\cap N}$.

Proof. If P is a projection, then

$$\langle P_M P_N x, y\rangle = \langle x, P_M P_N y\rangle\,, \quad \text{all } x, y \in X\,.$$

Moreover, since P_M and P_N are projections we also have

$$\langle P_M P_N x, y\rangle = \langle x, P_N P_M y\rangle\,, \quad \text{all } x, y \in X\,,$$

and consequently, we get

$$\langle x, P_M P_N y\rangle = \langle x, P_N P_M y\rangle\,, \quad \text{all } x, y \in X\,.$$

But this can only be true if $P_M P_N = P_N P_M$, and the necessity has been established.

Conversely, if P_N and P_M commute, then essentially reversing the above steps we get that $P = P_M P_N = P_N P_M$ satisfies

$$\langle Px, y\rangle = \langle x, Py\rangle, \quad \text{all } x, y \in X.$$

Moreover, since also

$$P^2 = (P_M P_N)(P_M P_N) = P_M(P_N P_M)P_N$$
$$= P_M(P_M P_N)P_N = P_M^2 P_N^2 = P_M P_N = P,$$

by Proposition 2.1 P is a projection. Furthermore, since

$$Px = P_M(P_N x) = P_N(P_M x) \in M \cap N, \quad \text{all } x \in X,$$

P projects X into $M \cap N$. On the other hand, if $x \in M \cap N$, then

$$Px = P_M(P_N x) = P_M x = x,$$

$P = P_{M \cap N}$, and P is the projection onto $M \cap N$. ∎

Before we consider the question of the difference of projections we need a preliminary result.

Lemma 2.5. Suppose X is a Hilbert space and let P_M, P_N be projections onto the subspaces M and N, respectively. Then the following four conditions are equivalent:

(i) $\langle P_M x, x\rangle \geq \langle P_N x, x\rangle$, all $x \in X$.

(ii) $M \supseteq N$.

(iii) $P_M P_N = P_N$.

(iv) $P_N P_M = P_N$.

Proof. (i) implies (ii). Let $x \in X$. Since $x - P_M x \in M^\perp$, we have

$$\|x\|^2 = \|P_M x\|^2 + \|x - P_M x\|^2.$$

Now, if $x \in N$, then $P_N x = x$, and $\langle P_N x, x\rangle = \|x\|^2$. Whence, by (i) we get

$$\|x\|^2 = \langle P_N x, x\rangle \leq \langle P_M x, x\rangle = \langle P_M x, P_M x\rangle = \|P_M x\|^2 \leq \|x\|^2.$$

Thus $\|x\| = \|P_M x\|$, and by (1.13) we conclude that $\|x - P_M x\| = 0$. Consequently, $x = P_M x \in M$, and (ii) holds.

(ii) implies (iii). Since $P_N x \in N \subseteq M$, we have $P_M P_N x = P_N x$.

(iii) implies (iv). By Proposition 2.4, since the composition $P_M P_N = P_N$ is a projection, the projections commute. Specifically, $P_N P_M = P_M P_N = P_N$, which is precisely (iv).

(iv) implies (i). It is a straightforward computation: For $x \in X$ we have

$$\langle P_N x, x\rangle = \|P_N x\|^2 = \|P_N P_M x\|^2 \leq \|P_M x\|^2 = \langle P_M x, x\rangle,$$

and (i) holds. ∎

Proposition 2.6. Suppose X is a Hilbert space, and let P_M, P_N be projections onto the subspaces M and N, respectively. The difference $P = P_M - P_N$ is a projection iff $N \subseteq M$. If this is the case, then $P = P_{M \ominus N}$.

Proof. Suppose that P is a projection. Then, since $P_M = P + P_N$ is also a projection, by Proposition 2.3 it follows that $PP_N = 0$. Whence

$$(P_M - P_N)P_N = P_M P_N - P_N = 0\,,$$

and (iii) of Lemma 2.4, and consequently, also (ii) in that lemma, are true.

Conversely, if $N \subseteq M$, it is clear that $P_N P_M = P_M P_N = P_N$, and

$$\begin{aligned}(P_M - P_N)^2 &= (P_M - P_N)(P_M - P_N)\\ &= P_M^2 - P_N P_M - P_M P_N + P_N^2\\ &= P_M - P_N - P_N + P_N = P_M - P_N\,.\end{aligned}$$

Also,

$$\langle Px, x\rangle = \langle P_M x, x\rangle - \langle P_N x, x\rangle = \langle x, P_M x\rangle - \langle x, P_N x\rangle = \langle x, Px\rangle\,,$$

and by Proposition 2.1, P is a projection.

Moreover, since as observed above $(P_M - P_N)P_N = 0$, by Proposition 2.3 the subspace Y of the projection $P_M - P_N$ satisfies $Y \oplus N = M$. Therefore $Y = M \ominus N$. ■

3. ORTHONORMAL SETS

A general question we address in this section is the following: Given a Hilbert space X and a subset Y of X, how can we best approximate elements of X by those of Y? A good measure of the approximation is given by the quantity

$$d(x, Y) = \inf_{y \in Y} \|x - y\|\,,$$

so we are naturally interested whether the inf above is actually achieved.

We begin by considering a simple example, namely, the case when Y is the (finite dimensional) subspace of X spanned by $\{x_1, \ldots, x_n\}$. Given $x \in X$, we seek to minimize the expression

$$\left\| x - \sum_{i=1}^{n} \lambda_i x_i \right\|\,, \quad \lambda_i \text{ scalars}\,. \tag{3.1}$$

Clearly we may assume the x_i's are linearly independent, and, by the Gram-Schmidt process, cf. 5.21 below, orthonormal. More precisely, we may assume that each x_i has norm 1, and that $x_i \perp x_j$ for $1 \leq i \neq j \leq n$. The x_i's are then said to constitute an orthonormal system, or ONS, in X.

A closer look at (3.1) and (1.13) suggests that we consider the projection of x onto Y, and in order to invoke the projection theorem we begin by showing that Y is closed.

Proposition 3.1. If M is a closed subspace of a normed space X and if $x \in X$, then the span $\{M, x\}$ of M and $\{x\}$ is also a closed subspace of X.

Proof. If $x \in M$, then the span $\{M, x\} = M$, and we are done. Otherwise, assume $x \notin M$ and suppose a sequence $\{m_n + \lambda_n x\}$ of elements of $\{M, x\}$ converges to an element $y \in X$; we must show that actually $y \in \{M, x\}$. First note that since the sequence converges it is bounded, i.e., there is a constant c such that

$$\|m_n + \lambda_n x\| \leq c, \quad \text{all } n.$$

We claim that $\{|\lambda_n|\}$ is also bounded. If this is not the case, there is a subsequence $n_k \to \infty$ such that $|\lambda_{n_k}| \to \infty$ as $k \to \infty$, and consequently,

$$\|\lambda_{n_k}^{-1} m_{n_k} + x\| \leq c / |\lambda_{n_k}| \to 0 \quad \text{as } k \to \infty.$$

Whence x belongs to the closure of M, which is M since M is closed, and this is a contradiction.

Now, since $\{|\lambda_n|\}$ is bounded, passing to a subsequence if necessary, we may assume that the λ_n's converge to a limit λ, say. But then

$$m_n = (m_n + \lambda_n x) - \lambda_n x \to y - \lambda x, \quad \text{as } m \to \infty,$$

and since M is closed, $y - \lambda x \in M$. Thus $y = (y - \lambda x) + \lambda x$ belongs to $\{M, x\}$. ■

Corollary 3.2. Let X be a normed space. Then the subspace Y spanned by $\{x_1, \ldots, x_n\}$ is closed.

Proof. Observe that $\{x_1\} = \{\lambda x_1 : \lambda \text{ is a scalar}\}$ is closed in X and apply Proposition 3.1 as many times as necessary. ■

The stage is now set to invoke the projection theorem, and to obtain the answer to our question: It is $P_Y x$. In fact, by the projection theorem we can find the λ_i's as follows: Since $x - P_Y x \in Y^\perp$, if $P_Y x = \sum_{i=1}^n \lambda_i x_i$, we have

$$\left\langle x - \sum_{i=1}^n \lambda_i x_i, x_k \right\rangle = 0\,, \quad 1 \le k \le n\,.$$

Moreoever, since $\{x_i\}$ is an ONS in X, it readily follows that

$$\langle x, x_k\rangle = \sum_{i=1}^n \lambda_i \langle x_i, x_k\rangle = \lambda_k\,, \quad 1 \le k \le n\,,$$

and consequently, the minimum value of the expression (3.1) equals

$$\left\| x - \sum_{i=1}^n \langle x, x_i\rangle x_i \right\| . \tag{3.2}$$

Furthermore, we may compute the exact value of the expression (3.2): Its square equals

$$\begin{aligned} \|x\|^2 - \left\langle \sum_{i=1}^n \langle x, x_i\rangle x_i, x\right\rangle - \left\langle x, \sum_{k=1}^n \langle x, x_k\rangle x_k\right\rangle + \sum_{i=1}^n |\langle x, x_i\rangle|^2 \\ = \|x\|^2 - \sum_{i=1}^n |\langle x, x_i\rangle|^2 . \end{aligned} \tag{3.3}$$

The minimizing values of the λ_k's, to wit, the scalars $\langle x, x_k\rangle$, are called the Fourier coefficients of x with respect to the ONS $\{x_k\}$.

In addition to providing a complete answer to the question posed above, (3.3) implies

Proposition 3.3 (Bessel's Inequality). Suppose $\{x_\alpha\}_{\alpha\in A}$ is an ONS in a Hilbert space X. Then Bessel's inequality holds, i.e.,

$$\sum_{\alpha\in A} |\langle x, x_\alpha\rangle|^2 \le \|x\|^2\,, \quad \text{all } x \in X\,. \tag{3.4}$$

In particular, for each x in X, all but an at most countable number of the Fourier coefficients $\langle x, x_\alpha\rangle$ of x with respect to the ONS $\{x_\alpha\}$ vanish.

Proof. Let $\alpha_1, \ldots, \alpha_n$ be a finite subset of A. Then, given $x \in X$, by (3.3) we have

$$\left\| x - \sum_{i=1}^n \langle x, x_{\alpha_i}\rangle x_{\alpha_i} \right\|^2 = \|x\|^2 - \sum_{i=1}^n |\langle x, x_{\alpha_i}\rangle|^2 \ge 0\,.$$

It then readily follows that

$$\sup_{\{\alpha_1,\ldots,\alpha_n\}\subseteq A} \sum_{i=1}^n |\langle x, x_{\alpha_i}\rangle|^2 \le \|x\|^2\,,$$

and (3.4) holds. ■

There is yet another way to interpret the inequality (3.4): Suppose A is endowed with the counting measure, and let T be the linear mapping from X into the space of sequences $(c_\alpha)_{\alpha\in A}$ which takes x into its sequence of Fourier coefficients,

$$Tx = (\langle x, x_\alpha \rangle)_{\alpha\in A}\,, \quad x \in X\,. \tag{3.5}$$

Bessel's inequality asserts that T is a bounded mapping from X into $\ell^2(A)$ with norm $\|T\| \le 1$. Since we are interested in describing X in terms of $\ell^2(A)$, we must decide when T is one-to-one and onto. We begin settling the "onto" question.

Theorem 3.4 (F. Riesz-Fischer). Suppose $\{x_\alpha\}_{\alpha\in A}$ is an ONS in a Hilbert space X, and let $(c_\alpha) \in \ell^2(A)$. If T is given by (3.5), then there exists $y \in X$ such that $Ty = (c_\alpha)_{\alpha\in A}$. More precisely, T is onto.

Proof. Since $(c_\alpha) \in \ell^2(A)$, there are at most countably many α's, $\alpha_1, \ldots, \alpha_i, \ldots$, say, so that $c_{\alpha_i} \ne 0$. Put now

$$y_n = \sum_{i=1}^{n} c_{\alpha_i} x_{\alpha_i}\,, \quad n = 1, 2 \ldots$$

We claim that the sequence $\{y_n\}$ is Cauchy in X, and consequently, it converges. This is not hard; indeed, let $n < m$, and note that by Proposition 1.4 we have

$$\|y_m - y_n\|^2 = \sum_{i=n+1}^{m} \|c_{\alpha_i} x_{\alpha_i}\|^2 = \sum_{i=n+1}^{m} |c_{\alpha_i}|^2\,.$$

Now, since $(c_\alpha) \in \ell^2(A)$, the right-hand side of the above equality is dominated by the tail of a convergent series, and it tends to 0 as $n \to \infty$. Whence the same is true of the left-hand side, and $\{y_n\}$ is Cauchy. Moreover, if $y \in X$ is the limit of the y_n's, from the continuity of the inner product we get

$$\langle y, x_\alpha \rangle = \lim_{n\to\infty} \langle y_n, x_\alpha \rangle = \begin{cases} 0 & \text{if } \alpha \ne \alpha_k, \text{ all } k \\ c_{\alpha_k} & \text{if } \alpha = \alpha_k. \end{cases}$$

Thus, $Ty = (c_\alpha)_{\alpha\in A}$. ■

Still the question as to whether T is one-to-one remains open; first a definition.

We say that an ONS $\{x_\alpha\}_{\alpha\in A}$ in a Hilbert space X is maximal, or complete, if no nonzero element can be added to it so that the resulting collection of elements is still an ONS in X.

Note that given a Hilbert space X, we can always find a maximal orthormal system in X; this is a simple consequence of Zorn's Lemma, cf. 5.23 below.

The stage is now set for

Theorem 3.5. Suppose $\{x_\alpha\}_{\alpha\in A}$ is an ONS in a Hilbert space X. Then the following properties are equivalent:

(i) $\{x_\alpha\}_{\alpha\in A}$ is maximal in X.

(ii) The collection of all finite linear combinations of the x_α's is dense in X.

(iii) (Plancherel's Equality) Equality holds in Bessel's inequality, i.e.,

$$\|x\|^2 = \sum_{\alpha\in A} |\langle x, x_\alpha\rangle|^2\,, \quad \text{all } x \in X\,. \tag{3.6}$$

(iv) (Parseval's Identity) For all $x, y \in X$, we have

$$\langle x, y\rangle = \sum_{\alpha\in A} \langle x, x_\alpha\rangle \overline{\langle y, x_\alpha\rangle}\,. \tag{3.7}$$

Proof. (i) implies (ii). Let M be the closure of the subspace of X consisting of all finite linear combinations of the x_α's; M is then a closed subspace of X. Now, if $X\backslash M \neq \emptyset$, by Theorem 1.7 we can find an element $x \in M^\perp$ with $\|x\| = 1$. In particular, we have

$$\langle x, x_\alpha\rangle = 0\,, \quad \text{all } \alpha \in A\,,$$

thus contradicting the assumed maximality of the x_α's.

(ii) implies (iii) By Bessel's inequality, it suffices to prove the "$\leq$" inequality in (3.6). Given $x \in X$ and $\varepsilon > 0$, there exists a finite subset $\{\alpha_1, \ldots, \alpha_n\}$ of A such that

$$\Big\| x - \sum\nolimits_{i=1}^n c_i x_{\alpha_i} \Big\| \leq \varepsilon\,. \tag{3.8}$$

Now, since the best approximation to x in the subspace spanned by $\{x_{\alpha_1}, \ldots, x_{\alpha_n}\}$ is given by $\sum_{i=1}^n \langle x, x_{\alpha_i}\rangle x_{\alpha_i}$, we also have

$$\Big\| x - \sum\nolimits_{i=1}^n \langle x, x_{\alpha_i}\rangle x_{\alpha_i} \Big\| \leq \varepsilon\,. \tag{3.9}$$

Moreover, since by Proposition 1.4 we have

$$\left\|\sum_{i=1}^{n}\langle x, x_{\alpha_i}\rangle x_{\alpha_i}\right\|^2 = \sum_{i=1}^{n}|\langle x, x_{\alpha_i}\rangle|^2 ,$$

from (3.9) it follows that

$$\begin{aligned}\|x\| &\le \left\|x - \sum_{i=1}^{n}\langle x, x_{\alpha_i}\rangle x_{\alpha_i}\right\| + \left\|\sum_{i=1}^{n}\langle x, x_{\alpha_i}\rangle x_{\alpha_i}\right\|\\ &\le \varepsilon + \left(\sum_{i=1}^{n}|\langle x, x_{\alpha_i}\rangle|^2\right)^{1/2}\\ &\le \varepsilon + \left(\sum_{\alpha\in A}|\langle x, x_{\alpha}\rangle|^2\right)^{1/2} . \end{aligned} \tag{3.10}$$

Thus, since $\varepsilon > 0$ is arbitrary, the inequality opposite to Bessel's inequality holds, and (3.6) is true.

(iii) implies (iv). Identity (3.7) is one of inner products and (3.6) one of norms; we derive the former from the latter via (1.5). Specifically, given $x, y \in X$, using (3.6) we compute $\langle x, y\rangle$ by evaluating the norms that appear on the right-hand side of (1.5). In fact, adding up the relation

$$\|x + y\|^2 = \sum_{\alpha\in A}|\langle x + y, x_\alpha\rangle|^2$$

to those corresponding to $x - y$ and $x \pm iy$, by (1.5) again, it readily follows that if $(\langle x, x_\alpha\rangle)$ and $(\langle y, x_\alpha\rangle)$ are the sequences of Fourier coefficients of x and y respectively, then

$$4\langle x, y\rangle = 4\sum_{\alpha\in A}\langle x, x_\alpha\rangle\overline{\langle y, x_\alpha\rangle} .$$

This is precisely (3.7). The argument involved in this step is known as "polarization."

(iv) implies (i). Suppose $\langle x, x_\alpha\rangle = 0$ for all $\alpha \in A$. Then, by (3.7) we get $\|x\|^2 = \langle x, x\rangle = \sum_{\alpha\in A}|\langle x, x_\alpha\rangle|^2 = 0$, and consequently, $x = 0$. Thus $\{x_\alpha\}$ is a maximal ONS in X. ■

Observe that, in particular, (iii) above implies that T is an isometry of X onto $\ell^2(A)$, and consequently, T is one-to-one. Thus T establishes an isomorphism between X and $\ell^2(A)$ provided $\{x_\alpha\}_{\alpha\in A}$ is a maximal ONS in X. We reiterate that, by Zorn's Lemma, all Hilbert spaces have a maximal ONS. The question is then, how to produce concrete examples of such systems. In the familiar case of $L^2(I)$, where I is an interval of the line, we construct such an example following Corollary 2.3 in Chapter XVII.

It is also apparent that (3.6) gives

$$x = \sum_{\alpha \in A} \langle x, x_\alpha \rangle x_\alpha\,, \quad \text{all } x \in X\,, \tag{3.11}$$

where the sum is understood to converge in the norm of X. Thus, in this setting, each element of X is represented by its Fourier series.

It is also customary to call a maximal ONS in X a basis. An interesting question we are able to settle at this time is when X has an at most countable basis.

Proposition 3.6. Suppose X is a Hilbert space. Then X has an at most countable basis iff X is separable, and in this case all bases are at most countable.

Proof. We do the sufficiency first. Let $\{x_n\}$ be an at most countable dense subset of X. First discard any x_n which is a linear combination of the x_i's with $1 \leq i \leq n-1$, and then, by the Gram-Schmidt process, orthonormalize the remaining elements. Call this ONS $\{x_n\}$ again, and note that its span coincides with that of the original ONS, and consequently, its closure is X. By (ii) in Theorem 3.6, the set $\{x_n\}$ is an at most countable basis for X.

Conversely, suppose $\{x_n\}$ is an ON basis for X, and note that by (ii) in Theorem 3.6, finite linear combinations of the x_n's are dense in X. Let now $\{\lambda_n\}$ be a countable dense subset of the field of scalars, and observe that

$$S = \left\{ \sum_{\text{finite sums}} \lambda_k y_k : y_k = x_n, \text{some } n \right\}$$

is a countable dense subset of X, and consequently, X is separable.

It only remains to show that any other basis $\{y_\alpha\}_{\alpha \in A}$ for X is also at most countable. For each integer n consider the set

$$A_n = \{\alpha \in A : \langle y_\alpha, x_n \rangle \neq 0\}\,, \quad n = 1, 2, \ldots$$

Since by (3.6)

$$\sum_{\alpha \in A} |\langle x_n, y_\alpha \rangle|^2 = \|x_n\|^2 = 1\,,$$

each A_n is at most countable. Thus $A = \bigcup_n A_n$ is also at most countable.

We claim that, unless $\alpha \in A$, $y_\alpha = 0$. For, if $y_\alpha \neq 0$, then by the maximality of the x_n's there exists an index n such that $\langle y_\alpha, x_n \rangle \neq 0$. In this case $\alpha \in A_n \subseteq A$, and we have finished. ■

4. SPECTRAL DECOMPOSITION OF COMPACT OPERATORS

As in the case of the Lebesgue L^2 spaces, there is a notion of weak convergence in Hilbert spaces. More precisely, inspired by the Riesz Representation Theorem, we say that a sequence (x_n) in a Hilbert space X converges weakly to $x \in X$, provided that

$$\lim_{n\to\infty} \langle x_n, y\rangle = \langle x, y\rangle\,, \quad \text{all } y \in X\,. \tag{4.1}$$

As before, convergent sequences are weakly convergent, but the converse is not true.

Also, bounded sequences have weakly convergent subsequences. Given a bounded sequence (y_n) in X, let $\{x_{\alpha_m}\}$ be the at most countable subset of a basis $\{x_\alpha\}_{\alpha\in A}$ for X such that the Fourier coefficients of the y_n's with respect to the ONS $\{x_\alpha\}_{\alpha\in A}$ do not vanish. Setting now

$$c_{n,m} = \langle x_{\alpha_m}, y_n\rangle, \quad \text{all } m, n,$$

we obtain a weakly convergent subsequence along the lines of the proof of Theorem 3.3 in Chapter XII.

How do weakly convergent sequences look?

Proposition 4.1. Let X be a Hilbert space, and suppose the sequence $\{x_n\} \subseteq X$ converges weakly to $x \in X$. Then

(i) The set $\{x_n\}$ is bounded.

(ii) The weak limit x lies in the subspace of X spanned by $\{x_1, x_2, \ldots\}$.

(iii) $\|x\| \le \liminf \|x_n\|$.

Proof. Let $y \in X$; since by assumption we have

$$\lim_{n\to\infty} L_y(x_n - x) = \lim_{n\to\infty} \langle x_n - x, y\rangle = 0\,,$$

it readily follows that for each $y \in X$ we have

$$\sup_n |L_y(x_n - x)| \le c_y < \infty\,.$$

Thus, by the Uniform boundedness principle we get

$$\sup_n \|x_n - x\| < \infty\,,$$

which gives (i) at once.

Next, if (ii) is not true, by Proposition 3.3 in Chapter XIV and Theorem 1.8 , there exists $y \in X$ such that

$$\langle x_n, y\rangle = 0\,, \quad \text{all } n\,, \quad \text{and} \quad \langle x, y\rangle \neq 0\,.$$

However, since the sequence $\{x_n\}$ converges weakly to x, this is not possible, and (ii) also holds.

Finally, let $\{x_{n_k}\}$ be a subsequence such that $\lim_{n_k\to\infty} \|x_{n_k}\| = \liminf \|x_n\|$. Since $\{x_{n_k}\}$ also converges weakly to x, we get

$$\begin{aligned} |\langle x, y\rangle| &= \lim_{n_k\to\infty} |\langle x_{n_k}, y\rangle| \leq \lim_{n_k\to\infty} (\,\|x_{n_k}\|\,\|y\|\,) \\ &\leq (\,\liminf \|x_n\|\,)\|y\|\,. \end{aligned}$$

and (iii) follows at once from Proposition 3.4 in Chapter XIV. ■

As in the case of the L^p spaces, this last result suggests under what conditions the notions of weak convergence and convergence coincide.

Proposition 4.2. Let X be a Hilbert space, and suppose $\{x_n\} \subseteq X$ converges weakly to $x \in X$. If in addition $\lim_{n\to\infty} \|x_n\| = \|x\|$, then $\lim_{n\to\infty} x_n = x$.

Proof. As usual,

$$\begin{aligned} \|x_n - x\|^2 &= \langle x_n - x, x_n - x\rangle \\ &= \|x_n\|^2 - \overline{\langle x_n, x\rangle} - \langle x_n - x\rangle + \|x\|^2\,. \end{aligned}$$

Whence, by assumption

$$\lim_{n\to\infty} \|x_n - x\|^2 = \|x\|^2 - \overline{\langle x, x\rangle} - \langle x, x\rangle + \|x\|^2 = 0\,. \quad ■$$

How does the notion of weak convergence fit into the theory of continuous linear operators? Well, if $T \in \mathcal{B}(X)$, then T also preserves weak convergence. Specifically, if the sequence $\{x_n\} \subseteq X$ converges weakly to $x \in X$, we also have that $\{Tx_n\}$ converges weakly to Tx. We prove this using the notion of adjoint mapping.

Given $y \in X$, consider the functional L on X defined by

$$Lx = \langle Tx, y\rangle\,, \quad \text{all } x \in X\,.$$

As before, cf. Proposition 1.9, it is clear that L is actually bounded, that the dependence on y is linear, and that if T^* denotes the correspondence between y and L, then T^* is also a bounded linear mapping on X with norm $\|T^*\| = \|T\|$, and

$$\langle Tx, y\rangle = \langle x, T^*y\rangle \,, \quad \text{all } x, y \in X \,. \tag{4.2}$$

T^* is called the adjoint of the operator T; some of its basic properties are discussed in 5.36-5.38 below.

Now, in our case, by (4.2) we have $\langle Tx_n, y\rangle = \langle x_n, T^*y\rangle$, and consequently,

$$\lim_{n\to\infty} \langle Tx_n, y\rangle = \langle x, T^*y\rangle = \langle Tx, y\rangle \,, \quad \text{all } y \in X \,,$$

which establishes the weak convergence of $\{Tx_n\}$ to Tx.

We can also consider the special case when an operator T maps weakly convergence sequences into convergent sequences. More precisely, we say that a linear mapping T from X into itself is completely continuous if whenever the sequence $\{x_n\} \subseteq X$ converges weakly to x, then we have

$$\lim_{n\to\infty} \|Tx_n - Tx\| = 0 \,.$$

Not all continuous mappings on X are completely continuous. Indeed, since as observed above there are sequences in ℓ^2 that converge weakly but not strongly, the identity map is one such operator. On the other hand, it is clear that completely continuous operators are also continuous.

Completely continuous operators are also called compact, and the following result explains why.

Proposition 4.3. Let X be a Hilbert space, and $T \in \mathcal{B}(X)$. Then, T is completely continuous iff given any bounded sequence $\{x_n\}$ of elements in X, there is a subsequence $\{Tx_{n_k}\} \subseteq X$ which converges.

Proof. Suppose first that T is completely continuous and that $\{x_n\} \subseteq X$ is bounded. Since as noted above $\{x_n\}$ has a weakly convergent subsequence, and since T is completely continuous, the necessity follows.

Conversely, assume the sequence $\{x_n\} \subseteq X$ is bounded. Passing to a subsequence if necessary we may assume that $\{x_n\}$ converges weakly to $x \in X$. Since T is bounded, $\{Tx_n\}$ is weakly convergent to Tx. Now, if the sequence $\{Tx_n\}$ does not converge to Tx, then there exist a subsequence $\{x_{n_k}\}$ and $\eta > 0$, such that

$$\|Tx_{n_k} - Tx\| \geq \eta \,, \quad k = 1, 2, \ldots$$

Note that the sequence $\{x_{n_k}\}$ is weakly convergent and therefore also bounded. By the assumptions on T there are yet another subsequence, which we call $\{x_{n_k}\}$ again, and $y \in X$ with such that $\lim_{k\to\infty} Tx_{n_k} = y$. But then $\{Tx_{n_k}\}$ is also weakly convergent to y, and since weak limits are unique, cf. 4.32 in Chapter XIV, it follows that $Tx = y$. Hence,

$$\lim_{k\to\infty} \|Tx_{n_k} - Tx\| = 0\,,$$

which is a contradiction. ■

What are some completely continuous operators? The following is a useful sufficient condition to identify them.

Proposition 4.4. Let X be a Hilbert space, $\{x_\alpha\}_{\alpha\in A}$ a basis for X, and $T \in \mathcal{B}(X)$. If

$$\sum_{\alpha,\beta\in A} |\langle Tx_\alpha, x_\beta\rangle|^2 < \infty\,, \tag{4.3}$$

then T is completely continuous.

Proof. By translations it suffices to show that if the sequence $\{y_n\} \subseteq X$ converges weakly to 0, then $\{Ty_n\}$ converges to 0.

First observe that by Theorem 3.5 there are at most countably many α's so that

$$\langle Ty_n, x_\alpha\rangle \neq 0\,, \quad \text{or} \quad \langle y_n, x_\alpha\rangle \neq 0\,,\ n = 1, 2, \ldots$$

We may, therefore, consider α as an integer index, and note that from (3.6) we obtain

$$\begin{aligned}\|Ty_n\|^2 &= \sum_{\alpha=1}^{N} |\langle Ty_n, x_\alpha\rangle|^2 + \sum_{\alpha=N+1}^{\infty} |\langle Ty_n, x_\alpha\rangle|^2 \\ &= A + B\,,\end{aligned}$$

say. Now, since

$$y_n = \sum_{\beta=1}^{\infty} \langle y_n, x_\beta\rangle x_\beta$$

and T is continuous, we get at once that

$$\langle Ty_n, x_\alpha\rangle = \sum_{\beta=1}^{\infty} \langle y_n, x_\beta\rangle \langle Tx_\beta, x_\alpha\rangle\,,$$

and consequently, by the Cauchy-Schwarz inequality we get

$$\begin{aligned} B &= \sum_{\alpha=N+1}^{\infty} \left| \sum_{\beta=1}^{\infty} \langle y_n, x_\beta \rangle \langle T x_\beta, x_\alpha \rangle \right|^2 \\ &\le \sum_{\alpha=N+1}^{\infty} \left(\sum_{\beta=1}^{\infty} |\langle y_n, x_\beta \rangle|^2 \right) \left(\sum_{\beta=1}^{\infty} |\langle T x_\beta, x_\alpha \rangle|^2 \right) \\ &\le \|y_n\|^2 \sum_{\alpha=N+1}^{\infty} \sum_{\beta} |\langle T x_\beta, x_\alpha \rangle|^2 . \end{aligned}$$

Furthermore, since $\{y_n\}$ is bounded, by (4.3) it follows that the right-hand side of the above inequality can be made as small as we wish, $\le \eta$ say, provided N is sufficiently large.

Once that N has been fixed it is not hard to estimate A. Indeed, since T is bounded and the sequence $\{y_n\}$ converges weakly to 0, then also $\{Ty_n\}$ converges weakly to 0, and consequently,

$$\lim_{n\to\infty} \langle T y_n, x_\alpha \rangle = 0\,, \quad \alpha = 1, \ldots, N\,.$$

Thus each summand in A goes to 0 with n, and we have $A \le \eta$, provided n is large enough. Finally, combining this estimate with that for B, we get

$$\|T y_n\|^2 \le 2\eta\,, \quad \text{all large } n\,. \quad \blacksquare$$

This result allows us to construct an interesting, and important, example of a compact operator. Let $I = [0,1]$, and consider the integral operator T on $L^2(I)$ given by

$$Tf(x) = \int_I k(x,t) f(t)\, dt\,, \quad x \in I\,,$$

with kernel $k(x,t) \in L^2(I \times I)$. We claim that T is a bounded, completely continuous mapping.

The boundedness follows from the Cauchy-Schwarz inequality. Indeed,

$$\begin{aligned} \|Tf\|^2 &= \int_I \left| \int_I k(x,t) f(t)\, dt \right|^2 dx \\ &\le \int_I \left(\int_I |k(x,y)|^2 dt \right) \left(\int_I |f(t)|^2 dt \right) dx \\ &= \left(\int_I \int_I |k(x,t)|^2 dt dx \right) \|f\|^2 . \end{aligned}$$

As for the compactness, let $\{\phi_n\}$ be a basis for $L^2(I)$, and consider the quantity

$$\sum_{m,n} |\langle T\phi_m, \phi_n\rangle|^2 ; \tag{4.4}$$

we must show it is finite. First note that T^* is given by

$$T^* f(x) = \int_I \overline{k(s,x)} f(s)\, ds\,, \quad x \in I\,.$$

By Tonelli's theorem, (4.4) can be evaluated by the iterated sum obtained by summing first over m and then over n. Now,

$$\begin{aligned}\sum_{m=1}^{\infty} |\langle T\phi_m, \phi_n\rangle|^2 &= \sum_{m=1}^{\infty} |\langle \phi_m, T^*\phi_n\rangle|^2 \\ &= \|T^*\phi_n\|^2\,, \quad n = 1, 2, \ldots\end{aligned}$$

Thus summing over n above we get

$$\begin{aligned}\sum_{m,n} |\langle T\phi_m, \phi_n\rangle|^2 &= \sum_{n=1}^{\infty} \|T^*\phi_n\|^2 = \sum_{n=1}^{\infty} \int_I \left| \int_I \overline{k(s,x)} \phi_n(s) ds \right|^2 dx \\ &= \int_I \sum_{n=1}^{\infty} |\langle \phi_n, k(\cdot, x)\rangle|^2 dx = \int_I \int_I |k(s,x)|^2 ds dx < \infty\,,\end{aligned}$$

(4.3) holds, and T is compact.

We are now ready to give a detailed description of a compact operator T with the additional property that $T^* = T$; any operator which satisfies this relation is said to be self-adjoint. For example, in the case of the integral operator described above, this corresponds to those kernels k that satisfy $k(x,t) = \overline{k(t,x)}$, $x, t \in I$. As we shall have the opportunity to verify in the case of compact operators, it turns out that the structure of such an operator is reminiscent to that of a symmetric matrix. And, as in the case of matrices, eigenvalues and related concepts play an important role in determining the properties of a compact self-adjoint operator.

An eigenvalue of a linear operator T defined on X is a scalar λ such that there exists $0 \neq x \in X$, with the property that

$$Tx = \lambda x\,. \tag{4.5}$$

An element $x \in X$ for which (4.5) holds is called an eigenvector associated, or corresponding, to the eigenvalue λ. The collection of all eigenvectors associated to an eigenvalue λ is a subspace X_λ of X, called the eigenspace corresponding to λ.

The relevant facts concerning a self-adjoint operator are included in our next result.

Proposition 4.5. Let X be a Hilbert space, and suppose T is a self-adjoint mapping defined on X. Then T satisfies the following properties:

(i) T is bounded.

(ii) For each $x \in X, \langle Tx, x\rangle$ is a real number.

(iii) The norm of T is given by

$$\|T\| = \sup_{\|x\|=1} |\langle Tx, x\rangle| . \tag{4.6}$$

(iv) All eigenvalues of T are real.

(v) Eigenspaces corresponding to different eigenvalues are orthogonal.

(vi) If P_λ denotes the projection onto the eigenspace X_λ corresponding to the eigenvalue λ of T, then

$$\lambda P_\lambda = TP_\lambda = P_\lambda T . \tag{4.7}$$

Proof. (i) Assume that $x_n \to x$, and that $Tx_n \to y$; by the Closed graph theorem it suffices to show that $x \in D(T)$ and $Tx = y$. Let $z \in X$, and observe that on the one hand,

$$\langle Tx_n, z\rangle = \langle x_n, Tz\rangle \to \langle x, Tz\rangle ,$$

and, on the other hand,

$$\langle Tx_n, z\rangle \to \langle y, z\rangle .$$

Thus, combining the above relations we have

$$\langle Tx, z\rangle = \langle x, Tz\rangle = \langle y, z\rangle \quad \text{all } z \in X ,$$

which, since z is arbitrary, means that $x \in D(T)$ and that $Tx = y$.
(ii) Since for $x \in X$ we have $\langle Tx, x\rangle = \langle x, Tx\rangle = \overline{\langle Tx, x\rangle}$, as anticipated, $\langle Tx, x\rangle$ is real.
(iii) Let $\eta = \sup_{\|x\|=1} |\langle Tx, x\rangle|$. Since for $\|x\| = 1$ we have $|\langle Tx, x\rangle| \leq \|Tx\|\,\|x\| \leq \|T\|$, it follows that $\eta \leq \|T\|$. On the other hand, since by a simple computation it readily follows that

$$4\Re\langle Tx, y\rangle = \langle T(x+y), x+y\rangle - \langle T(x-y), x-y\rangle ,$$

by putting $x_1 = (1/\|x+y\|)(x+y)$, $x_2 = (1/\|x-y\|)(x-y)$, $\|x_1\| = \|x_2\| = 1$, we get that

$$\begin{aligned} 4\Re\langle Tx, y\rangle &= \|x+y\|^2\langle Tx_1, x_1\rangle - \|x-y\|^2\langle Tx_2, x_2\rangle \\ &\leq \eta\|x+y\|^2 + \eta\|x-y\|^2 = 2\eta\left(\|x\|^2 + \|y\|^2\right) . \end{aligned} \tag{4.8}$$

Pick now $x \in X$ such that $Tx \neq 0$ and $\|x\| = 1$, and put $y = (1/\|Tx\|)Tx$, $\|y\| = 1$, in (4.8). That inequality then becomes

$$4\Re\langle Tx, Tx\rangle/\|Tx\| \leq 4\eta\,,$$

or, equivalently, $\|Tx\| \leq \eta$. This gives $\|T\| \leq \eta$, and (4.6) holds.
(iv) If λ is an eigenvalue of T and x is an eigenvector corresponding to λ with $\|x\| = 1$, we have

$$\lambda = \lambda\langle x, x\rangle = \langle Tx, x\rangle\,,$$

which by (ii) above is real.
(v) Let $\lambda \neq \mu$ be eigenvalues of T, and let X_λ and X_μ denote the eigenspaces corresponding to λ and μ, respectively. Now, assume $x \in X_\lambda$, $y \in X_\mu$, and note that since the eigenvalues are real we get

$$\lambda\langle x, y\rangle = \langle Tx, y\rangle = \langle x, Ty\rangle = \mu\langle x, y\rangle\,,$$

which can only be true if $\langle x, y\rangle = 0$.
(vi) Since $P_\lambda x \in X_\lambda$ for all $x \in X$, it readily follows that

$$TP_\lambda x = \lambda P_\lambda x\,, \quad x \in X\,.$$

As for the commutation relation in (4.7), observe that for $x, y \in X$ we have

$$\begin{aligned}\langle TP_\lambda x, y\rangle &= \langle \lambda P_\lambda x, y\rangle = \langle x, \lambda P_\lambda y\rangle \\ &= \langle x, TP_\lambda y\rangle = \langle Tx, P_\lambda y\rangle = \langle P_\lambda Tx, y\rangle\,.\end{aligned}$$

Since x and y are arbitrary, this can only happen if $TP_\lambda = P_\lambda T$. ■

Referring to the finite dimensional case, it is possible to represent a self-adjoint mapping $T: C^n \to C^n$ in terms of projections. First choose a basis for C^n and represent T by a Hermitian matrix, which we also denote by T; for simplicity assume that the matrix T has n different, necessarily real, eigenvalues $\lambda_1 < \ldots < \lambda_n$, say. Then T has an ONS of n eigenvectors $x_1, \ldots, x_n$, say, where x_i corresponds to the eigenvalue λ_i, $1 \leq i \leq n$. But this is a basis for C^n, so that

$$x = \sum_{i=1}^{n} \langle x, x_i\rangle x_i\,, \quad \text{all } x \in C^n\,.$$

Consequently, Tx is equal to

$$\sum_{i=1}^{n} \langle x, x_i\rangle Tx_i = \sum_{i=1}^{n} \lambda_i \langle x, x_i\rangle\,, \quad x \in C^n\,.$$

Thus, if $P_i: C^n \to \{x_i\}$ denotes the projection of C^n onto the eigenspace of T corresponding to the eigenvalue λ_i, $1 \le i \le n$, the above expression becomes

$$T = \sum_{i=1}^{n} \lambda_i P_i .$$

This is a representation of T in terms of projections; a natural question is whether a similar representation holds for compact self-adjoint operators on a Hilbert space.

The first step is to show that such operators do have eigenvalues.

Lemma 4.6. Let X be a Hilbert space and T a compact self-adjoint mapping defined on X. Then T has a nonzero eigenvalue.

Proof. Since the conclusion of the theorem is obvious when $T = 0$, we assume that $T \neq 0$. Let

$$\mu = \inf_{\|x\|=1} \langle Tx, x\rangle , \quad \text{and} \quad \eta = \sup_{\|x\|=1} \langle Tx, x\rangle ;$$

by (iii) in Theorem 4.4, we have $\|T\| = \max\{|\mu|, \eta\}$. We claim that

$$\lambda = \begin{cases} \mu & \text{if } \|T\| = |\mu| \\ \eta & \text{if } \|T\| = \eta, \end{cases}$$

is an eigenvalue of T.

Consider, for instance, the case when $\|T\| = |\mu| > 0$. By the definition of μ there exists a sequence $\{x_n\}$ such that

$$\lim_{n\to\infty} \langle Tx_n, x_n\rangle = \mu , \quad \|x_n\| = 1 , \text{all } n .$$

Since T is compact we can choose a subsequence, which we call $\{x_n\}$ again, such that $\lim_{n\to\infty} Tx_n = y$, say. Moreover, since

$$\begin{aligned} \|Tx_n - \mu x_n\|^2 &= \|Tx_n\|^2 - 2\mu\langle Tx_n, x_n\rangle + \mu^2 \\ &\le \|T\|^2 - 2\mu\langle Tx_n, x_n\rangle + \mu^2 , \end{aligned}$$

and the right-hand side above tends to 0 as $n \to \infty$, we get that

$$\lim_{n\to\infty} Tx_n - \mu x_n = 0 .$$

But then, writing

$$x_n = \frac{1}{\mu}(Tx_n - (Tx_n - \mu x_n)) , \quad n = 1, 2, \ldots$$

it follows that $\lim_{n\to\infty} x_n = (1/\mu)y$. In this case we have

$$y = \lim_{n\to\infty} Tx_n = (1/\mu)Ty\,,$$

and consequently, μ is an eigenvalue for T, and y is an eigenvector corresponding to μ. ■

It is interesting to point out that the assumption that T is self-adjoint is necessary for the validity of Lemma 4.6. Indeed, let $X = \ell^2$ and suppose $T: X \to X$ is given by

$$T((x_1, \ldots, x_n, \ldots)) = (0, x_1/1, \ldots, x_n/n, \ldots)\,.$$

T is compact, but not self-adjoint. Also, as the reader can readily verify, T has no nonzero eigenvalues.

Corollary 4.7. Let X be a Hilbert space, and suppose T is a compact self-adjoint operator defined on X. If T has no nonzero eigenvalues, then $T = 0$.

Observe that the eigenvalue we just found in Lemma 4.6 is one with largest absolute value. Indeed, if λ is an eigenvalue of T, and x is an eigenvector corresponding to λ with $\|x\| = 1$, then we have

$$|\lambda| = |\lambda|\langle x, x\rangle = |\langle Tx, x\rangle| \le \|T\| = |\mu|\,,$$

which is precisely our remark.

The stage is now set for the description of the action of a compact self-adjoint mapping in terms of projections.

Theorem 4.8 (Spectral Decomposition of Compact Operators). Let X be a Hilbert space, and T a compact self-adjoint mapping defined on X. Then, the set of distinct eigenvalues $\{\lambda_n\}$ of T is at most countable, and if P_{λ_n} denotes the projection of X onto the eigenspace X_{λ_n} corresponding to the eigenvalue λ_n, we have

$$T = \sum\nolimits_n \lambda_n P_{\lambda_n}\,. \tag{4.9}$$

The convergence of the series in (4.9) is understood to be in the sense of the norm of $\mathcal{B}(X)$, i.e.,

$$\lim_{m\to\infty} \Big\| T - \sum\nolimits_{n=1}^{m} \lambda_n P_{\lambda_n} \Big\| = 0\,.$$

Proof. Unless $T = 0$, by Lemma 4.6, T has an eigenvalue λ_1, say, with the largest absolute value. If $T_1 = T$ and if P_{λ_1} denotes the projection of X onto X_{λ_1}, the eigenspace corresponding to the eigenvalue λ_1, consider the mapping T_2 given by $T_2 = T_1 - \lambda_1 P_{\lambda_1}$. In view of (vi) in Proposition 4.4 we can rewrite T_2 as

$$T_2 = T_1 - T_1 P_{\lambda_1} = T_1(I - P_{\lambda_1}) = (I - P_{\lambda_1})T_1 . \tag{4.10}$$

Since T_1 is compact and self-adjoint, by 5.48 below, also T_2 is compact and self-adjoint. Moreover, since $I - P_{\lambda_1}$ is a projection, and as such $\|I - P_{\lambda_1}\| \leq 1$, we have

$$\|T_2\| \leq \|I - P_{\lambda_1}\| \, \|T_1\| \leq \|T_1\| . \tag{4.11}$$

Whence applying Lemma 4.6 to T_2 now, unless $T_2 = 0$ we can find an eigenvalue $0 \neq \lambda_2$ of T_2 with largest absolute value. By (4.11), $|\lambda_1| \geq |\lambda_2|$.

Moreover, we claim that the following is also true: λ_1 is not an eigenvalue of T_2, and every eigenvalue of T_2 is at the same time an eigenvalue of T_1, and the corresponding eigenspaces coincide.

First we show that λ_1 is not an eigenvalue of T_2. For if it were, let $0 \neq x$ be an eigenvector corresponding to λ_1, and note that by (4.10) and (4.7) we have

$$T_1 x - \lambda_1 P_{\lambda_1} x = \lambda_1 x . \tag{4.12}$$

Now, applying P_{λ_1} to both sides, again by (4.7), we get

$$\lambda_1 P_{\lambda_1} x - \lambda_1 P_{\lambda_1} x = \lambda_1 P_{\lambda_1} x = 0 , \tag{4.13}$$

which substituted into (4.12) gives $T_1 x = \lambda_1 x$, i.e., $x \in X_{\lambda_1}$. But if this is the case, then by (4.13) we also have $x = P_{\lambda_1} x = 0$, which gives the desired contradiction.

We now show that every nonzero eigenvalue of T_2 is at the same time an eigenvalue of T_1, and the corresponding eigenspaces coincide. In fact, let $\lambda \neq 0$ be an eigenvalue of T_2, and $x \neq 0$ an eigenvector of T_2 corresponding to λ. By the definition of T_2 we have

$$T_1(I - P_{\lambda_1})x = \lambda x , \tag{4.14}$$

and consequently, it follows that

$$(I - P_{\lambda_1})T_1(I - P_{\lambda_1})x = \lambda(I - P_{\lambda_1})x . \tag{4.15}$$

Moreover, since T_1 commutes with $(I - P_{\lambda_1})$, the left-hand side of (4.15) equals $T_1(I - P_{\lambda_1})$, which happens to be the left-hand side of (4.14).

Whence equating the right-hand sides of (4.14) and (4.15), and since $\lambda \neq 0$, we get $x = (I - P_{\lambda_1})x$, which gives

$$T_1 x = T_1(I - P_{\lambda_1})x = T_2 x = \lambda x\,.$$

Thus, λ is also an eigenvalue of T_1 with eigenvector x, and consequently the eigenspace corresponding to λ as an eigenvector of T_2 is contained or equal to that of λ as an eigenvalue of T_1.

Now, since $\lambda \neq \lambda_1$, by (v) in Proposition 4.5 it follows that the eigenspaces X_λ and X_{λ_1}, corresponding to T_1, are orthogonal. Whence if y is an eigenvector of T_1 corresponding to λ it follows that $P_{\lambda_1} y = 0$, and by (4.10) we have

$$T_2 y = T_1 y - \lambda_1 P_{\lambda_1} y = T_1 y = \lambda y\,.$$

Thus, y is also an eigenvector of T_2 corresponding to λ, and the eigenspaces coincide.

Repeating this process we construct compact self-adjoint operators $T_1 = T, T_2, \ldots, T_n$, and eigenvalues $\lambda_1, \ldots, \lambda_n$ of these operators, such that

$$T_{k+1} = T_k - \lambda_k P_{\lambda_k} = T - \sum\nolimits_{i=1}^{k} \lambda_i P_{\lambda_i}\,, \quad k = 1, \ldots, n-1\,,$$

and

$$|\lambda_1| \geq \ldots \geq |\lambda_n|\,, \quad \|T_k\| = |\lambda_k|\,, \quad k = 1, \ldots, n\,.$$

Further, by what has been shown above, the λ_k's are distinct eigenvalues of T_1.

Now, if for some n we have $T_n = 0$, then the sum in (4.9) is finite, and the conclusion follows. However, if $T_n \neq 0$ for every n, then the process described above leads to a sequence $\{T_n\}$ of compact self-adjoint operators and a corresponding sequence $\{\lambda_n\}$ of eigenvalues. Next we show that in this case $\lambda_n \to 0$ as $n \to \infty$. Suppose not, then it follows that

$$|\lambda_n| \geq \varepsilon > 0\,, \quad \text{all } n\,.$$

Choose now an ONS $\{x_n\}$ consisting of eigenvectors associated to the λ_n's. By the Pythagorean theorem we get

$$\begin{aligned}\|Tx_m - Tx_n\|^2 &= \|\lambda_m x_m - \lambda_n x_n\|^2 \\ &= |\lambda_m|^2 + |\lambda_n|^2 \geq 2\varepsilon^2\,, \quad m \neq n\,,\end{aligned}$$

and consequently, neither the sequence $\{Tx_n\}$ nor any of its subsequences converges, thus contradicting the fact that T is compact.

Moreover, since $\|T_n\| = |\lambda_n|$ for all n, we also have $\lim_{n\to\infty} \|T_n\| = 0$, and (4.9) holds.

There is yet a last detail to be checked, namely, that T has no nonzero eigenvalues apart from the λ_n's. For, if $\lambda \neq 0$ is such an eigenvalue and if $x \neq 0$ is an eigenvector corresponding to λ, then by (4.9) we get

$$\lambda x = Tx = \sum_n \lambda_n P_{\lambda_n} x \,. \tag{4.16}$$

Now, by (v) in Proposition 4.5 the elements $P_{\lambda_n} x$, $n = 1, 2, \ldots$ are pairwise orthogonal. Hence by (4.16) it follows that

$$\lambda P_{\lambda_m} x = P_{\lambda_m} \left(\sum_n \lambda_n P_{\lambda_n} x \right) = \lambda_m P_{\lambda_m} x \,, \quad \text{all } m \,,$$

and since $\lambda \neq \lambda_m$, we have that $P_{\lambda_m} x = 0$ for all m. Again by (4.16) we get that $\lambda x = 0$, which, in turn, implies that $x = 0$, a contradiction. ∎

Theorem 4.8 has many important applications, including the development of a functional calculus for compact self-adjoint operators. For instance, if T is such an operator, to represent T^2 observe that on account of (4.9) we have

$$T^2 x = \sum_n \lambda_n P_{\lambda_n} Tx = \sum_n \lambda_n^2 P_{\lambda_n} x \,, \quad \text{all } x \in X \,.$$

It is then apparent that for polynomials p, we also have

$$p(T)x = \sum_n p(\lambda_n) P_{\lambda_n} x \,, \quad \text{all } x \in X \,,$$

where the notation $p(T)$ is self-explanatory. Moreover, since functions f that are continuous on $[\mu, \eta]$ are uniform limits of polynomials, we prove this in Corollary 2.3 in Chapter XVII, we also have

$$f(T)x = \sum_n f(\lambda_n) P_{\lambda_n} x \,, \quad \text{all } x \in X \,.$$

In fact, the expression on the right-hand side above defines $f(T)$.

There is yet another way to express the identity in (4.9). Consider the operators $E_\lambda = \sum_{\lambda_n < \lambda} P_{\lambda_n}$, $\lambda \in R$. It is not hard to check that the family $\{E\lambda\}_{\lambda\in R}$, has the following property: For $x, y \in X$, put $\phi(\lambda) = \langle E_\lambda x, y \rangle$;

ϕ is a right-continuous function that vanishes for $\lambda < \mu$, and that is constant for $\lambda > \eta$. We then have

$$\langle Tx, y\rangle = \mu\phi(\mu) + \int_{\mu}^{\eta} \lambda \, d\phi(\lambda),$$

where the integral is an ordinary Riemann-Stieltjes integral. A similar representation is true for arbitrary self-adjoint operators, not necessarily compact; we will not discuss it here.

5. PROBLEMS AND QUESTIONS

5.1 Suppose X is a real inner product space and that the elements $x, y \in X$ satisfy $\|x + y\|^2 = \|x\|^2 + \|y\|^2$. Show that $x \perp y$. Is the result true for complex inner product spaces?

5.2 Suppose X is an inner product space and show that the elements $x, y \in X$ satisfy $\|x + y\| = \|x\| + \|y\|$ iff there exists a scalar $\lambda > 0$ such that $x = \lambda y$.

5.3 If X is a real inner product space and $x, y \in X$ satisfy $\|x\| = \|y\|$, then $(x - y) \perp (x + y)$. What does this mean geometrically? What does the assumption imply in case X is a complex inner product space?

5.4 Suppose x, y, z are elements of an inner product space X. Show that Appolonius' identity holds, to wit,

$$\|z - x\|^2 + \|z - y\|^2 = \|x - y\|^2/2 + 2\|z - (x + y)/2\|^2 .$$

5.5 Can we obtain the norm $\|z\| = |z_1| + |z_2|$ in C^2 from an inner product?

5.6 Discuss under what conditions equality holds in the Cauchy-Schwarz inequality.

5.7 Show that if y, x, x_n are elements of an inner product space X, $n = 1, 2, \ldots$, and if $y \perp x_n$ for all n, and $\lim_{n\to\infty} x_n = x$, then $y \perp x$.

5.8 Show that in an inner product space X, $x \perp y$ iff $\|x + \lambda y\| = \|x - \lambda y\|$ for all scalars λ. Further, $x \perp y$ iff $\|x + \lambda y\| \geq \|x\|$ for all scalars λ.

5.9 Prove that the span of a subset M of a Hilbert space X is dense in X iff $M^{\perp} = \{0\}$.

5.10 Let $M_1 \supseteq M_2$ be nonempty subsets of an inner product space X. Show that (a) $M_1 \subseteq M_1^{\perp\perp}$, (b) $M_1^{\perp} \subseteq M_2^{\perp}$, and (c) $M_1^{\perp\perp\perp} = M_1^{\perp}$. Further, show that a subspace Y of a Hilbert space X is closed iff $Y = Y^{\perp\perp}$.

5.11 Suppose M is a closed subspace of a Hilbert space X and let x be an element of X. Prove that

$$\min\{\|x-y\| : y \in M\} = \min\{|\langle x,y\rangle| : y \in M^{\perp}, \|y\| = 1\}.$$

5.12 Let $I = [0,1]$. Compute $\max \int_I x^3 f\, dx$, and $\min_{a,b} \int_I (x^5 - a - bx)^2 dx$, subject to the conditions

$$\int_I x^k f(x)\, dx = 0\,, k = 0,1,2, \quad \text{and} \quad \int_I |f(x)|^2 dx = 1\,.$$

5.13 The following extension of Theorem 1.8 is called the Lax-Milgram lemma. Let X be a Hilbert space. Let $B(x,y)$ be a complex-valued functional defined on $X \times X$ which satisfies the following four conditions:

(i) $B(x_1 + \lambda x_2, y) = B(x_1, y) + \lambda B(x_2, y)$ for all x_1, x_2, y in X and scalars λ.

(ii) $B(x, y_1 + \lambda y_2) = B(x, y_1) + \overline{\lambda} B(x, y_2)$ for all x, y_1, y_2 in X and scalars λ.

(iii) There is a positive constant k such that $|B(x,y)| \leq k\|x\|\,\|y\|$ for all x, y in X.

(iv) There exists a positive constant c such that $|B(x,x)| \geq c\|x\|^2$ for all $x \in X$.

Then for every $L \in X^*$ there exists a unique element $y \in X$ such that $Lx = B(x,y)$ for all $x \in X$. More precisely, there exists a uniquely determined bounded linear operator T with a bounded inverse T^{-1} such that $\langle x,y\rangle = B(x,Ty)$ for all $x, y \in X$, and $\|T\| \leq 1/c$, $\|T^{-1}\| \leq k$.

5.14 Let X be a complex inner product space. A complex-valued functional $H(x,y)$ defined on $X \times X$ is said to be a Hermitian form provided that the following two conditions hold:

(i) $H(x_1 + \lambda x_2, y) = H(x_1, y) + \lambda H(x_2, y)$ for all x_1, x_2, y in X, and scalars λ.

(ii) $H(x,y) = \overline{H(y,x)}$ for all $x, y \in X$.

H is said to be positive semidefinite if

(iii) $H(x,x) \geq 0$ for all $x \in X$.

Show that if (i), (ii) and (iii) hold, then H satisfies the following Cauchy-Schwarz inequality

$$|H(x,y)| \leq H(x,x)H(y,y), \quad \text{all } x, y \in X.$$

Further, show that $p(x) = \sqrt{H(x,x)}$ defines a seminorm on X.

5.15 Let X, Y be inner product spaces and $T: X \to Y$ be a bounded linear operator. Then, (a) $T = 0$ iff $\langle Tx, y \rangle = 0$ for all x in X and y in Y, and, (b) If $X = Y$ is a complex linear space, then $T = 0$ iff $\langle Tx, x \rangle = 0$ for all x in X.

5.16 Suppose X is a Hilbert space, and let $T \in \mathcal{B}(X)$. Show that $\|T\| = \sup_{\|x\|=\|y\|=1} |\langle Tx, y \rangle|$.

5.17 Suppose $P: X \to X$ is a linear map that satisfies (2.1) in Proposition 2.1 and so that P^2 is a projection. Is P a projection?

5.18 Let $I = [0,1]$ and consider the linear mapping $T: L^2(I) \to L^2(I)$ given by $Tf(x) = a(x)f(x)$, $f \in L^2(I)$. Find necessary and sufficient conditions on $a(x)$ for T to be a projection.

5.19 Let X be a Hilbert space and suppose $P_1, \ldots, P_n$ are projections on X. Find necessary and sufficient conditions for $P = P_1 + \ldots + P_n$ to be a projection. What does the subspace of X onto which P projects look like?

5.20 Let M be a closed subspace of an infinite-dimensional Hilbert space X. Show that the projection P_M is compact iff M is finite dimensional.

5.21 (Gram-Schmidt) Given an arbitrary linearly independent sequence of elements $\{y_n\}$ in an inner product space X there is an ONS of elements $\{x_n\}$ in X such that

$$\operatorname{span}\{y_1, \ldots, y_n\} = \operatorname{span}\{x_1, \ldots, x_n\} \quad \text{for every } n.$$

5.22 Suppose $\|x\| = 1$. Show that at most $1/\varepsilon^2$ of the Fourier coefficients of x with respect to any ONS in X exceed, in modulus, any $\varepsilon > 0$.

5.23 Suppose X is a Hilbert space, show that X has an orthonormal basis. Moreover, in case X is separable, the existence of such a basis may be established without invoking Zorn's Lemma.

5.24 Suppose $\{x_n\}$ is a basis for a Hilbert space X, and let $y_n = x_n - x_{n+1}$, $n \geq 1$. Show that the system $\{y_n\}$, although not orthonormal, is nevertheless complete in X.

5.25 Suppose $\{x_n\}$ is a basis for a Hilbert space X, and $\{y_n\} \subseteq X$ is such that $\sum_n \|x_n - y_n\|^2 < 1$. Show that $\{y_n\}$ is complete in X. Is the same conclusion true if $\sum_n \|x_n - y_n\|^2 < \infty$ instead?

5.26 (Paley-Wiener) Suppose $\{x_n\}$ is a basis for a Hilbert space X, and suppose the sequence $\{y_n\} \subseteq X$ satisfies

$$\left\| \sum_{m=1}^{n} \lambda_m (x_m - y_m) \right\| \leq A \left\| \sum_{m=1}^{n} \lambda_m x_m \right\| ,$$

for any scalars λ_m, $m \geq 1$, and a constant A, $0 \leq A < 1$, independent of the choice of scalars and n. Show that each $x \in X$ can be written as $x = \sum_{n=1}^{\infty} c_n y_n$, where the c_n's are scalars and the sum converges in X.

5.27 Suppose $I = [a,b]$ is an interval of the line. Show that an ONS $\{\phi_n\}$ is complete in $L^2(I)$ iff $\sum_{n=1}^{\infty} \left(\int_{[a,x]} \phi_n \right)^2 = x - a$ for all $x \in I$.

5.28 Let $\{\phi_n\}$ be an orthonormal sequence in $L^2(I)$, where $I = [0,1]$. Show that if there is a constant M such that $|\phi_n(x)| < M$ a.e. for all n, and if $\sum_{n=1}^{\infty} \lambda_n \phi_n(x)$ converges a.e., then $\lim_{n\to\infty} \lambda_n = 0$.

5.29 Suppose μ is a Borel measure on $I = (-\pi,\pi]$ with the property that if $a_0 = \frac{1}{\pi} \int_I d\mu$, $a_n = \frac{1}{2\pi} \int_I \cos nx \, d\mu(x)$ and $b_n = \frac{1}{2\pi} \int_I \sin nx \, d\mu(x)$, $n = 1,2,\ldots$, then $a_0^2 + \sum_{n=1\infty} (a_n^2 + b_n^2) < \infty$. Show that μ is absolutely continuous with respect to the restriction λ of the Lebesgue measure to I, and that $d\mu/d\lambda \in L^2(I)$.

5.30 Referring to 5.13, if the Hermitian form there is bounded, i.e., if $|H(x,y)| \leq k\|x\| \, \|y\|$ for all $x, y \in X$, prove that there exists a self-adjoint operator T such that $H(x,y) = \langle Tx, y \rangle$ for all $x, y \in X$.

5.31 If $H(x,y)$ is a Hermitian form, then $H(x) = H(x,x)$ is called a quadratic form. Prove that Hermitian and quadratic forms are related by

$$4H(x,y) = H(x+y) + H(x-y) + iH(x+iy) - iH(x-iy) .$$

5.32 Prove, without invoking the Continuum Hypothesis, that the dimension of any infinite-dimensional Hilbert space is at least c.

5.33 Let X be a Hilbert space, and $T, T_n \in \mathcal{B}(X), n = 1,2,\ldots$ Show that if $\lim_{n\to\infty} \langle T_n x, y \rangle = \langle Tx, y \rangle$ uniformly for $y \in X$, $\|y\| = 1$, then $\lim_{n\to\infty} \|T_n x - Tx\| = 0$ for each $x \in X$. The assumption concerning the uniform convergence means that the sequence $\{\langle Tx_n, y \rangle\}$

converges to $\langle Tx, y\rangle$, uniformly over those $y \in X$ with $\|y\| = 1$ for each $x \in X$. Moreover, if $\lim_{n\to\infty} \|T_n x - Tx\| = 0$ uniformly for $x \in X, \|x\| = 1$, then $\lim_{n\to\infty} \|T_n - T\| = 0$ (in $\mathcal{B}(X)$).

5.34 Let $I = [0,1]$ and $X = L^2(I)$. Show that the linear operator T on X given by $Tf(x) = \int_{[0,x]} f(y)\,dy$, $0 \le x \le 1$, is bounded, and that $\|T\| \le 1$. Furthermore, show that $T^* = -T + P$, where P is the orthogonal projection onto the subspace of constant functions.

5.35 Let $I = [0,1]$ and consider the linear operator T defined for f in $L^2(I)$ by $Tf(x) = f(x) + 2\int_{[0,x]} e^{(x-t)} f(t)\,dt$, $0 \le x \le 1$. Show that $T \in \mathcal{B}(L^2(I))$, and that if $\phi(x) = e^x$ and $\psi(x) = e^{-x}$, the following properties hold: (a) $T\phi = \psi$, (b) If $\langle f, \phi\rangle = 0$, then $\langle Tf, \psi\rangle = 0$ and $\|Tf\|_2 = \|f\|_2$, (c) If $\langle g, \psi\rangle = 0$, then there exists $f \in L^2(I)$ such that $\langle f, \phi\rangle = 0$ and $g = Tf$. Conclude from these observations that T is one-to-one, onto, and that its inverse is continuous.

5.36 Let X be a Hilbert space, and $T, T_1 \in \mathcal{B}(X)$. Show that the adjoint mapping satisfies the following properties: (a) $\|T^*T\| = \|T\|^2$, (b) $(\lambda T + \lambda_1 T_1)^* = \overline{\lambda} T^* + \overline{\lambda}_1 T_1^*$, (c) $(TT_1)^* = T_1^* T^*$, and, (d) $T^{**} = T$.

5.37 Let X be a Hilbert space, and $T \in \mathcal{B}(X)$. If as usual $R(T)$ and $N(T)$ denote the range and the nullspace of T respectively, show that: (a) $R(T)^\perp = N(T^*)$, and, (b) $N(T)^\perp = \overline{R(T^*)}$.

5.38 Show that if X is a complex Hilbert space and T is a bounded linear operator on X such that $\langle Tx, x\rangle$ is real for all x in X, then T is self-adjoint. This result is generally false if X is a real Hilbert space.

5.39 Suppose T_1, T_2 are linear mappings defined on X that, in addition, satisfy $\langle T_1 x, y\rangle = \langle x, T_2 y\rangle$ for all $x, y \in X$. Show that $T_1 \in \mathcal{B}(X)$, and that $T_2 = T_1^*$.

5.40 Define a relation among the self-adjoint transformations on a Hilbert space X by writing $T_1 \le T_2$, or $T_2 \ge T_1$, when $\langle T_2 x, x\rangle \ge \langle T_1 x, x\rangle$ for all $x \in X$. Show that this relation is a partial ordering.
A mapping $T \ge 0$ is said to be positive. Prove that positive self-adjoint mappings satisfy the following generalized Cauchy-Schwarz inequality: $|\langle Tx, y\rangle|^2 \le \langle Tx, x\rangle\langle Ty, y\rangle$.

5.41 Prove that if the sequence $\{T_n\}$ of self-adjoint operators on a Hilbert space X satisfies $0 \le T_1 \le \ldots \le T_n \le \ldots \le I$, then there exists a bounded self-adjoint mapping T on X such that $\lim_{n\to\infty} \|T_n - T\| = 0$.

5.42 Let $I = [-1,1]$. Is there a bounded linear functional L on $L^2(I)$ such that $Lf = f(0)$ for every $f \in C^1(I)$? Is there a bounded linear functional L_1 on $L^2(I)$ such that $L_1 f = f'(0)$ for every $f \in C^1(I)$?

5.43 Let X be a Hilbert space, and $T, T_1, T_2, \ldots$ be bounded linear operators on X. Assume that $\lim_{n\to\infty}\langle T_n x, y\rangle = \langle Tx, y\rangle$ for all $x, y \in X$. Show that for some constant c, $\|T_n\| \le c$ for all n.

5.44 Let $I = [0,1]$ and suppose X is a linear subspace of $C(I)$ which is closed with respect to the norm of $L^2(I)$. Prove that X is also closed in $C(I)$, and, as a consequence of this, show that that there is a constant c such that $\|f\|_2 \le \|f\|_\infty \le c\|f\|_2$ for every $f \in X$. Furthermore, show that there is a function $k(\cdot, y) \in L^2(I)$ such that

$$f(y) = \int_I k(x,y) f(x)\, dx\,, \quad \text{all } f \in X\,, \quad y \in I\,.$$

Finally, show that the dimension of X cannot exceed c^2.

5.45 Assume that $\{x_n\}$ is a sequence of elements of a Hilbert space X such that $\|x_n\| = 1$ for $n = 1, 2, \ldots$ If the x_n's converge weakly to an element $x \in X$, does it necessarily follow that $\lim_{n\to\infty}\|x_n - x\| = 0$? Similarly, if the sequence $\{x_n\}$ converges weakly to $x \in X$, and if the convergence is uniform over $y \in X$ with $\|y\| = 1$, does it follow that $\lim_{n\to\infty}\|x - x_n\| = 0$? The assumption concerning the uniform convergence means that the sequence $(\langle x_n,y\rangle)$ converges to $\langle x,y\rangle$, uniformly over those $y \in X$ with $\|y\| = 1$.

5.46 Let X be a Hilbert space and suppose that $\{x_n\} \subseteq X$ converges weakly to $x \in X$. Prove that there is a subsequence $\{x_{n_k}\}$ such that its arithmetic means $(x_{n_1} + \cdots + x_{n_k})/k$ converge to x in the norm of X.

5.47 Suppose $\{x_n\}$ is a basis for a Hilbert space X, and $\lambda_n \to 0$ is a sequence of scalars. Prove that the mapping T on X given by $Tx = \sum_{n=1}^{\infty} \lambda_n \langle x, x_n\rangle x_n$ is compact.

5.48 Prove that if T_1, T_2 are compact operators on a Hilbert space X and λ is a scalar, then T_1T_2 and $T_1 + \lambda T_2$ are also compact. Further, if T is bounded, then also TT_1 and T_1T are compact.

5.49 Show that if $\{T_n\} \subseteq \mathcal{B}(X)$ is a sequence of compact operators on a Hilbert space X and $\lim_{n\to\infty} T_n = T$ (in $\mathcal{B}(X)$), then T is also compact.

5.50 Prove that if T is a compact mapping on a Hilbert space X, then so is its adjoint T^*. Is the converse true?

CHAPTER **XVII**

Fourier Series

In this section we discuss how the different results we have covered thus far fit into the theory of Fourier series.

1. THE DIRICHLET KERNEL

A trigonometric polynomial $p(x)$ is an expression of the form

$$p(x) = \sum_{|k|\le n} c_k e^{ikx}, \quad |c_{-n}| + |c_n| \ne 0\,. \tag{1.1}$$

n is the degree of p and the c_k's are (possibly) complex constants. Thus p is a continuous function of period 2π and is therefore determined by its values on $T = (-\pi, \pi]$ or any other interval of length 2π for that matter. Given a trigonometric polynomial p of degree $\le n$, we can easily compute the constants c_k by means of

$$c_k = \frac{1}{2\pi}\int_T p(x)e^{-ikx}\,dx\,, \quad |k| \le n\,.$$

This observation follows at once from the fact that

$$\frac{1}{2\pi}\int_T e^{ikx}\,dx = \begin{cases} 0 & \text{if } k \ne 0 \\ 1 & \text{if } k = 0. \end{cases}$$

A trigonometric series is an expression of the form

$$\sum_{k=-\infty}^{\infty} c_k e^{ikx}\,. \tag{1.2}$$

Since we make no assumption concerning the convergence of this series, (1.2) only formally represents a function of period 2π.

A Fourier series is a trigonometric series for which there is a periodic Lebesgue integrable function f such that

$$c_k = c_k(f) = \frac{1}{2\pi}\int_T f(x)e^{-ikx}\,dx\,, \quad \text{all } k\,. \tag{1.3}$$

In this case we call the constants c_k the Fourier coefficients of f and denote this correspondence by

$$f \sim \sum_k c_k e^{ikx}\,. \tag{1.4}$$

Still in the case of Fourier series no assumption concerning the convergence of the series (1.4) is made. More precisely, if $s_n(f,x)$ denotes the trigonometric polynomial of degree $\leq n$ corresponding to the symmetric partial sums of (1.4) of order n, i.e.,

$$s_n(f,x) = \sum_{|k|\leq n} c_k e^{ikx}\,, \tag{1.5}$$

then nothing is known or assumed concerning the existence of the $\lim_{n\to\infty} s_n(f,x)$ for any x in T.

Now, by (1.3) it is clear that $|c_k(f)| \leq \|f\|_1$ for all k, and there is more we can say in this direction.

Theorem 1.1 (Riemann-Lebesgue). Let f be a periodic integrable function on T. Then $c_k(f) \to 0$ as $|k| \to \infty$.

Proof. We invoke the fact that trigonometric polynomials are dense in $L(T)$; a proof of this is given in Corollary 2.3 below. Now, given $\varepsilon > 0$, we show that $|c_k| \leq \varepsilon$ provided $|k| > k_0$ is large enough. Let p be a trigonometric polynomial such that $\|f - p\|_1 \leq \varepsilon$, and let $k_0 =$ degree of p. Then for $|k| > k_0$ we have

$$c_k(f) = \frac{1}{2\pi}\int_T (f(x) - p(x))e^{-ikx}\,dx\,,$$

and consequently, $|c_k(f)| \leq \|f - p\|_1 \leq \varepsilon$. ∎

Now that there is some hope that the Fourier series of an integrable function converges, we take a closer look at $s_n(f,x)$. It can also be written as

$$\sum_{|k|\le n}\left(\frac{1}{2\pi}\int_T f(t)e^{-ikt}\,dt\right)e^{ikx}=\frac{1}{\pi}\int_T f(t)\left(\frac{1}{2}\sum_{|k|\le n}e^{ik(x-t)}\right)dt$$
$$=\frac{1}{\pi}\int_T f(t)D_n(x-t)\,dt\,, \qquad (1.6)$$

where we have denoted by

$$D_n(x)=\frac{1}{2}\sum_{|k|\le n}e^{ikx}\,,\quad n=0,1,\ldots \qquad (1.7)$$

the Dirichlet kernel of order n. In other words, $s_n(f,x)$ is essentially the convolution of f with the Dirichlet kernel D_n of order n.

We list some properties of these kernels. In the first place by summing the geometric series in (1.7) we get that $D_n(x)$ equals

$$\frac{1}{2}e^{-inx}\frac{\left(e^{i(2n+1)x}-1\right)}{\left(e^{ix}-1\right)}=\frac{1}{2}\frac{e^{i(n+1)x}-e^{-inx}}{e^{ix/2}\left(e^{ix/2}-e^{-ix/2}\right)}$$
$$=\frac{1}{2}\frac{e^{i(n+1/2)x}-e^{-i(n+1/2)x}}{e^{ix/2}-e^{-ix/2}}$$
$$=\frac{1}{2}\frac{\sin((n+1/2)x)}{\sin(x/2)}\,,\quad n=0,1,\ldots \qquad (1.8)$$

Thus D_n is an even function, and integrating (1.7) over T it readily follows that

$$\frac{1}{\pi}\int_T D_n(x)\,dx=\frac{2}{\pi}\int_{[0,\pi]}D_n(x)\,dx=1\,,\quad \text{all } n\,. \qquad (1.9)$$

It is also possible to estimate D_n. In fact, by (1.8),

$$|D_n(x)|\le\frac{1}{2}\sum_{|k|\le n}|e^{ikx}|=\frac{2n+1}{2}=n+\frac{1}{2}\,,\quad \text{all } n\,. \qquad (1.10)$$

Moreover, since as is readily seen

$$1/(2\sin(x/2))\le\pi/2x\,,\quad \text{for } 0<x<\pi\,, \qquad (1.11)$$

by (1.7) it follows that

$$|D_n(x)|\le\pi/2|x|\,,\quad 0<|x|<\pi\,,\quad \text{all } n\,. \qquad (1.12)$$

Because of these properties, the D_n's look like an approximate identity parametrized by n rather than ε.

The natural question is, then, how closely does the behaviour of the Dirichlet kernels resemble that of an approximate identity. In particular, do the convolutions in (1.6) tend to f a.e.? Or, in other words, do the partial sums $s_n(f,\cdot)$ tend to f a.e.? We begin by considering the question of the convergence for continuous functions f, and to do so we introduce the Lebesgue constants L_n. They are given by

$$L_n = \frac{1}{\pi}\int_T |D_n(x)|\,dx\,, \quad n = 0,1,\ldots$$

This is one reason why: In Chapter XIII we have seen that in addition to (1.9) a uniform bound on the L_n's is an important ingredient in establishing the convergence of the approximate identities to the function in question. There is yet another way to see this: From (1.6) it is clear that

$$|s_n(f,0)| \le \frac{1}{\pi}\int_T |f(t)|\,|D_n(t)|\,dt = \|f\|_\infty\, L_n\,.$$

Moreover, setting $f_n(x) = \operatorname{sgn} D_n(x)$, we readily see that $\|f_n\|_\infty = 1$ for all n and consequently,

$$\sup_{f\in L^\infty(T),\|f\|_\infty\le 1} |s_n(f,0)| = L_n\,, \quad \text{all } n\,.$$

Further, since the function f_n is real-valued and discontinuous at a finite number of points, it is easy to modify its values in small neighbourhoods of those points to obtain that also for continuous functions

$$\sup_{f\in C(T),\|f\|_\infty\le 1} |s_n(f,0)| = L_n\,. \tag{1.13}$$

Proposition 1.2. $L_n \sim (4/\pi^2)\ln n$, as $n\to\infty$.

Proof. Since D_n is even and $\sin(t/2) > 0$ for $0 < t < \pi$, we have that

$$\begin{aligned} L_n =& \frac{2}{\pi}\int_{[0,\pi]} |\sin((n+1/2)t)|\left(\frac{1}{2\sin(t/2)} - \frac{1}{t}\right)\,dt \\ &+ \frac{2}{\pi}\int_{[0,\pi]} |\sin((n+1/2)t)|\,\frac{dt}{t} = A_n + B_n\,, \end{aligned}$$

say. It is clear that the integrand in A_n is bounded, independently of n, and consequently the A_n's are bounded. As for the B_n's, the change of variables $(n+1/2)t = s$ gives

$$B_n = \frac{2}{\pi}\int_{[0,(n+1/2)\pi]} |\sin s|\,\frac{ds}{s}$$
$$= \frac{2}{\pi}\left(\int_{[0,n\pi]} + \int_{(n\pi,(n+1/2)\pi]}\right) |\sin s|\,\frac{ds}{s} = B'_n + B''_n\,,$$

say. Clearly $B''_n \le k$ for all n. Thus we will be done once we show that $B'_n \sim (4/\pi^2)\ln n$ for large n. To see this rewrite

$$B'_n = \frac{2}{\pi}\sum_{k=0}^{n-1}\int_{[k\pi,(k+1)\pi]} \frac{|\sin s|}{s}\,ds$$
$$= \frac{2}{\pi}\sum_{k=0}^{n-1}\int_{[0,\pi]} \frac{|\sin(k\pi+t)|}{k\pi+t}\,dt$$
$$= \frac{2}{\pi}\int_{[0,\pi]} \sin t\left\{\sum_{k=0}^{n-1}\frac{1}{k\pi+t}\right\}dt\,.$$

The expression in $\{\cdot\}$ in the above integral can be estimated below and above, uniformly for $t \in (0,\pi]$, by

$$\frac{1}{\pi}\sum_{k=0}^{n-1}\frac{1}{k+1} = \frac{1}{\pi}\sum_{k=1}^{n}\frac{1}{k} - \frac{1}{\pi}\,, \quad \text{and} \quad \frac{1}{\pi}\sum_{k=1}^{n-1}\frac{1}{k}\,,$$

respectively. Since $\int_{[0,\pi]}\sin t\,dt = 2$, these estimates show that

$$B'_n \sim \frac{4}{\pi^2}\sum_{k=1}^{n}\frac{1}{k} \sim \frac{4}{\pi^2}\ln n\,, \quad \text{large } n\,. \quad \blacksquare$$

As a consequence of this result we now know that, on account of (1.13), for each (large) n there is a continuous function f, $|f(x)| \le 1$ for all x in T, such that $|s_n(f,0)| \sim (4/\pi^2)\ln n$. But, do there exist continuous functions whose Fourier series have large partial sums at 0? Assuming that no such functions exist we will reach a contradiction. Indeed, in the notation of Theorem 3.2 in Chapter XV, let $X = C(T)$, $Y = C$, and put

$$T_n f = s_n(f,0) = \frac{1}{\pi}\int_T f(x)D_n(x)\,dx\,, \quad n = 0,1,2,\ldots \tag{1.14}$$

It is clear that the T_n's form a sequence of bounded linear mappings from X into Y, and by Proposition 1.2 and (1.13) it follows that

$$\|T_n\| = \sup_{f\in C(T),\|f\|_\infty \le 1} |T_n f| \sim \frac{4}{\pi^2}\ln n\,, \quad \text{all } n\,. \tag{1.15}$$

Suppose now that the Fourier series of every continuous function converges at 0; in particular, the partial sums will be bounded there, i.e., $|T_n| = |s_n(f,0)| \le c_f < \infty$ for each $f \in C(T)$ and all n. Then by the Uniform Boundedness Principle there is a constant c such that

$$\|T_n\| \sim (4/\pi^2)\ln n \le c\,, \quad \text{all } n\,,$$

which is impossible. In fact, by (3.3) in Chapter XV, the set of those continuous functions f whose Fourier series converges at the origin is of first category in $C(T)$.

As for explicit examples, in 1876 du Bois-Reymond (1831–1889) constructed a continuous function with a nonconvergent Fourier series at a single point. This led to an example where there is no convergence at each point of a dense set of points in T; this set nevertheless is of Lebesgue measure 0. In 1926 Kolmogorov produced an example of an integrable function whose Fourier series diverges a.e. in T. Until 1966 it was not known whether or not there exists a continuous function with that property. In that year Carleson proved that "Lusin's conjecture" is true, to wit, the Fourier series of every $L^2(T)$ function converges a.e. Shortly after, Hunt extended Carleson's argument to prove that the Fourier series of $L^p(T)$ functions converge a.e. in T provided $p > 1$.

The Principle of Condensation of Singularities allows us to construct functions whose Fourier series do not converge at many points.

Example 1.3. There exists a real-valued continuous function $f(x)$ of period 2π such that the partial sums $s_n(f,x)$ of its Fourier series expansion satisfy the condition

$$\limsup_{n\to\infty} |s_n(f,x)| = \infty\,,$$

for those x's on a set $B \subset T$ which may be taken to contain any sequence $\{x_m\} \subset T$.

Indeed, from (1.7) is clear that for a given fixed $x \in T$, $s_n(f,x)$ is a bounded linear functional on $C(T)$ with norm $L_n \to \infty$ as $n \to \infty$. Therefore, if we take a sequence $\{x_m\} \subset T$, by Theorem 3.3 in Chapter XV, the set

$$\mathcal{B} = \{f \in C(T) \colon \limsup_{n\to\infty} |s_n(f,x)| = \infty \text{ for } x = x_1, x_2, \ldots\}$$

is of second category in $C(T)$. Hence, by the completeness of $C(T)$, the (bad) set $\mathcal{B}$ is nonempty.

In fact, there is more we can say about functions in $\mathcal{B}$: If we take $\{x_m\}$ to be a dense sequence of points in T, then for any $f \in \mathcal{B}$ the set

$$B = \{x \in T : \limsup_{n\to\infty} |s_n(f,x)| = \infty\}$$

is uncountable. To see this, given $f \in \mathcal{B}$, let

$$F_{m,n} = \{x \in T : |s_n(f,x)| \le m\}\,, \quad F_m = \bigcap_{n=1}^{\infty} F_{m,n}\,.$$

By the continuity of f it readily follows that the $F_{m,n}$'s and hence F_m are closed subsets of T. If we can show that $\bigcup_{m=1}^{\infty} F_m$ is of first category in T, then the set

$$B = T \setminus \textstyle\bigcup_{m=1}^{\infty} F_m \supseteq \{x_1, x_2, \ldots\}$$

would be of second category in T, and so uncountable. So, finally we prove that each F_m is of first category in T. Suppose some F_{m_0} is of second category in T. Then the closed set F_{m_0} must contain a closed subinterval $[a,b]$ of T. This implies that

$$\sup_{n\ge1} |s_n(f,x)| \le m_0\,, \quad \text{all } x \in [a,b]\,,$$

contradicting the fact that P contains a dense subset of T.

2. CESÀRO SUMMABILITY

In the previous section we saw that the notion of pointwise convergence is not the ideal one for dealing with Fourier series of continuous, or integrable, functions. Now we address the convergence question from the more general point of view of the arithmetic means. We begin by defining the notion of Cesàro $(C,1)$ summability.

Given a sequence of complex numbers $\{c_k\}$, we say that it converges to L in the Cesàro $(C,1)$ sense, and we write

$$\lim_{k\to\infty} c_k = L(C,1)$$

provided that $\lim_{k\to\infty} C_k = L$, where C_k is the average $(c_1 + \cdots + c_k)/k$. It would be reassuring to know that convergent sequences also converge $(C,1)$ to the same limit.

Proposition 2.1. Let $\lim_{k\to\infty} c_k = L$; then $\lim_{k\to\infty} c_k = L(C,1)$.

Proof. We may assume that $L = 0$ by replacing c_k by $c_k - L$ if necessary. Observe that the c_k's have the following properties:

(i) $|c_k| \le M$, all k.

(ii) Given $\varepsilon > 0$, there is a k_0 such that $|c_k| \le \varepsilon$ provided $k \ge k_0$.

It is now a simple matter to estimate the C_k's. Indeed,

$$|C_k| \le \frac{|c_1| + \cdots |c_{k_0-1}| + \cdots + |c_k|}{k}$$
$$\le \frac{k_0 M}{k} + \varepsilon\frac{(k-k_0)}{k} \le \frac{k_0 M}{k} + \varepsilon\,, \quad k \ge k_0\,.$$

Therefore, by first picking ε and thus fixing k_0, and then letting $k \to \infty$ we see that $|C_k|$ can be made arbitrarily small for k large. ■

Note that the oscillating sequence $c_k = (-1)^k$ has limit 0 in the $(C,1)$ sense.

In a similar vein we define the Cesàro summability of series. Given a sequence (c_k) of complex numbers, put

$$s_n = \sum_{k=1}^{n} c_k\,, \quad \text{and} \quad \sigma_n = \frac{1}{n}\sum_{k=1}^{n} s_k\,.$$

If $\lim_{n\to\infty} \sigma_n = s$, then we say that the series $\sum c_k$ is C_1-summable to s, and we write $\sum c_k = s(C,1)$.

By Proposition 2.1 it follows that if $\lim_{n\to\infty} s_n = s$, then $\sum c_k = s(C,1)$. On the other hand, if $c_k = z^k$, $z \ne 1$ a complex number with $|z| = 1$, then $\sum z^k$ does not converge, yet

$$\sum_{k=0}^{\infty} z^k = \frac{1}{1-z}(C,1)\,.$$

We pass now to explore how the notion of Cesàro summability applies to Fourier series. As before we begin by determining the integral, or convolution, representation of the Cesàro means $\sigma_n(f,x)$ corresponding to (the Fourier series of) f. More precisely, given $f \sim \sum c_k e^{ikx}$, let

$$\sigma_n(f,x) = \frac{s_0(f,x) + \cdots + s_n(f,x)}{n+1}\,, \quad n = 0, 1, \ldots \tag{2.1}$$

It is fairly easy to obtain an explicit expression of $\sigma_n(f,x)$. Indeed, note that the numerator of (2.1) is the sum of

$$\begin{array}{c} c_0 \\ c_{-1}e^{-ix} + c_0 + c_1 e^{ix} \\ \vdots \\ c_{-n}e^{-inx} + \cdots + c_0 + \cdots + c_n e^{inx} \end{array}$$

which equals $c_{-n}e^{-inx} + 2c_{-n+1}e^{-i(n-1)x} + \cdots + (n+1)c_0 + \cdots + c_n e^{inx}$. Thus, by dividing by $(n+1)$ we get that

$$\sigma_n(f,x) = \sum_{|k|\le n} \left(1 - \frac{|k|}{n+1}\right) c_k e^{ikx} . \tag{2.2}$$

As for the integral representation alluded to above, by (1.6) we readily see that

$$\begin{aligned} \sigma_n(f,x) &= \frac{f * 2D_0(x) + \cdots + f * 2D_n(x)}{n+1} \\ &= \frac{f * (2\sum_{k=1}^n D_k)(x)}{n+1} = f * 2K_n(x) , \end{aligned}$$

say, where

$$K_n(x) = \frac{\sum_{k=1}^n D_k(x)}{n+1} \tag{2.3}$$

is the Fejér kernel of order n. It is not hard to compute $K_n(x)$. In the first place, the numerator in (2.3) equals

$$\sum_{k=1}^n \frac{\sin((k+1/2)x)}{2\sin(x/2)} = \frac{1}{2\sin(x/2)} \Im \left(\sum_{k=1}^n e^{i(k+1/2)x}\right) .$$

By summing the geometric series we see that the imaginary part of the above sum equals

$$\begin{aligned} &\Im\left(e^{ix/2}\frac{(1-e^{i(n+1)x})}{(1-e^{ix})}\right) \\ &\quad = \Im\left(e^{i(n+1)x/2}\frac{(e^{-i(n+1)x/2} - e^{i(n+1)x/2})}{(e^{-ix/2} - e^{ix/2})}\right) \\ &\quad = \frac{\sin((n+1)x/2)}{\sin(x/2)} \Im(e^{i(n+1)x/2}) = \frac{\sin^2((n+1)x/2)}{\sin(x/2)} . \end{aligned}$$

Whence

$$K_n(x) = \frac{1}{2(n+1)} \left(\frac{\sin((n+1)x/2)}{\sin(x/2)} \right)^2 . \tag{2.4}$$

The following properties of $K_n(x)$ are readily verified: It is a positive even function and by (2.3) above

$$\frac{1}{\pi} \int_T K_n(x)\, dx = \frac{2}{\pi} \int_{[0,\pi]} K_n(x)\, dx = 1 . \tag{2.5}$$

Thus, in contrast to the Dirichlet kernels $D_n(x)$, the Fejér kernels have uniformly bounded L^1 norms.

It is also immediate to estimate $K_n(x)$. By (1.10),

$$\begin{aligned} K_n(x) &\le \frac{1}{n+1} \sum_{k=1}^{n} |D_k(x)| \le \frac{1}{n+1} \sum_{k=0}^{n} (k+1/2) \\ &= \frac{1}{n+1} \left(\frac{n(n+1)}{2} + \frac{n+1}{2} \right) = \frac{n+1}{2} . \end{aligned} \tag{2.6}$$

Similarly, by (1.11),

$$K_n(x) \le \frac{2\pi}{2(n+1)\, x^2} , \qquad 0 < |x| < \pi . \tag{2.7}$$

This is all we need to know about these kernels.

Concerning the representation of $\sigma_n(f,x)$, since $K_n(x)$ is even we have that $f * 2K_n(x)$ equals either

$$\frac{1}{\pi} \int_T \left(\frac{f(x+t) + f(x-t)}{2} \right) K_n(t)\, dt , \tag{2.8}$$

or,

$$\frac{1}{\pi} \int_{[0,\pi]} (f(x+t) + f(x-t))\, K_n(t)\, dt . \tag{2.9}$$

The first result we consider is how the convolution with the Fejér kernel behaves at the points of continuity of a function. In fact, a slightly more general result is

Theorem 2.2 (Fejér). suppose f is a periodic integrable function defined on T, $f \sim \sum c_k e^{ikx}$. If the limits $f(x \pm 0)$ exist, then

$$\sum c_k e^{ikx} = \left(\frac{f(x+0) + f(x-0)}{2} \right) (C,1) . \tag{2.10}$$

In particular, if f is continuous at every point of an interval $I \subseteq T$, then the convergence is uniform over I.

Proof. We may, and do, assume that $f(x) = (f(x+0)+f(x-0))/2$. On account of (2.9) and (2.5) we have

$$\sigma_n(f,x)-f(x) = \frac{2}{\pi}\int_{[0,\pi]}\left(\frac{f(x+t)+f(x-t)}{2} - f(x)\right) K_n(t)\,dt\,. \quad (2.11)$$

Thus, for $0 < \eta < \pi$, we see that

$$\begin{aligned} &|\sigma_n(f,x) - f(x)| \\ &\quad \le \frac{2}{\pi}\left(\int_{[0,\eta)} + \int_{[\eta,\pi]}\right)\left|\frac{f(x+t)+f(x-t)}{2} - f(x)\right| K_n(t)\,dt \\ &\quad = A + B\,, \end{aligned} \quad (2.12)$$

say. We estimate A first. By assumption, given $\varepsilon > 0$, there exists $\delta > 0$ such that

$$\left|\frac{f(x+t)+f(x-t)}{2} - f(x)\right| \le \varepsilon$$

provided $0 \le t < \delta$. We set $\eta = \delta$ in (2.12) and note that

$$A \le \varepsilon \frac{2}{\pi}\int_{[0,\delta]} K_n(t)\,dt \le \varepsilon\,.$$

To bound B we introduce the quantity $M_n(\delta) = \sup_{\delta\le t\le\pi} K_n(t)$. Then, by (2.7), $M_n(\delta) \le c/(N+1)\delta^2$. Thus $\lim_{n\to\infty} M_n(\delta) = 0$ for each $\delta > 0$, and consequently

$$\begin{aligned} B &\le M_n(\delta)\frac{2}{\pi}\int_{[0,\pi]}(|f(x+t)| + |f(x-t)| + 2|f(x)|)\,dt \\ &\le c_f M_n(\delta) \to 0\,, \quad \text{as } n \to \infty\,. \end{aligned} \quad (2.13)$$

This completes the proof of the first assertion. If now f is continuous on a closed interval $I \subseteq T$, then (2.13) holds uniformly over I, and the theorem is completely proved. ∎

An interesting application of Fejér's theorem is the familiar Weierstrass theorem that asserts that continuous functions are the uniform limit of polynomials.

Corollary 2.3 (Weierstrass). Let $f \in C(T)$. Then, given $\varepsilon > 0$, there is a trigonometric polynomial p such that $|f(x) - p(x)| \le \varepsilon$ for all $x \in T$.

Proof. The natural candidate for $p(x)$ is $\sigma_n(f,x)$, with n sufficiently large. By Fejér's theorem this choice works. ■

As for the convergence in norm, we have

Theorem 2.4. Let $1 \le p < \infty$. Then

$$\lim_{n\to\infty} \|\sigma_n(f) - f\|_p = 0\,, \quad \text{all } f \in L^p(T)\,. \tag{2.14}$$

Proof. Let

$$F(t) = \left(\frac{1}{2\pi}\int_T |f(x+t) - f(x)|^p dx\right)^{1/p};$$

by (v) in Section 1 of Chapter XII, $F(t)$ is continuous at $t = 0$. Thus Theorem 2.2 gives that $\lim_{n\to\infty} \sigma_n(F,0) = F(0) = 0$. Now, by (2.5) and (2.8), we have

$$\sigma_n(f,x) - f(x) = \frac{1}{\pi}\int_T (f(x+t) - f(x))K_n(t)\,dt\,,$$

and consequently, by Minkowski's integral inequality, 4.23 in Chapter XIII, we get

$$\begin{aligned}\|\sigma_n(f) - f\|_p &\le \frac{1}{\pi}\int_T \|f(\cdot + t) - f(\cdot)\|_p K_n(t)\,dt \\ &= \frac{1}{\pi}\int_T F(t)K_n(t)\,dt = \sigma_n(F,0)\,.\end{aligned}$$

Whence $\limsup \|\sigma_n(f) - f\|_p \le \limsup \sigma_n(F,0) = 0$. ■

In particular, referring to a question left open in Chapter XVI, the trigonometric polynomials are dense in $L^2(T)$, and consequently, the ONS $\left\{\frac{1}{\sqrt{\pi}}e^{ikx}\right\}$ is complete in $L^2(T)$.

The stronger version of (2.14) is also true, to wit,

$$\lim_{n\to\infty} \|s_n(f) - f\|_p = 0\,, \quad \text{all } f \in L^p(T)\,, 1 < p < \infty\,.$$

This result is known as the M.Riesz theorem and it is much harder to prove.

Another consequence of Theorem 2.2 is the uniqueness of the Fourier expansion.

Corollary 2.5. Let $f \in L(T)$, $f \sim \sum c_k e^{ikx}$. If $c_k = 0$ for all k, then $f = 0$ a.e.

Proof. Note that $\sigma_n(f,x) = 0$ for all n, and then apply Theorem 2.4 . ■

3. POINTWISE CONVERGENCE

We close our discussion on a positive note, namely, we show that the Cesàro means of the Fourier series of an integrable function converge to the function a.e. This result was proved by Lebesgue, and it represents one of the early successes of the new concept of Lebesgue integral.

Theorem 3.1. Suppose $f \in L(T)$. Then $\lim_{n\to\infty} \sigma_n(f,x) = f(x)$ at each point x in the Lebesgue set of f. In particular this statement is true a.e. in T.

Proof. We proceed as in the proof of Theorem 2.2, but in a more deliberate fashion. By (2.12) of that theorem

$$\begin{aligned}
&|\sigma_n(f,x) - f(x)| \\
&\quad \le \frac{2}{\pi}\left(\int_{[0,\eta)} + \int_{[\eta,\pi]}\right)\left|\frac{f(x+t)+f(x-t)}{2} - f(x)\right| K_n(t)\,dt \\
&\quad = A + B\,,
\end{aligned}$$

say. Further, by (2.7)

$$B \le \frac{c}{n+1}\int_{[\eta,\pi]}\left|\frac{f(x+t)+f(x-t)}{2} - f(x)\right|\frac{dt}{t^2} \le \frac{c}{(n+1)\,\eta^2}\,c_f\,,$$

where c_f depends, of course, on f. So this term can be made small, but there must be a balance between η and n. Choosing $\eta = 1/n^{1/4}$, for instance, we get at once that B can be made arbitrarily small for sufficiently large n.

To bound A is a more delicate pursuit. We split A into two integrals,

$$\frac{2}{\pi}\left(\int_{[0,1/n]} + \int_{[1/n,1/n^{1/4})}\right) = A_1 + A_2\,,$$

say. Now, by estimate (2.6) it readily follows that

$$A_1 \leq cn \int_{[0,1/n)} \left| \frac{f(x+t)+f(x-t)}{2} - f(x) \right| dt \,,$$

Further, by 3.24 in Chapter VIII, the right-hand side of the above estimate tends to 0 as $n \to \infty$ at each point x in the Lebesgue set of f. Thus, it only remains to deal with A_2. Again, by (2.7) it is dominated by

$$\frac{c}{n} \int_{[1/n,1/n^{1/4})} \left| \frac{f(x+t)+f(x-t)}{2} - f(x) \right| \frac{dt}{t^2} \,. \tag{3.1}$$

Let now

$$F_x(t) = F(t) = \int_{[0,t)} \left| \frac{f(x+s)+f(x-s)}{2} - f(x) \right| ds \,.$$

As F is absolutely continuous, we may rewrite (3.1)

$$\frac{c}{n} \int_{[1/n,1/n^{1/4})} F'(t) \frac{dt}{t^2} \,.$$

Whence integrating by parts we see that

$$A_2 \leq \frac{c}{n} \frac{F(t)}{t^2} \Bigg]_{1/n}^{1/n^{1/4}} + \frac{2c}{n} \int_{[1/n,1/n^{1/4}]} F(t) \frac{dt}{t^3} \,. \tag{3.2}$$

Since by 3.24 in Chapter VIII $\lim_{t\to 0} F(t)/t = 0$ at each Lebesgue point x of f, the integrated term in (3.2) above tends to 0 as $n \to \infty$. The same is true for the integral in (3.2) since, given $\varepsilon > 0$, we may first choose n large enough so that $F(t)/t \leq \varepsilon$ in $[1/n, 1/n^{1/4}]$, and observe that the integral there does not exceed

$$\frac{c\varepsilon}{n} \int_{[1/n,\infty)} \frac{dt}{t^2} = c\varepsilon \,.$$

The combination of all these estimates leads to the desired conclusion and we have finished. ■

CHAPTER XVIII

Remarks on Problems and Questions

'DON'T PANIC!'

These remarks consist of hints and comments to some of the problems and questions included in the text.

CHAPTER I

5.17 A contains a subset A_n of n elements for each n; consider the subset $B = \bigcup_n A_n$.

5.25 A continuous real-valued function on [0,1] is determined by the values it takes on $Q \cap [0,1]$.

5.29 A subset of R^2 consisting of a disk and any part of its boundary is convex.

CHAPTER II

5.10 Suppose not and consider the first element of the nonempty set $\{m \in M : P(m) \text{ is not true}\}$.

5.14 Given a set A, let f be a selection function for $\mathcal{P}(A) \setminus \{\emptyset\}$. We may now construct a well-ordering on A as follows: Put $a_1 = f(A)$, $a_2 = f(A \setminus \{a_1\})$, $a_3 = f(A \setminus \{a_1, a_2\})$, and in general, define by transfinite induction $a_b = f(A \setminus \{\bigcup_{d<b} a_d\})$. The construction stops as soon as we exhaust all the elements of A.

5.17 Suppose A is a set with card $A = a$ and let $\mathcal{F}$ denote the collection of functions $f: X \times X \to X$, $X \subseteq A$, such that f sets up a one-to-one correspondence from $X \times X$ onto X; by 5.18 in Chapter I, $\mathcal{F} \neq \emptyset$. It is not hard to see that $\mathcal{F}$ is partially ordered by restriction, and that all the conditions of Zorn's Lemma are satisfied. Consider now a maximal element.

5.18 Let B_1, B_2 be Hamel bases for R, and given $x \in B_1$, let $B_2(x) = \{\text{finite collection of } y_i\text{'s in } B_2 : x = r_1y_1 + \cdots + r_ny_n,\ r_1 \ldots, r_n \text{ rational}\}$; similarly for $y \in B_2$ introduce the sets $B_1(y) \subset B_1$. Consider now the following assertion: $x \in B_2(y)$ iff $y \in B_1(x)$.

5.19 Let $(M, \prec)$ be a nonempty partially ordered set and assume that every chain in M has an upper bound; we must find a maximal element of M. By assumption there is a listing $m_0, m_1, \ldots$ of the elements of M. Now, by transfinite induction, let $m'_0 = m_0$, and for an ordinal $a > 0$, let $m'_a = m_b$, where b is the least ordinal such that m_b is an upper bound of the chain $C = \{m'_d : d < a\}$ and $m_b \notin C$. Note that $\{m'_d : d < a\}$ is always a chain, and that m_b exists unless m'_{a-1} is a maximal element of M. This construction eventually comes to an end, and we obtain a maximal element of M.

5.28 Pick $y \in \overline{f^{-1}(\{0\})} \setminus f^{-1}(\{0\})$ and a sequence $\{y_n\} \subseteq f^{-1}(\{0\})$ such that $\lim_{n\to\infty} y_n = y$. Now, given $x \in R$, write it as

$$x = (x - (f(x)/f(y))(y - y_n)) + (f(x)/f(y))(y - y_n) = x_1 + x_2\,,$$

say, and observe that $f(x_1) = 0$ and $x_2 \to 0$.

5.33 Suppose B is an at most countable subset of A. Then the set $B^* = \bigcup_{b\in B} I_b$ is also an at most countable, and consequently, proper subset of A. In fact, B^* is an initial segment I_a of A, say, and a is an upper bound of B.

CHAPTER III

4.2 Imagine a rising sun on the x-axis at ∞. Then $\{(x,y) \in R^2 : y \geq f^*(x)\}$ is illuminated by the sun whereas $\{(x,y) \in R^2 : y < f^*(x)\}$ lies in the shadow.

4.4 If $D = \{r_1, r_2, \ldots\}$ is a countable subset of $[a,b]$, define functions $f_1, \ldots, f_n, \ldots$ by

$$f_n(x) = \begin{cases} -1/n^2 & \text{if } x < r_n \\ 0 & \text{if } x = r_n \\ 1/n^2 & \text{if } x > r_n\,, \end{cases}$$

and let $f(x) = \sum_{n=1}^{\infty} f_n(x)$. Note that since $|f_n(x)| \leq 1/n^2$ for all x, the series converges uniformly. Further, since each f_n is increasing, so is f. Let now $x \notin D$. Then each function $f_1 + \cdots + f_n$ is continuous at x and so is the uniform limit, f. If, on the other hand, $x \in D$, then $x = x_N$ for some N and $f = f_N + \sum_{n \neq N} f_n$. By the above argument the sum is continuous at x, but f_N is not.

4.14 Choose a partition $\mathcal{P}_1$ consisting of points $a = x_0 < \ldots < x_n = x$ so that $\sum_{\text{over } \mathcal{P}_1} |\Delta_k f| > V(x) - \varepsilon$ and using the uniform continuity of f on $[a,x]$ find $\delta > 0$ with the property that $|f(x) - f(x')| < \varepsilon/4(n+1)$ if $|x - x'| < \delta$. Show now that any partition $\mathcal{P}$ with norm less than or equal to $\eta = \wedge(\delta, \min \Delta_k x)$ will do.

4.16 By the mean value theorem we get $\sum_{j=1}^{n} |f(x_j) - f(x_{j-1})| = \sum_{j=1}^{n} |f'(t_j)| \Delta_j x$ for appropriate $t_j \in (x_{j-1}, x_j)$. Hence by 4.15 and Theorem 2.6 we have

$$V(x) = \lim_{\text{norm}(\mathcal{P}) \to 0} \sum_{j=1}^{n} |f'(t_j)| \Delta_j x = \int_a^x |f'(t)|\, dt\,.$$

Moreover, by 4.15,

$$2P(x) = V(x) + (f(x) - f(a)) = \int_a^x |f'(t)|\, dt + \int_a^x f'(t)\, dt\,.$$

4.23 If $f(b) = f(a)$ the identity clearly holds. Otherwise, if $f(b) \neq f(a)$, since all upper and lower sums lie between $m(f(b) - f(a))$ and $M(f(b) - f(a))$, so does the integral. Thus, the quotient $c = \int_a^b g\, df / \int_a^b df$ lies between m and M.

4.25 Note that $f_1(x) - f_1(y) = f_2(x) - f_2(y)$ whenever $x, y \in D$. Let $\varepsilon > 0$ be given. Since g is uniformly continuous on I, there is a $\delta > 0$ such that $|g(x) - g(y)| \leq \varepsilon/(f_1(b) - f_1(a) + 1)$ whenever $|x - y| < \delta$. Because D is dense in R there is a partition $\mathcal{P}$ of I consisting of points in D so that the distance between no two succesive points exceeds δ. The stage is now set to show that $| \int_a^b g\, df_1 - \int_a^b g\, df_2| \leq \varepsilon$.

4.27 There are two ways to go about this problem: Either integrate by parts and invoke 4.23 or note that by 4.22 the function $\int_a^b g\, df - g(a) \int_a^x df - g(b) \int_x^b df$ is continuous and invoke the mean value theorem.

CHAPTER IV

4.14 Let $X = \{a_1, \ldots, a_6\}$, $A_1 = \{a_1, a_2, a_3\}$ and $A_2 = \{a_3, a_4, a_5\}$; $\mathcal{S}(\{A_1, A_2\})$ consists of exactly 16 sets.

4.16 Partially order $\mathcal{A}$ by set inclusion; by Zorn's Lemma there exists a maximal chain $\mathcal{C}$, say, in $\mathcal{A}$. Furthermore, since $\mathcal{A}$ is infinite, so is $\mathcal{C}$. Indeed, suppose there are only finitely many elements $C_0 = \emptyset \subset C_1 \ldots \subset C_n = X$, say, in $\mathcal{C}$. Then, let $A \in \mathcal{A} \setminus \mathcal{C}$, and let n_0 be the largest integer less than n such that $A \setminus C_{n_0} \neq \emptyset$. Since $A \neq C_{n_0+1}$ it follows that $\emptyset \subset \ldots \subset C_{n_0} \subset (C_{n_0} \cup ((A \setminus C_{n_0}) \cap C_{n_0+1})) \subset C_{n_0+1} \subset \ldots \subset C_n$, is a chain of elements of $\mathcal{A}$ that properly contains $\mathcal{C}$, contrary to its assumed maximality. So, $\mathcal{C}$ is an infinite maximal chain. If $\mathcal{C}$ has no "first" nonempty element, then there exists an infinite sequence $\{A_k\} \subseteq \mathcal{C}$ such that $A_1 \supset \ldots \supset A_n \supset \ldots \supset \emptyset$. Otherwise designate the first nonempty element of $\mathcal{C}$ by B_1 and take a look at $\mathcal{C} \setminus \{B_1\}$. If this family has no first nonempty element, pick as before an infinite decreasing sequence, and if it has a first element B_2, say, note that $B_1 \subset B_2$, and consider $\mathcal{C} \setminus \{B_1, B_2\}$. Proceeding in this fashion we either obtain an infinite decreasing sequence of nonempty elements of $\mathcal{C}$, or else a sequence $B_1 \subset \ldots \subset B_n \subset \ldots$ In the latter case, by looking instead at $X \setminus B_1 \supset \ldots \supset X \setminus B_n \supset \ldots$ we also obtain an infinite decreasing sequence of elements of $\mathcal{C}$. Thus, in any case, there exists an infinite decreasing sequence $\{A_k\} \subseteq \mathcal{C}$, say, and the sequence $A'_k = A_k \setminus A_{k+1}$, $k = 1, \ldots,$ is composed of infinitely many pairwise disjoint nonempty subsets of $\mathcal{A}$.

4.17 Divide R^n into a mesh of congruent nonoverlapping closed intervals each of measure 1. Next discard any interval in the mesh which does not intersect the open set $\mathcal{O}$ and separate those intervals that are totally contained in $\mathcal{O}$. Next subdivide each of the remaining intervals into 2^n congruent nonoverlapping closed intervals by bisecting the sides. Once again discard any interval in this new mesh which does not intersect $\mathcal{O}$ and separate those intervals which are totally contained in $\mathcal{O}$. As for those intervals that are left, they intersect both $\mathcal{O}$ and $R^n \setminus \mathcal{O}$. Keep going.

4.29 The restriction that μ_2 is a finite measure is not necessary. Indeed, when $\mu_2(X) = \infty$ we may define

$$\mu_3(E) = \sup\{\mu_1(B) - \mu_2(B) : B \subset E, \mu_2(B) < \infty\}, \quad E \in \mathcal{M}.$$

By the way, μ_3 is uniquely determined if μ_2 is σ-finite.

4.31 If X is uncountable and $\mu(E) = 0$ if $E \subset X$ is at most countable and $\mu(E) = \infty$ if $E \subseteq X$ is uncountable, then μ is not semifinite.

4.32 Let $A \in \mathcal{M}$, with $\mu(A) = \infty$. If the assertion is false, then

$$0 < \sup\{\mu(B) : B \subset A, B \in \mathcal{M}, \mu(B) < \infty\} = \eta < \infty .$$

Since η is finite we can find a nondecreasing sequence $\{B_n\} \subseteq \mathcal{M}$ such that $\mu(B_{n+1}) \geq \mu(B_n) > \eta - 1/n$. Let $B = \bigcup_{n=1}^{\infty} B_n$; $B \subset A$, $B \in \mathcal{M}$ and $\mu(B) = \eta$. Since $\mu(A) = \infty$, also $\mu(A \setminus B) = \infty$. The fact that μ is semifinite now leads to a contradiction.

4.33 Given $A \in \mathcal{M}$, let

$$\mu_1(A) = \sup\{\mu(B) : B \subset A, B \in \mathcal{M}, \mu(B) < \infty\} ,$$

and

$$\mu_2(A) = \sup\{\mu(B) - \mu_1(B) : B \subset A, B \in \mathcal{M}, \mu_1(B) < \infty\} .$$

CHAPTER V

3.15 Consider the function $f(x) = 10^{-N(x)}$, where $N(x)$ is the number of zeros and nines after the decimal point in the expansion of x; the convention $10^{-\infty} = 0$ is in effect.

3.24 Suppose that $E \subseteq [0,1]$ and that, to the contrary, $E \cap (E + r) = \emptyset$ for every $r \in [0,1]$. Then note that the sets $E + r_n$, $r_n \in [0,1] \cap Q$, are measurable, pairwise disjoint and contained in [0,2].

3.25 3.24 is useful here.

3.26 Let B_1, B_2 be the sets of real numbers

$$B_1 = \{r \in R : r = c_0 + c_2/2^2 + c_4/2^4 + \cdots\}$$

and

$$B_2 = \{r \in R : r = c_1/2 + c_3/2^3 + \cdots\} ,$$

where c_0 is an arbitrary integer and $c_k = 0$ or 1 for all k. Show now that $|B_1| = |B_2| = 0$. Furthermore, $B_1 \cup B_2$ is also null, and it contains a Hamel basis for R.

3.28 Let $D = C \times C \subset R^2$ and observe that for any number $-1 \leq r \leq 1$, the line $y = x + r$ meets D in at least one point. Indeed, since C is obtained by a sequence of removals of "middle thirds", D can be thought of as the intersection of a countable family $D_1, D_2, \ldots$ of closed subsets of $[0,1] \times [0,1]$ obtained as follows: The set D_1 consists of four 1/3 by 1/3 closed squares located in the corners of $[0,1] \times [0,1]$, the set D_2 consists of of sixteen 1/9 by 1/9 squares located by fours in the corners of the squares of D_1, and so on. For any given $r \in [-1,1]$, the line $y = x + r$ meets at least one of the four squares in D_1, at least one of the four squares of D_2 that lies in D_1 and so on. Now note that there is a point (x,y), say, that belongs to every square of this sequence. This point therefore belongs to D, and consequently, we have $y = x + r$, with $x, y \in C$.

3.30 Let $r_1, r_2, \ldots$ denote the sequence of all the rational numbers in $(0,1)$. Start out as in the construction of the Cantor set but extend the open interval to be removed in the first step so that its endpoints are irrational and so that r_1 belongs to this interval. At the second step remove from each of the two remaining closed intervals an open "middle third" so that the endpoints are irrational and r_2 is removed. If this process is carried out a Cantor-like set consisting entirely of irrational numbers remains.

3.34 Let A_1 be a Cantor subset of $[0,1]$ of measure 1/2. In each interval contiguous with A_1, i.e., in every connected component of $[0,1] \setminus A_1$ in $[0,1]$, construct a Cantor set the measure of which is half that of this interval, and denote by A_2 the union of these compact sets; clearly $|A_2| = 1/4$. Once $A_1, \ldots, A_n$ have been defined, A_{n+1} is constructed as follows: Choose in every interval contiguous to the compact set $K_n = A_1 \cup \ldots \cup A_n$ a Cantor set of measure half of that interval, and let A_{n+1} denote the union of those compact sets. Note that $|A_{n+1}| = 1/2^{n+1}$. Finally put $E = \bigcup_{n=0}^{\infty} A_{2n+1}$, and show that for every nondegenerate interval $I \subseteq [0,1]$ we have $0 < |E \cap I| < |I|$.

CHAPTER VI

4.30 By an argument analogous to Proposition 3.3, there exists a subsequence $\{f_{n_k}\}$ which is Cauchy μ-a.e. This subsequence determines a measurable function f such that $\lim_{k\to\infty} f_{n_k} = f$ μ-a.e., and consequently also $\lim_{k\to\infty} f_{n_k} = f$ in probability. To complete the argument observe that given $\varepsilon > 0$, it follows that $\{|f_n - f| > \varepsilon\} \subseteq \{|f_n - f_{n_k}| > \varepsilon/2\} \cup \{|f_{n_k} - f| > \varepsilon/2\}$. The first set on the right-hand side above may be handled because the sequence is Cauchy in measure and the second because the subsequence converges μ-a.e.

4.35 Choose b_n such that $\mu(X \setminus A_n) \le 2^{-n}$, where $A_n = \{|f_n| < b_n\}$, put $a_n = 1/nb_n$, and consider the set $\liminf A_n$.

4.36 A proof by pictures works. By Lusin's theorem there is a closed subset K of [0,1] such that $|B| = |[0,1] \setminus K| < \varepsilon$ and $f|K$ is continuous on K in the relative topology of K. Now, B is open and it can be written as an at most countable union of disjoint open intervals $B = \bigcup_n (a_n, b_n)$, say. Since $a_n, b_n \in K$ for all n, it is now clear how to extend f linearly to a continuous function F defined on [0,1] with the desired properties.

4.38 Let $A_1, A_2, \ldots$ be the sequence of nonmeasurable sets constructed in Chapter V, put $B_n = \bigcup_{i=n}^{\infty} A_i$, and let $\{f_n\}$ be the sequence of bounded functions on [0,1] given by $f_n = \chi_{B_n}$, $n = 1, 2, \ldots$ The conclusion of Egorov's theorem fails for this sequence.

CHAPTER VII

4.5 It is false. However, show that for each $\eta > 0$ it is true that $|\{|x| > \lambda, |f| > \eta\}| \to 0$ as $\lambda \to \infty$.

4.10 Feel free to use Proposition 2.4 in Chapter XIII.

4.22 Suppose not. Then, passing to a subsequence if necessary, we may assume there exists $\varepsilon > 0$ such that $\mu(E_n) \geq \varepsilon$ for all n. Let $G_k = \{0 < f < 1/k\}$, $k = 1, 2, \ldots$ By assumption $\lim G_k = \emptyset$, and since $\mu(G_1) < \infty$, it readily follows that there exists an index k_0 such that $\mu(G_{k_0}) \leq \varepsilon/2$, and consequently, $\mu(E_n \setminus G_{k_0}) \geq \varepsilon/2$ for all n. Thus, $\int_{E_n} f\,d\mu \geq \int_{E_n \setminus G_{k_0}} f\,d\mu \geq \mu(E_n \setminus G_{k_0})/k_0 \geq \varepsilon/2k_0$ for all n, contradicting the assumption that $\lim_{n\to\infty} \int_{E_n} f\,d\mu = 0$.

4.24 Since $|f_n - f| \leq 2|f_n| + 2|f|$, we have $2|f| + 2|f_n| - |f_n - f| \geq 0$. Thus by Fatou's Lemma we get

$$\int_X (2|f|+2|f|)\,d\mu \leq 2\int_X |f|\,d\mu + 2\liminf \int_X |f_n|\,d\mu - \limsup \int_X |f_n - f|\,d\mu\,.$$

Since $f \in L(\mu)$, any expression involving $\int_X |f|\,d\mu$ may be cancelled, and the conclusion obtains. Note that the weaker assumption $\liminf \int_X |f_n|\,d\mu \leq \int_X |f|\,d\mu$ is all that is needed to complete the argument. The idea of this proof is due to Novinger, *Proc. Amer. Math. Soc.* **34** (1972), 627-628. Also note that in $(R, \mathcal{L}, |\cdot|)$, the sequence $f_n = \chi_{[-n,n]}$, $n = 1, 2, \ldots$, tends to 1 everywhere, $\lim \int_R f_n = \int_R 1 = \infty$ but $\int_R |f_n - 1| \not\to 0$.

4.30 and **4.32** You may want to use 4.46 to do these problems.

4.36 No, but try with the additional assumption that $\lim_{n\to\infty} g_n = g$ μ-a.e.

4.37 The inequality $|a-b|/(1+|a-b|) \le |a-c|/(1+|a-c|) + |b-c|/(1+|b-c|)$ is useful in this context.

4.38 Let N be the set of points $x \in X$ for which either $|f(x)| > g(x)$ or $|f_n(x)| > g(x)$ for some n; by assumption $\mu(N) = 0$. Now, by LDCT, we get $\lim_{n\to\infty} \int_{X\setminus N} |f_n - f|\, d\mu = 0$, and consequently by Chebychev's inequality, $\lim_{n\to\infty} \mu((X \setminus N) \cap \{|f_n - f| > \lambda\}) = 0$ for all $\lambda > 0$.

4.40 Given $\varepsilon > 0$, we must find a measurable set $B \subset X$ such that $\mu(B) < \varepsilon$ and so that the sequence $\{f_n\}$ converges to 0 uniformly on $X \setminus B$. Let $E_k = \{g \ge 1/k\}$, $k = 1, 2, \ldots$ Since $g \in L(X)$, $\mu(E_k) < \infty$, and also $E_k \subseteq E_{k+1}$ for all k. By Egorov's theorem there exist measurable sets $B_k \subset E_k$ such that $\mu(B_k) < \varepsilon/2^k$ and the f_n's converge uniformly to 0 on $E_k \setminus B_k$. It only remains to check that if $B = \bigcup_{k=1}^\infty B_k$, $\mu(B) < \varepsilon$, then the f_n's converge uniformly to 0 on $X \setminus B$.

4.41 Suppose $|f|, |f_n| \le \phi \in L(\mu)$, $n = 1, 2, \ldots$ Given $\varepsilon > 0$, note that

$$\begin{aligned}\int_X |f_n - f|\, d\mu &= \left(\int_{\{|f_n-f|>\varepsilon\phi\}} + \int_{\{|f_n-f|\le\varepsilon\phi\}}\right) |f_n - f|\, d\mu \\ &\le 2\int_{\{|f_n-f|>\varepsilon\phi\}} \phi\, d\mu + \varepsilon \int_{\{|f_n-f|\le\varepsilon\phi\}} \phi\, d\mu \\ &\le 2\int_{\{\phi\le\eta\}} \phi\, d\mu + 2\int_{\{|f_n-f|>\varepsilon\eta\}} \phi\, d\mu + \varepsilon\|\phi\|_1 ,\end{aligned}$$

where η is an arbitrary positive real number, independent of n. Thus, since the f_n's converge to f in measure, we get that $\limsup \int_X |f_n - f|\, d\mu \le 2\int_{\{\phi\le\eta\}} \phi\, d\mu + \varepsilon\|\phi\|_1 = I + J$, say. The desired conclusion now follows since, by LDCT, $\lim_{\eta\to 0} I = 0$, and ε is arbitrary. In fact, note that we have proved a stronger statement, to wit, $\lim_{n\to\infty} \int_X |f - f_n|\, d\mu = 0$.

4.43 The integral of $(f - f^2)^2$ over X vanishes. By the way, the reader may be interested in proving that if f is nonnegative, a similar statement is true for any consecutive three integers. How about a similar statement for any three integers?

4.46 For simplicity assume that $I = [-1,1] \times \cdots \times [-1,1]$, and note that with the notation $rI = [-r,r] \times \cdots \times [-r,r]$, we can write I as the pairwise disjoin union $I = \bigcup_{k=0}^\infty (2^{-k}I \setminus 2^{-(k+1)}I) = \bigcup_{k=1}^\infty J_k$, say. Now, on each "ring" J_k we have $|x|^{-\eta} \sim 2^{k\eta}$, where the constant for $\sim$ is a dimensional constant independent of k. Also $|J_k| \sim 2^{-kn}$, with a dimensional constant independent of k. Thus,

$$\int_I |x|^{-\eta}\,dx = \sum_{k=0}^{\infty} \int_{J_k} |x|^{-\eta}\,dx \sim c \sum_{k=0}^{\infty} 2^{k(\eta-n)}\,,$$

which converges iff $\eta < n$.

CHAPTER VIII

3.1 Given $a, b > 0$ and ε, it is apparent that

$$ab = \frac{1}{2} 2(\varepsilon a)\left(\frac{b}{\varepsilon}\right) \le \frac{1}{2}\left(\varepsilon^2 a^2 + \frac{b^2}{\varepsilon^2}\right).$$

Therefore,

$$\int_X |fg|\,d\mu \le \frac{1}{2}\left(\varepsilon^2 \int_X f^2\,d\mu + \frac{1}{\varepsilon^2}\int_X g^2\,d\mu\right).$$

Now, unless $\int_X f^2 d\mu = 0$, and in this case $f = 0$ μ-a.e. and (3.1) holds trivially, we may choose $\varepsilon = \left(\int_X g^2\,d\mu / \int_X f^2\,d\mu\right)^{1/4}$. This choice of ε gives (3.1) at once.

3.4 Let $r_1, r_2, \ldots$ be an enumeration of the rational numbers in $I =$ [0,1] and put

$$f(x) = \sum_{n=1}^{\infty} \frac{1}{2^n} \frac{\chi_{[-1+r_n, 1+r_n]}(x)}{\sqrt{|x - r_n|}}\,, \quad x \in I\,.$$

By Theorem 1.6, and 4.6 and 4.46 in Chapter VII, it follows that $f \in L(I)$ and consequently, f is finite a.e. This function f works for the interval I, to obtain a function that gives the example in R, just put together functions that look like f. By the way, not only is f discontinuous at every point of I and unbounded on I, but it remains so after modifying its values on any Lebesgue null subset of I. Also, f^2 is finite a.e. on I, but it is not integrable there.

3.7 Let $\{f_n\}$ be the sequence of simple integrable functions constructed in Theorem 1.12 in Chapter VI. Then for each $E \in \mathcal{M}$ we have $\int_E |f|\,d\mu \le \int_E |f - f_n|\,d\mu + \int_E |f_n|\,d\mu = I + J$, say. By LDCT, $\lim_{n\to\infty} I =$ 0, uniformly for $E \in \mathcal{M}$. So, fix n large enough so that $I < \varepsilon/2$, and since

f_n is bounded by c, say, observe that $\int_E |f_n|\,d\mu \le c\mu(E) < \varepsilon/2$ provided that $\mu(E) < \delta = \varepsilon/2c$.

3.8 If $F(n) = \int_{\{|f|>n\}} |f|\,d\mu$ for $n = 1, 2, \dots$, it follows that $\int_X |f|\,d\mu \ge \sum_{n=1}^{\infty} n(F(n) - F(n+1))$. What we need now is to "sum by parts," and this is achieved through the following relation, known as Abel's transformation: $\sum_{n=1}^{k} u_n v_n = \sum_{n=1}^{k-1} U_n(v_n - v_{n+1}) + U_k v_k$, where $U_n = u_1 + \cdots + u_n$ for $n = 1, 2, \dots, k$.

3.10 Since as is readily seen $\int_X |f_n|\,d\mu \le c$ for all n, by Fatou's Lemma, f is integrable. Put $h_n = |f - f_n| \ge 0$. Then the sequence $\{h_n\}$ is also uniformly integrable and $\lim_{n\to\infty} h_n = 0$ μ-a.e. Further, if $h_{n,r} = h_n \chi_{\{h_n \le r\}}$ denotes the sequence of cut-offs of the h_n's at level r, by LDCT we have $\lim_{n\to\infty} \int_X h_{n,r}\,d\mu = 0$. Now, $\int_X h_n\,d\mu = \int_{\{h_n > r\}} h_n\,d\mu + \int_X h_{n,r}\,d\mu = I + J$, say. Given $\varepsilon > 0$, choose first r so that $I \le \varepsilon/2$ for all n, and then pick N so that $J \le \varepsilon/2$ for all $n \ge N$.

3.15 Assume first that f vanishes off a bounded interval I of R^n. In this case $f(y) + f(y+h)$ equals either $f(y)$, $f(y+h)$ or zero provided that $|h|$ is sufficiently large, the exact value depends on I. For these values of h, and by 4.6 in Chapter VII, we have $\int_{R^n} |f(y+h) + f(y)|\,dy = \int_{R^n} |f(y+h)|\,dy + \int_{R^n} |f(y)|\,dy = 2\int_{R^n} |f(y)|\,dy$, and the limit in question equals this last quantity. For arbitrary integrable functions consider a limiting argument working with truncates.

3.16 By passing to a subset if necessary, we may assume that $0 < |A| < \infty$. Let now $\phi(x) = \int_{R^n} \chi_A(y)\chi_A(x+y)\,dy$, $x \in R^n$. By the continuity of the translates of integrable functions it follows that ϕ is continuous, and since $\phi(0) = |A| > 0$, there exists $\delta > 0$ such that $\phi(x) > 0$ provided that $|x| < \delta$. This in particular means that the integrand of the integral defining $\phi(x)$ does not vanish identically, and consequently, there is $y \in A$ such that $x + y \in A$, or $x \in A - A$.

3.17 We show first that f is bounded in a neighbourhood of the origin. Indeed, let k be so large that $|A| = |\{x \in R : |f(x)| \le k\}| > 0$. Then, by 3.16, $A - A$ contains a neighbourhood of the origin, and this gives our assertion. Next note that if f is bounded on $(-\varepsilon, \varepsilon)$, say, we have $\int_{[0,\varepsilon]} f(x+y)\,dy = \varepsilon f(x) + \int_{[0,\varepsilon]} f(y)\,dy$. But this implies that $\varepsilon f(x) = \int_{[x,x+\varepsilon]} f(y)\,dy - \int_{[0,\varepsilon]} f(y)\,dy = \int_{[0,x+\varepsilon]} f(y)\,dy - \int_{[0,x]} f(y)\,dy - \int_{[0,\varepsilon]} f(y)\,dy = \int_{[\varepsilon,x+\varepsilon]} f(y)\,dy - \int_{[0,x]} f(y)\,dy = \int_{[0,x]} (f(y+\varepsilon) - f(y))\,dy = f(\varepsilon)x$. It is also possible to avoid the first part of the argument by considering $\cos(f(x))$ instead; in this case the integral relation is a bit more complicated.

3.22 The Wiener covering argument works with balls in place of intervals.

3.23 We may assume that $f \in L(R^n)$ is nonnegative since replacing f by $|f|$ does not change Mf. Consider next a sequence $\{f_k\}$ of nonnegative integrable functions that vanish off a compact set of R^n with the following two properties: (i) $f_k \le f$ a.e., and, (ii) The f_k's increase to f a.e. By Theorem 2.1 there is a constant c independent of $\lambda > 0$ and k such that $\lambda|\{Mf_k > \lambda\}| \le c\int_{R^n} f_k\,dy \le c\int_{R^n} f\,dy$. Now, since the f_k's increase to f it follows that Mf_k increases to Mf everywhere and consequently, $\lim_{k\to\infty} |\{Mf_k > \lambda\}| = |\{Mf > \lambda\}|$.

3.24 Let $r_1, r_2, \ldots$ be an enumeration of the rational numbers, and for each r_k let N_k be the subset of those points x in R^n where the relation

$$\lim_{r\to 0} \frac{1}{|I(x,r)|}\int_{I(x,r)} |f(y) - r_k|\,dy = |f(x) - r_k|$$

fails to hold. Since $|f(y)-r_k|$ is integrable on bounded sets, by (a straightforward variant of) Theorem 2.2, $|N_k| = 0$ for all k. Let $N = \bigcup_k N_k$; N is also a null subset of R^n. Now, for any $x, r,$ and r_k we have

$$\begin{aligned}
&\frac{1}{|I(x,r)|}\int_{I(x,r)} |f(y) - f(x)|\,dy \\
&\qquad\le \frac{1}{|I(x,r)|}\int_{I(x,r)} |f(y) - r_k|\,dy + \frac{1}{|I(x,r)|}\int_{I(x,r)} |f(x) - r_k|\,dy \\
&\qquad= \frac{1}{|I(x,r)|}\int_{I(x,r)} |f(y) - r_k|\,dy + |f(x) - r_k|\,.
\end{aligned}$$

Therefore, if $x \notin N$ and if $f(x)$ is finite,

$$\limsup_{r\to 0} \frac{1}{|I(x,r)|}\int_{I(x,r)} |f(y) - f(x)|\,dy \le 2|f(x) - r_k|\,,$$

where the right-hand side of the above inequality can be made arbitrarily small since r_k is any rational number.

3.29 Divide I into two nonoverlapping closed intervals, each of equal length. If J denotes one of these intervals, then either $\frac{1}{|J|}\int_J f\,dy \le \lambda$, or else $\frac{1}{|J|}\int_J f\,dy > \lambda$. In the latter case we separate J and it becomes one of the I_k's in the conclusion. Then (i) clearly holds since $\frac{1}{|J|}\int_J f\,dy \le \frac{2}{2|J|}\int_I f\,dy \le 2\lambda$. On the other hand, in the former case we proceed with subdividing J, and repeat this process until we are in the second case; if we are never forced into the second case we just keep subdividing. We denote by $\bigcup_k I_k$ the union of the pairwise possibly nonoverlapping closed intervals obtained from the second case, and we claim that (ii) holds. Indeed, by the Lebesgue differentiation theorem, for almost every point x of $I \setminus \bigcup_k I_k$ we have $f(x) = \lim_{|J|\to 0} \frac{1}{|J|}\int_J f\,dy$, where the limit is taken over those J's which satisfy the second case and which contain x; thus (ii) also holds. Finally, $|\bigcup_k I_k| = \sum_k |I_k| \le \frac{1}{\lambda}\sum_k \int_{I_k} f\,dy = \frac{1}{\lambda}\int_{\bigcup_k I_k} f\,dy$.

3.31 If f is integrable and $\lambda > 0$ is given, decompose R into a sequence of nonoverlapping closed intervals I each of equal length such that $\frac{1}{|I|}\int_I f\,dy \le \lambda$ for each I. Now apply 3.29 to each I separately.

3.32 We begin by decomposing R^n into a mesh of nonoverlapping closed intervals I, each of equal size, and whose diameter is so large that $\frac{1}{|I|}\int_I f\,dy \le \lambda$ for every I in this mesh. We consider now one interval at a time, as follows: If I is in the mesh we divide it into 2^n congruent nonoverlapping closed intervals by bisecting each of the sides of I. Let J be one of these new intervals and note that either $\frac{1}{|J|}\int_J f\,dy \le \lambda$, or else $\frac{1}{|J|}\int_J f\,dy > \lambda$. In the latter case we separate J and note that $\frac{1}{|J|}\int_J f\,dy \le \frac{2^n}{2^n|J|}\int_I f\,dy \le 2^n\lambda$; in the former case, we subdivide. Keep going.

CHAPTER IX

3.1 Given a family $\mathcal{C}$ of subsets of X, let $\mathcal{C}^*$ denote the class of subsets of X consisting of the at most countable unions of differences of sets in $\mathcal{C}$, i.e., $\mathcal{C}^* = \{\bigcup_n (A_n \setminus B_n) : A_n, B_n \in \mathcal{C}\}$. It is clear that if $\mathcal{C} \subseteq \mathcal{S}(\mathcal{A})$, then $\mathcal{C}^* \subseteq \mathcal{S}(\mathcal{A})$. Also, if $\emptyset$ and X are in $\mathcal{C}$, then $\emptyset$ and X are in $\mathcal{C}^*$, and $\mathcal{C} \subseteq \mathcal{C}^*$. In addition, by 5.30 and 5.31 in Chapter I, if $\operatorname{card}\mathcal{C} \le c$, then $\operatorname{card}\mathcal{C}^* \le c$. Let now $\mathcal{C}$ be the countable family consisting of open intervals in the line with rational endpoints, and intervals with one endpoint ∞ or $-\infty$ and the other enpoint rational, and $(-\infty,\infty)$; clearly $\emptyset$ and $(-\infty,\infty)$ are in $\mathcal{C}$. Let A be the well-ordered uncountable set with ordinal Ω constructed in Chapter II; we now show that to each ordinal $\alpha \in A$, there corresponds a class $\mathcal{C}_\alpha$ of subsets of R which satisfies the following two properties: (a) If the ordinals $\alpha, \beta \in A$ satisfy $\alpha < \beta$, then $\mathcal{C} \subseteq \mathcal{C}_\alpha \subseteq \mathcal{C}_\beta \subseteq \mathcal{S}(\mathcal{C})$, and (b) $\operatorname{card}\mathcal{C}_\alpha \le c$ for all α. This is how we go about it: To the first element 1 of A we associate the class $\mathcal{C}_1 = \mathcal{C}$. Assuming now that $\mathcal{C}_\alpha$ has been assigned to each ordinal $\alpha < \beta$ and that (a) and (b) hold, let $\mathcal{C}_\beta = (\bigcup_{\alpha<\beta}\mathcal{C}_\alpha)^*$. Then $\mathcal{C}_\beta$ also satisfies (a) and (b) above, and by the principle of transfinite induction, cf. 5.10 in Chapter II, a class $\mathcal{C}_\alpha$ exists for each $\alpha \in A$ satisfying (a) and (b) above. Furthermore, we claim that $\mathcal{M} = \bigcup_{\alpha\in A} \mathcal{C}_\alpha$ is a σ-algebra that coincides with $\mathcal{S}(\mathcal{C})$. To see this, let $\{A_n\}$ be a sequence of sets in $\mathcal{M}$ and observe that to each n there corresponds a class $\mathcal{C}_{\alpha_n}$ so that $A_n \in \mathcal{C}_{\alpha_n}$. Let $\alpha = \sup \alpha_n$; by 5.33 in Chapter II, $\alpha \in A$. Since $\mathcal{C}_{\alpha_n} \subseteq \mathcal{C}_\alpha$ for all n, we have $\{A_n\} \subseteq \mathcal{C}_\alpha$ and $\bigcup_n A_n = \bigcup_n (A_n \setminus \emptyset) \in \mathcal{C}_{\alpha+1} \subseteq \mathcal{M}$. In a similar fashion we show that if $A_1, A_2 \in \mathcal{M}$, then $A_1 \setminus A_2 \in \mathcal{M}$. Finally, since $\operatorname{card} \mathcal{C}_\alpha \le c$ for all α and $\operatorname{card} A \le c$, it follows that $\operatorname{card} \mathcal{B}_1 = \operatorname{card} \mathcal{S}(\mathcal{C}) = \operatorname{card} \mathcal{M} \le c$. That $\operatorname{card} \mathcal{B}_1 = c$ follows combining this result with 4.16 in Chapter IV. Also, in the construction following Theorem 1.7 in Chapter VI, the set A there is Lebesgue but not Borel measurable.

3.5 Since μ is regular and $\mu(\{x\}) = 0$ for each x, given $\varepsilon > 0$, there exists $\delta(x) > 0$ such that the open ball $B(x, \delta(x))$ satisfies $\mu(B(x, \delta(x)) < \varepsilon/2$. Now, by the Heine-Borel Theorem there exists a finite sequence $x_1, \ldots, x_N$, say, such that $X \subseteq \bigcup_{n=1}^N B(x_n, \delta(x_n))$. The choice $\delta = \min\{\delta(x_1), \ldots, \delta(x_N)\}/2$ will do.

3.7 Theorem 2.3 in Chapter IV is useful here.

3.14 The set $A = \{x : \mu(\{x\}) > 0\}$ is at most countable; let $x_1, x_2, \ldots$ be an enumeration of A. If $A = \emptyset$ put $\mu_d = 0$ and $\mu_c = \mu$. Otherwise put $\mu_d = \sum_k \mu(\{x_k\})\delta_{x_k}$, and note that since $\mu_d(E) = \sum_{x\in E} \mu(\{x\}) \le \mu(E)$ for each Borel set E, then by 4.29 in Chapter IV there exists a unique measure μ_c, say, such that $\mu_c(E) = \mu(E) - \mu_d(E)$ for all $E \in \mathcal{B}$.

3.15 Let $x_1 < y_1$ satisfy $0 < \mu((-\infty, x_1]) < \eta < \mu((-\infty, y_1]) < \infty$. Having chosen points $x_1 \le \ldots \le x_{n-1} < y_{n-1} \le \ldots \le y_1$ such that $\mu((-\infty, x_{n-1}]) < \eta < \mu((-\infty, y_{n-1}])$, divide the interval $[x_{n-1}, y_{n-1}]$ into two equal nonoverlapping closed subintervals $[x_{n-1}, r]$ and $[r, y_{n-1}]$. If $\mu((-\infty, r]) = \eta$ we are done. Otherwise, if $\mu((-\infty, r]) < \eta$ set $x_n = r$ and $y_n = y_{n-1}$, and if $\mu((-\infty, r]) > \eta$ set $x_n = x_{n-1}$ and $y_n = r$. Keep going.

3.16 Let $\mathcal{V} = \{V : V \text{ is open and } \mu(V) = 0\}$, and put $G = \bigcup_{V\in\mathcal{V}} V$; G is open. We show next that $\mu(G) = 0$. To see this, let K be a compact subset G. Then there exist $V_1, \ldots, V_m$ in $\mathcal{V}$ such that $K \subset \bigcup_{k=1}^m V_k$, and consequently, $\mu(K) = 0$; by regularity, also $\mu(G) = 0$. Now put $C = R^n \setminus G$; clearly (a) holds. Also, if $\mathcal{O}$ is an open set such that $C \cap \mathcal{O} \ne \emptyset$, and if $\mu(C \cap \mathcal{O}) = 0$, then we also have $\mu(\mathcal{O}) = \mu(\mathcal{O} \cap E) + \mu(\mathcal{O} \cap (R^n \setminus E)) = 0$, implying that $\mathcal{O} \subseteq (R^n \setminus C)$, contrary to the fact that $C \cap \mathcal{O} \ne \emptyset$. Finally, if C_1 is another closed subset that satisfies (a) and (b), then from (a) it follows that $R^n \setminus C_1 \subseteq G$, and so $C = R^n \setminus G \subseteq C_1$. Moreover, since $\mu(G \cap C_1) = 0$, from (b) it follows that $G \cap C_1 = \emptyset$. Hence, $C_1 \subseteq (R^n \setminus G) =$

C. Finally, if K is a compact subset of R, let $x_1, x_2, \ldots$ be a dense subset of K, put $\mu = \sum_n 2^{-n}\delta_{x_n}$, and verify that $\operatorname{supp}\mu = K$. What if K is only closed?

3.21 To compare the integrals we need to integrate by parts. Since it is not apparent how to do this, we sum by parts instead using Abel's transformation, cf. 3.8 above.

3.25 To prove the necessity, choose continuity points $x_0 < \ldots < x_k$ of F such that $F(x_0) < \varepsilon$, $F(x_k) > 1-\varepsilon$, and $x_i - x_{i-1} < \varepsilon$ for $i = 1, \ldots, k$. If n is sufficiently large, then $|F(x_i) - F_n(x_i)| < \varepsilon/2$ for all i. Suppose now that $x_{i-1} \le x \le x_i$. Then $F_n(x) \le F_n(x_i) \le F(x_i) + \varepsilon/2 \le F(x+\varepsilon) + \varepsilon/2$; the inequality going the other direction is established along similar lines.

3.31 (i) implies (ii) follows from the usual convergence results once we change variables, cf. 3.24. Also, if $f = \chi_E$, then the set D_f of discontinuities of f is ∂E, and from $\mu(\partial E) = 0$ and (i) it follows that $\mu_n(E) = \int_R f\,d\mu_n \to \int_R f\,d\mu = \mu(E)$. Thus, (i) also implies (iii). Furthermore, since $\partial(-\infty, x] = \{x\}$, (iii) implies (i). To deduce (i) from (ii), consider the corresponding distribution functions F_n, F, assume that $x < y$, and let $f(t)$ be the function equal to 1 for $t \le x$, 0 for $t \ge y$ and linear on $[x,y]$, i.e., $f(t) = (y-t)/(y-x)$ for $x \le t \le y$. Since $F_n(x) \le \int_R f\,d\mu_n$ and $\int_R f\,d\mu \le F(y)$, it follows from (ii) that $\limsup F_n(x) \le F(y)$; whence letting y decrease to x we get $\limsup F_n(x) \le F(x)$. Similarly we show that $\lim_{u\to x, u<x} F(u) \le \liminf F_n(x)$, and this implies the convergence at continuity points.

3.32 The relations $2f(x/3) = f(x)$ and $2f(2/3 + x/3) - 1 = f(x)$ are useful here.

CHAPTER X

4.9 By considering if necessary the functions $f_k - f_k(a)$ we may assume that the f_k's are nonnegative. Thus $f = \sum_k f_k$ is nonnegative and nondecreasing and, by Theorem 2.1, f' exists a.e. in I. Consider next the partial sums $s_n = f_1 + \cdots + f_n$, and let N be a null subset of I so that f' and f_n' exist off N, $n = 1, 2, \ldots$ For any $x \in (a,b)$ and $h > 0$ with $x + h \in (a,b)$, we have

$$\frac{f(x+h) - f(x)}{h} \ge \frac{s_n(x+h) - s_n(x)}{h},$$

and consequently, $s_n'(x) \le f'(x)$ for $x \in I \setminus N$. Moreover, since $s_n'(x) \le s_{n+1}'(x) \le f'(x)$, $\lim_{n\to\infty} s_n'(x)$ exists for $x \in I \setminus N$ and it does not exceed

$f'(x)$. But, is it equal to $f'(x)$? To show that this is the case it suffices to find a subsequence s'_{n_k} that converges to f' a.e. Let $n_1 < n_2 < \ldots < n_k < \ldots$ be so that $\sum_{k=1}^{\infty}(f(b) - s_{n_k}(b)) < \infty$. This implies that for each n_k and all $x \in (a,b)$ we have $0 \le f(x) - s_{n_k}(x) \le f(b) - s_{n_k}(b)$ and consequently the series $\sum_k (f(x) - s_{n_k}(x))$ converges. Since the terms of this series are monotone functions that have finite derivatives a.e., the above argument gives that $\sum_k (f'(x) - s'_{n_k}(x))$ converges a.e. and $\lim_{k\to\infty}(f'(x) - s'_{n_k}(x)) = 0$ a.e.

4.12 Let f denote the Cantor-Lebesgue function on $[0,1]$, and extend it to be 0 for $x \le 0$ and 1 for $x \ge 1$. Let $\{[a_n,b_n]\}$ be an enumeration of the family of closed subintervals of $[0,1]$ with rational endpoints, and put $f_n(x) = f((x - a_n)/(b_n - a_n))$. Then $g(x) = \sum_{n=1}^{\infty} 2^{-n} f_n(x)$ is continuous and strictly increasing on $[0,1]$ and, by 4.8, $g' = 0$ a.e.

4.22 Let $\{\mathcal{O}_n\}$ be a decreasing sequence of open sets containing N such that $|\mathcal{O}_n| < 1/2^n$, $n = 1,2,\ldots$, let $\phi_n = \sum_{k=1}^{n} \chi_{\mathcal{O}_n}$, and put $\phi = \lim_{n\to\infty} \phi_n$. It is not hard to verify that $\phi \in L(I)$ and that $\int_I \phi < 1$. Further, put $\psi_n(x) = \int_{[0,x]} \phi_n(y)\,dy$ and $\psi(x) = \int_{[0,x]} \phi(y)\,dy$. Now, if $x \in I$ and $[x, x+h] \subset \mathcal{O}_n$, we have

$$\frac{\psi(x+h) - \psi(x)}{h} \ge \frac{1}{h}\int_{[x,x+h]} \phi_n(y)\,dy = n\,.$$

Thus the right-hand side derivative of ψ at x is ∞; similarly for the left-hand side derivative.

4.23 If E is the set constructed in 3.34 in Chapter V, consider the function $f(x) = \int_{[0,x]}(\chi_E - \chi_{I\setminus E})\,dy$, $0 < x < 1$.

CHAPTER XI

3.4 The Closed Graph Theorem, which is discussed in Chapter XV, is relevant to answer the question.

3.6 We observe first that sets E with $\nu(E) > -\infty$ enjoy the following property: Given $\varepsilon > 0$, there is a measurable subset B of E such that (a) $\nu(B) \ge \nu(E)$, and, (b) $\nu(B') > -\varepsilon$ for each measurable subset B' of B. Indeed, if $\nu(E) > -\varepsilon$ and the same is true for each of its measurable subsets, we are done. Otherwise let B_1 be a measurable subset of E with $\nu(B_1) \le -\varepsilon$. Since $\nu(E)$ is finite, we have $\nu(E \setminus B_1) \ge \nu(E)$, and we may repeat the above argument with $E\setminus B_1$ in place of E. If $\nu(E\setminus B_1) > -\varepsilon$ and the same is true for each of its measurable subsets, we are done. Otherwise let $B_2 \subseteq E \backslash B_1$ be such that $\nu(B_2) \leq -\varepsilon$; note that B_2 is disjoint with B_1 and that $\nu(B_2) \geq \nu(E \backslash B_1) \geq \nu(E)$. We proceed in this fashion recursively and either stop after a finite number of steps after obtainig a set B_k with the desired properties, or else we get a pairwise disjoint family $B_1, B_2, \ldots$ of measurable subsets of E such that $\nu(B_k) \leq -\varepsilon$ for all k. In the latter case put $B = \bigcup_k B_k$, and note that $\nu(E \setminus B) = \nu(E) - \sum_k \nu(B_k) = \infty$, which contradicts our assumption on ν. To complete the proof, put $E_1 = E$, and define E_n inductively as follows: Having defined $E_1, \ldots, E_{n-1}$, by the above remark we can find a measurable subset E_n of E_{n-1} such that $\nu(E_n) \geq \nu(E_{n-1} \geq \nu(E)$ and $\nu(B) \geq -1/n$ for each measurable subset B of E_n. Finally put $A = \bigcap_n E_n \subseteq E_1 = E$, and verify that A is positive with respect to ν. By the way, this result can be used to show that arbitrary signed measures ν on $(X, \mathcal{M})$ admit a Hahn decomposition. The proof runs along these lines: We may assume that ν does not take the value ∞ for otherwise we work with $-\nu$. Put $\eta = \sup\{\nu(A) : A$ is positive with respect to $\nu\} \geq 0$, let $\{A_n\}$ be a sequence of positive sets such that $\lim_{n\to\infty} \nu(A_n) = \eta$, and set $A = \bigcup_n A_n$. A is also positive with respect to η and $\nu(A) = \eta < \infty$. Put now $B = X \setminus A$; we claim that B is negative with respect to ν. For, if this is not the case, then B contains a measurable subset of measure $\varepsilon > 0$, say, and consequently, also a subset A_1 with $\nu(A_1) \geq \varepsilon$ which is positive with respect to ν. In particular, A_1 and A are disjoint, and $\eta \geq \nu(A \cup A_1) = \nu(A) + \nu(A_1) = \eta + \varepsilon$.

3.33 By considering either $\mu_m - \mu$ or $\mu - \mu_m$, we may assume that $\mu_m(A)$ decreases to 0 for each $A \in \mathcal{B}_n$. Thus for each interval I we have $\mu_m(I)/|I| \geq \mu_{m+1}(I)/|I|$, and consequently $D\mu_m \geq D\mu_{m+1}$ and $\{D\mu_m\}$ is a nonincreasing sequence of function. Let $A_\eta = \{x \in R^n : D\mu_m(x) \geq \eta$ for all $n\}$. By 3.32 it follows that $\mu_m(A_\eta) \geq \eta|A_\eta|$. If we let $n \to \infty$ we get that $0 \geq \eta|A_\eta|$ and so A_η is null. Now we must show that $D\mu_m \to 0$ a.e. Indeed, if $A = \{x \in R^n : \lim_{m\to\infty} D\mu_m(x) > 0\}$, then $A = \bigcup_k A_{1/k}$, and hence $|A| = 0$.

CHAPTER XII

4.11 To show the necessity, let $f \in L^p(X) \setminus L^q(X)$, and set $E_n = \{|f| \geq n\}$. Thus, $\mu(E_n) < \infty$ for all n, and $\lim_{n\to\infty} \mu(E_n) = 0$. Furthermore, if $\mu(E_n) = 0$ for some n, then $f \in L^\infty(X)$, and by 4.6 it follows that $f \in L^q(X)$, a contradiction. Conversely, let us first show that there exists a pairwise disjoint sequence $\{E_n\}$ of measurable sets such that $0 < \mu(E_n) < 1/2^n$. By assumption there exists a sequence $\{A_n\}$ of measurable sets such that $0 < \mu(A_n) \leq 1/2^n$ and $\mu(A_{n+1}) \leq \mu(A_n)/4$; the sequence $E_n = A_n \setminus \bigcup_{k=n+1}^{\infty} A_k$ has the desired properties. If $q < \infty$ put $f = \sum_{n=1}^{\infty} \mu(E_n)^{-1/q} \chi_{E_n}$, and if $q = \infty$ put $f = \sum_{n=1}^{\infty} n^{1/p} \chi_{E_n}$.

4.16 Note that $|f_n - f|^p \le 2^p |f_n|^p + 2^p |f|^p$, and proceed as in 4.24 in Chapter VII.

4.19 To prove the necessity of (a), given $\varepsilon > 0$, choose an integer n_0 so that $\|f - f_n\|_p < \varepsilon$ for all $n \ge n_0$, and then pick $B_\varepsilon, C_\varepsilon \in \mathcal{M}$ of finite measure such that $\int_{X \setminus B_\varepsilon} |f|^p \, d\mu < \varepsilon$ and $\int_{X \setminus C_\varepsilon} |f_n|^p \, d\mu < \varepsilon$ for $n = 1, \ldots, n_0$; now put $A_\varepsilon = B_\varepsilon \cup C_\varepsilon$. (b) follows along similar lines using 3.7 in Chapter VIII. As for the sufficiency, by (a), Fatou's Lemma and Theorem 1.2, we may reduce the problem to one where μ is a finite measure. Given $\varepsilon > 0$, let $\delta > 0$ be chosen so that (b) holds. By Egorov's theorem there is a measurable subset B with $\mu(B) < \delta$ such that f_n converges uniformly to f on $X \setminus B$, and by Fatou's Lemma it follows that $\int_B |f|^p \, d\mu < \varepsilon$. By Theorem 1.2 we then see that $\int_X |f - f_n|^p \, d\mu < 3^p \varepsilon$ for all sufficiently large n. Thus we conclude that $f = (f - f_n) + f_n \in L^p(\mu)$, and $\|f - f_n\|_p \to 0$ as $n \to \infty$.

4.20 If f is a nonnegative simple function, $f = \sum_{n=1}^{N} c_n \chi_{A_n}$, $\mu(A_n) = a_n < \infty$ for all n, and $c_1 > c_2 > \ldots > c_n$, then, by the properties established in 4.11-4.12 of Chapter VI, it follows that

$$\int_0^\infty p\, t^{p-1} \mu(\{f > t\}) \, dt = a_1 \int_{c_1}^{c_2} p\, t^{p-1} \, dt + (a_1 + a_2) \int_{c_3}^{c_2} p t^{p-1} \, dt + \cdots$$
$$+ (a_1 + \cdots + a_N) \int_0^{c_N} p\, t^{p-1} \, dt = a_1 c_1^p + \cdots + a_N c_N^p = \|f\|_p^p .$$

For arbitrary L^p functions, take limits. A simpler proof follows from Fubini's theorem, to be discussed in Chapter XIII, and the identity $|f(x)|^p = \int_{[0, |f(x)|]} p\, t^{p-1} \, dt$.

4.22 If $\mathcal{O}_t = \{|f| > t\}$, try $g = f \chi_{\mathcal{O}_t}$.

4.24 Given $f \in L^p(R^n)$ and $t > 0$, let $\mathcal{O}_t$ be as in 4.22 and write $f = f_1 + f_\infty$, where $f_1 = f \chi_{\mathcal{O}_{t/2}}$, and $f_\infty \in L^\infty(R^n)$, $\|f_\infty\|_\infty \le t/2$. Note that since $\{Mf > t\} \subseteq \{Mf_1 > t/2\} \cup \{Mf_\infty > t/2\}$ and $\|Mf_\infty\|_\infty \le \|f_\infty\| \le t/2$, by the Hardy-Littlewood maximal theorem it readily follows that

$$|\{Mf > t\}| \le |\{Mf_1 > t/2\}| \le c(t/2)^{-1} \|f_1\|_1$$
$$= c t_1 \int_{\{|f| > t/2\}} |f| \, dx .$$

Whence from 4.20 and Fubini's theorem, to be discussed in Chapter XIII, we get that

$$\|Mf\|_p^p \le c \int_{[0,\infty)} p\,t^{p-1} t^{-1} \int_{\{|f|>t/2\}} |f(x)|\,dx dt$$

$$\le c \int_{R^n} |f(x)| \left(\int_{[0,2|f(x)|]} p\,t^{p-2}\,dt \right) dx = c \int_{R^n} |f(x)|\,|f(x)|^{p-1}\,dx\,.$$

The constant c above depends, of course, on n and p. It is surprising that if $M_1 f(x)$ denotes the Hardy-Littlewood maximal function with respect to balls introduced in 3.22 in Chapter VIII, then it is true that $\|M_1 f\|_p \le c\|f\|_p$, where c depends only on p and is independent of n. The proof of this interesting result has been given by Stein and Strömberg, cf. *Ark. Mat.* **21** (1983), 259-269.

4.27 To show the sufficiency observe that $\nu(E) > 0$ implies $\mu(E) > 0$, and consequently, $\nu \ll \mu$. Let $f = d\mu/d\nu$, and for a sequence (b_n) of nonnegative real numbers that increases from 0 to ∞, let $E_n = \{b_n \le f < b_{n+1}\}$. Then the E_n's form a measurable partition of X, and, since $\nu(E_n) \ge b_n \mu(E_n)$, the condition gives $\|f\|_p^p \le c$.

4.28 Let q be the conjugate index to p, and $\varepsilon > 0$. By Young's inequality we have

$$f_n \le \frac{\varepsilon^p f_n^p}{p} + \frac{1}{q\varepsilon^q}\,, \quad \text{all } n\,.$$

4.32 We have $\|f_n g_n - fg\|_p \le \|(f_n - f)g_n\|_p + \|f(g_n - g)\|_p = I + J$, say. By assumption, $\lim_{n\to\infty} I = 0$. Now, given $\varepsilon > 0$, let δ be chosen so that $\int_E |f|^p\,d\mu < \varepsilon$ provided that $\mu(E) < \delta$, and pick a measurable set A, $\mu(A) < \infty$, such that $\int_{X\setminus A} |f|^p\,d\mu < \varepsilon$. Since f is finite μ-a.e on A, by (an extension of) Proposition 2.1 in Chapter VI, there exist a measurable subset B of A and a constant M such that $\mu(B) \le \wedge\{\mu(A)/2, \delta\}$ and $|f| \le M$ on $A \setminus B$. If k denotes the uniform bound for the g_n's and g, then $J^p \le 2k \int_{X\setminus A} |f|^p\,d\mu + 2k \int_B |f|^p\,d\mu + M \int_{A\setminus B} |g_n - g|^p\,d\mu$.

4.33 Part (a) of the necessity follows from the Uniform Boundedness Principle, to be discussed in Chapter XV; also cf. Proposition 4.1 in Chapter XVI. As for the sufficiency, from (b) it follows that $\lim_{n\to\infty} \int_X f_n \phi\,d\mu = \int_X f\phi\,d\mu$ for all ϕ's in the dense subset of $L^p(\mu)$ consisting of the simple functions. The result for arbitrary $\phi \in L^{p'}(\mu)$, $1/p + 1/p' = 1$, follows by a limiting argument.

4.34 If $\liminf \|f_n\|_p \le k$, by Fatou's Lemma $\|f\|_p \le k$ as well. Thus, by Hölder's inequality, $|\int_X fg\,d\mu|, \limsup |\int_X f_n g\,d\mu| < \infty$. Now, given

$\varepsilon > 0$ and $g \in L^q$, $1/p + 1/q = 1$, by Young's inequality we have

$$f_n g \le \varepsilon |f_n| \frac{|g|}{\varepsilon} \le \frac{\varepsilon^p |f_n|^p}{p} + \frac{|g|^q}{q\varepsilon^q},$$

and consequently, $\varepsilon^p |f_n|^p/p + |g|^q/q\varepsilon^q - f_n g \ge 0$. Applying Fatou's Lemma to this nonnegative function it follows that

$$\frac{\varepsilon^p \|f\|_p^p}{p} + \frac{\|g\|_q^q}{q\varepsilon^q} - \int_X fg\, d\mu \le \frac{\varepsilon^p k^p}{p} k + \frac{\|g\|_q^q}{q\varepsilon^q} - \limsup \int_X f_n g\, d\mu .$$

Unraveling this expression we get $\limsup \int_X f_n g\, d\mu \le \int_X fg\, d\mu + \varepsilon^p k/p$, which implies, since ε is arbitrary, that $\limsup \int_X f_n g\, d\mu \le \int_X fg\, d\mu$. The opposite inequality follows from the estimate $\int_X fg\, d\mu \le \liminf \int_X f_n g\, d\mu$, obtained by considering the sequence $\{-f_n\}$.

4.35 Assume that the sequence f_n does not converge weakly to f, choose $g \in L^{p'}$ such that $\limsup | \int_X f_n g\, d\mu - \int_X fg\, d\mu| = \eta > 0$, and find a subsequence $n_1 < n_2 < \dots$ such that $\lim_{k\to\infty} | \int_X (f_{n_k} - f) g\, d\mu| = \eta$. Next invoke (an appropriate version of) Proposition 3.3 in Chapter VI to conclude that there is a further subsequence of the n_k's, which we call n_k again for simplicity, such that $\lim_{k\to\infty} f_{n_k} = f$ μ-a.e. This leads to a contradiction.

4.36 We begin by observing that given $0 < p < \infty$ and $0 < \varepsilon < 1$, there exist constants c, independent of ε, and c_ε such that

$$\big| |a+b|^p - |a|^p \big| \le c\varepsilon |a|^p + c_\varepsilon |b|^p \,, \text{all real } a, b . \tag{1.1}$$

We may prove this by considering two cases, to wit, when a, b are of the same sign, and when they are not. For instance, in the former case, suppose that $a, b > 0$, and note that when $b \le \varepsilon a$, by the mean value theorem, the left-hand side of (1.1) is dominated by $a^p((1+\varepsilon)^p - 1) \le p\varepsilon a^p$, and when $b > \varepsilon a$, by $((b/\varepsilon) + b)^p = ((1/\varepsilon) + 1)^p b^p$. With (1.1) out of the way, note that by Fatou's Lemma $f \in L^p(\mu)$, and that

$$\begin{aligned} \big| |f_n - f|^p - |f_n|^p - |f|^p \big| &\le \big| |f_n - f|^p - |f_n|^p \big| + |f|^p \\ &\le c\varepsilon |f_n|^p + c_\varepsilon |f|^p + |f|^p = c\varepsilon |f_n|^p + c_\varepsilon |f|^p . \end{aligned}$$

Whence, $c\varepsilon |f_n|^p + c_\varepsilon |f|^p - \big| |f_n - f|^p - |f_n|^p - |f|^p \big| \ge 0$, and, by applying Fatou's Lemma to this sequence of nonnegative functions we get

$$\begin{aligned} (c\varepsilon + c_\varepsilon) \int_X |f|^p\, d\mu &\le c\varepsilon \liminf \int_X |f_n|^p\, d\mu \\ &\quad + c_\varepsilon \int_X |f|^p\, d\mu - \limsup \int_X \big| |f_n - f|^p - |f_n|^p - |f|^p \big|\, d\mu . \end{aligned}$$

Thus, $\limsup \int_X \big| |f_n - f|^p - |f_n|^p - |f|^p \big|\, d\mu \le \varepsilon c k$, and the conclusion follows since ε is arbitrary. This result is due to Brezis and Lieb, *Proc. Amer. Math. Soc.* **88** (1983), 486-490.

CHAPTER XIII

4.2 Consider $\sum_{k=-m^2}^{m^2} f((k-1)/m, y)\chi_{[(k-1)/m,k/m]}(x)$.

4.9 First, since $\delta(y) \to \delta(x)$ as $y \to x$, we get that $M_\lambda(x,f) = \infty$ if $x \notin F$. Next, note that since $\delta = 0$ in F, the integral defining M_λ can be restricted to $I \setminus F$. Thus, by Tonelli's theorem, $\int_F M_\lambda(x,f)\,dx = \int_{I\setminus F}\int_F (\delta^\lambda(y)/|x-y|^{1+\lambda})\,dx\,dy$. In order to estimate the integral over F, fix $y \in I \setminus F$, and note that for $x \in F$, $|x-y| \ge \delta(y) > 0$. Thus, the integral in question is dominated by $\int_{|x-y|\ge\delta(y)} (1/|x-y|^{1+\lambda})\,dx = 2\int_{[\delta(y),\infty)} t^{-(1+\lambda)}dt = 2\lambda^{-1}|I \setminus F|$. This implies that $\int_F M_\lambda(x,f)\,dx \le 2\lambda^{-1}|I \setminus F|$.

4.13 Hölder's inequality and the continuity of translates of $L^p(R^n)$ functions should do the job.

4.14 Suppose $f \in L^p(R^n)$, $g \in L^q(R^n)$, and $\phi \in L^{r'}(R^n)$, where $1/r + 1/r' = 1$, are nonnegative. Young's convolution theorem follows then from the estimate

$$\int_{R^n}\int_{R^n} \phi(x) f(x-y) g(y)\,dx\,dy \le \|f\|_p \|g\|_q \|\phi\|_{r'}\,.$$

If α_1, α_2 are nonnegative numbers such that $\alpha_1 + \alpha_2 = 1$, and similarly for β_1, β_2 and γ_1, γ_2, then the left-hand side of the above inequality may be rewritten $\int_{R^n}\int_{R^n} \phi(x)^{\alpha_1+\alpha_2} f(x-y)^{\beta_1+\beta_2} g(y)^{\gamma_1+\gamma_2}\,dx\,dy$. By 4.4 in Chapter XII, if $1/s + 1/t + 1/u = 1$, this expression is dominated by the product of the three integrals $\left(\int_{R^n}\int_{R^n} \phi(x)^{\alpha_1 s} f(x-y)^{\beta_1 s} dx dy\right)^{1/s}$, $\left(\int_{R^n}\int_{R^n} f(x-y)^{\beta_2 t} g(y)^{\gamma_1 t} dx dy\right)^{1/t}$ and $\left(\int_{R^n}\int_{R^n} \phi(x)^{\alpha_2 u} g(y)^{\gamma_2 u} dx dy\right)^{1/u}$. Clearly we will be done once we choose our parameters so that $\alpha_1 s = r'$, $\beta_1 s = p$, $\beta_2 t = p$, $\gamma_1 t = q$, $\alpha_2 u = r'$, and $\gamma_2 u = q$. It is a simple arithmetic task to verify that if $1/p + 1/p' = 1$, and $1/q + 1/q' = 1$, then the choice $s = q'$, $t = r$, and $u = p'$ works.

4.15 We do the case $p = 1$. We look carefully at the proof of Theorem 3.7 and note that the term A there is also estimated by η. As for the B term, it does not exceed

$$\int_{\{|y|\ge M\}} |f(x-y)|\phi_\varepsilon(y)\,dy + |f(x)| \int_{\{|y|\ge M\}} \phi_\varepsilon(y)\,dy = B_1 + B_2\,,$$

say. By (2.10), $\lim_{\varepsilon\to 0} B_2 = 0$. Write $\phi(y) = \psi(y)/|y|^n$, where $\lim \psi(y) = 0$ as $|y| \to \infty$. Thus,

$$B_2 = \int_{\{|y|\ge M\}} |f(x-y)|\psi(y/\varepsilon)|y|^{-n} dy \le M^{-n} \left(\sup_{\{|y|\ge M\}} \psi(y/\varepsilon) \right) \|f\|_1\,,$$

where the sup above goes to 0 with ε.

4.17 The formula to be proved reduces to a familiar formula of integration by parts when $n = 1$. Writing $f = f^+ - f^-$, we note that F is the difference of two bounded increasing functions; hence F is BV on $[r, R]$, and the integral $\int_r^R \phi(\rho)\, dF(\rho)$ is well-defined. Assume now that f is nonnegative, let $I = \int_{\{r \le |x| \le R\}} f(x)\phi(|x|)\, dx$, and let $\{r = \rho_0 < \ldots < \rho_k = R\}$ be a partition of $[r, R]$. If m_i and M_i denote the minimum and maximum values of ϕ on $[\rho_{i-1}, \rho_i]$ respectively, $1 \le i \le n$, it readily follows that

$$\sum_{i=1}^{k} m_i(F(\rho_i) - F(\rho_{i-1})) \le I \le \sum_{i=1}^{k} M_i(F(\rho_i) - F(\rho_{i-1})),$$

where the extreme terms above converge to $\int_r^R \phi(\rho)\, dF(\rho)$.

4.18 We do the case $p = 1$. Let x_0 be a Lebesgue point of f, by considering the function $f(x_0 + x)$ we may assume that $x_0 = 0$. Now, if f is continuous at 0, we are done by 4.11. Hence, subtracting from f a continuous function that vanishes outside a bounded interval which equals $f(0)$ at 0 we may also suppose that $f(0) = 0$. The assumptions on ϕ can be combined into a single estimate, to wit, $\phi(x) \le c/(1 + |x|)^{n+\lambda}$. Therefore,

$$|f_\varepsilon(0)| \le c \int_{R^n} |f(x)| \frac{\varepsilon^\lambda}{(\varepsilon + |x|)^{n+\lambda}} dx, \tag{1.2}$$

and it suffices to show that the integral on the right-hand side above tends to 0. With the notation of 4.17, let $F(\rho) = \int_{\{|x| \le \rho\}} |f(x)|\, dx$. Our assumptions imply that given $\eta > 0$, there exists $\delta > 0$ such that $F(\rho) < \eta\rho^n$ provided that $\rho \le \delta$. Writing the integral on the right-hand side of (1.2) as a sum of integrals $A + B$, say, where A extends over $|x| \le \delta$ and B extends over $|x| > \delta$, with $\phi(\rho) = \varepsilon^\lambda/(\varepsilon + \rho)^{n+\lambda}$, with some work, by 4.13, it follows that $A = \int_0^\delta (\varepsilon^\lambda/(\varepsilon + \rho)^{n+\lambda})\, dF(\rho)$. Integration by parts gives now that $\limsup_{\varepsilon \to 0} A \le c\eta$. As for B, note that if $|x| > \delta$, then $\varepsilon + |x| > \delta$, so that B does not exceed $\varepsilon^\lambda \|f\|_1/\delta^{n+\lambda}$. Hence $\lim_{\varepsilon \to 0} B = 0$. The result now follows since η is arbitrary.

4.23 Use Theorem 2.4 in Chapter XII.

CHAPTER XIV

4.6 Since M is a closed set properly contained in X there is an element $x \in X$ such that $d(x, M) = \eta > 0$. Also, there exists $y \in M$

such that $\|x - y\| < \eta/(1-\varepsilon)$. Put $x_0 = \frac{1}{\|x-y\|}(x - y)$, and verify that $\|x_0\| = 1$ and $\|x_0 - x\| \geq 1 - \varepsilon$. Note that M has to be closed. Indeed, if $I = [0,1]$, and $X = C(I)$ and M is the subspace of polynomials on I, then $\overline{M} = X$ and the conclusion does not follow. The element x_0 is "almost perpendicular" to M, and Riesz's lemma does not always assure that a perpendicular element may be found. To see this consider the following setting: Let $I = [0,1]$, let X be the closed subspace of $C(I)$ consisting of those functions f such that $f(0) = 0$, and consider $M = \{f \in X : \int_I f\, dy = 0\}$; M is a proper closed subspace of X. Suppose there is an element $g \in X \setminus M$ such that $\inf_{f \in M} \|g - f\| \geq 1$. If $h \in X \setminus M$ and $a(h) = \int_I g / \int_I h$, then $\int_I (g(x) - a(h)h(x))\, dx = 0$ and $g - a(h)h \in M$. Therefore, we have $\|g - (g - a(h)h)\| = \|a(h)h\| \geq 1$. Let $h_n(x) = x^{1/n}$, $n = 1, 2, \ldots$; $h_n \in X \setminus M$ and hence $\|a(h_n)h_n\| \geq 1$. But $a(h_n) = ((n+1)/n) \int_I g$, and since $\|h_n\| = 1$ it follows that $\left|\int_I g\right| \geq ((n+1)/n)$ for each positive integer n. This in particular means that $\left|\int_I g\right| \geq 1$. Now, since $g(0) = 0$ and $\|g\| = 1$, $\left|\int_I g\right| < 1$, which is a contradiction.

4.8 Let $\eta = |\lambda_1| + \cdots |\lambda_n|$; if $\eta = 0$ there is nothing to prove. Otherwise, dividing both sides of the inequality by η, it follows that it is equivalent to prove that for some constant $c > 0$, we have

$$\|\mu_1 x_1 + \cdots \mu_n x_n\| \geq c, \quad \text{whenever} \quad \sum_{i=1}^{n} |\mu_i| = 1\,.$$

Suppose this last statement is false. Then for each integer k there exists a choice of constants $\mu_{1,k}, \ldots, \mu_{n,k}$ such that $\|\mu_{1,k}x_1 + \cdots + \mu_{n,k}x_n\| \leq 1/k$. Now, by the Bolzano-Weierstrass Theorem, we can find a subsequence $k_m \to \infty$ with the property that $\lim_{m\to\infty} \mu_{i,k_m} = \mu_i^*$ exists for $1 \leq i \leq n$. It now readily follows that not all the μ_i^*'s are zero, and that $\mu_1^* x_1 + \cdots + \mu_n^* x_n = 0$, contradicting the linear independence of the x_i's.

4.10 All the norm properties for the intersection are obvious except for the completeness. If $\{f_n\}$ is Cauchy in $L^p(\mu) \cap L^q(\mu)$, then it is also Cauchy in $L^p(\mu)$ and in $L^q(\mu)$, and if $\lim f_n = f$ (in $L^p(\mu)$), and $\lim f_n = g$ (in $L^q(\mu)$), then $f = g$ μ-a.e., and $\lim f_n = f$ (in $L^p(\mu) \cap L^q(\mu)$). For the sum there are two properties that require some thought: (a) That the norm of $f = 0$ implies $f = 0$ μ-a.e., and, (b) Completeness. To show (a) note that there are sequences $\{g_n\} \subseteq L^p(\mu)$ and $\{h_n\} \subseteq L^q(\mu)$ such that $f = g_n + h_n$ for all n, and $\lim \|g_n\|_p = \lim \|h_n\|_q = 0$. Then $g_n + h_n = g_1 + h_1$, and $g_n - g_1 = h_1 - h_n$ converges to $-g_1$ (in $L^p(\mu)$) and to h_1 (in $L^q(\mu)$); this implies that $f = g_1 + h_1 = 0$. To show (b) it suffices to check that if $\sum \|f_n\| < \infty$, then $\lim_{N\to\infty} \sum_{n=1}^{N} f_n$ exists in

$L^p(\mu)+L^q(\mu)$. To characterize the functionals on these spaces it is helpful to think about the relationship between the norm in $L^p(\mu)\cap L^q(\mu)$ and $\int_{[0,\infty)}\mu(\{|f|>\lambda\})\,d\max(\lambda^p,\lambda^q)$, and the relationship between the norm in $L^p(\mu)+L^q(\mu)$ and $\int_{[0,\infty)}\mu(\{|f|>\lambda\})\,d\min(\lambda^p,\lambda^q)$.

4.11 A Banach space B need not have a Schauder basis. If B is a Banach space over the reals and $\{x_n\}$ is a Schauder basis for B, then the at most countable set $\sum_{n=1}^N r_n x_n$, where the r_n's are arbitrary rational numbers, is dense in B.

4.16 If $x_0\in X$, $\|x_0\|\le 1$, then also $e^{i\vartheta}x_0\in X$, $\|e^{i\vartheta}x_0\|=\|x_0\|$ and $L(e^{i\vartheta}x_0)=e^{i\vartheta}Lx_0$.

4.23 To prove the sufficiency observe that if a linear combination $\lambda_1x_1+\cdots+\lambda_nx_n=0$, then also $\sum\lambda_kL_0x_k=0$. For an arbitrary element $x=\sum\lambda_kx_k$ in the span of Y put $Lx=\sum\lambda_kL_0x_k$. Our previous observation implies that, in spite of the possible multiplicity of the representations of x as an element in the span of Y, the value of the functional Lx is uniquely determined. The linearity and boundedness of L on the span of Y are readily established. To extend L to all of X, invoke the Hahn-Banach Theorem.

4.24 Given intervals I,J of R^n, let $f=\frac{1}{|I|}\chi_I-\frac{1}{|J|}\chi_J$. Then $f\in D$, and consequently, $\frac{1}{|I|}\int_I g\,dx=\frac{1}{|J|}\int_J g\,dx$. By the Lebesgue Differentiation Theorem it readily follows that g coincides with a constant c, say, a.e. Further, since $g\in L^q(R^n)$, c must be 0. Next, suppose that D is not dense in $L^p(R^n)$. Then, by Proposition 3.1, there exist a function $\phi\in L^p(R^n)\backslash\overline{D}$ and a bounded linear functional L on $L^p(R^n)$ such that $L\phi\neq 0$ and $Lf=0$ for all $f\in D$. Whence, by Riesz's theorem, there exists $g\in L^q(R^n)$ such that $Lf=\int_{R^n}fg\,dx$ for all $f\in L^p(R^n)$, and by the first part of the argument $g=0$ a.e. Thus L is the zero functional, a contradiction. As for $L^p(I)$, it is not hard to see that in this case D is a closed proper subspace of $L^p(I)$. For instance, $g(x)=1$ for all $x\in I$ is a nonzero element of $L^q(I)$ that satisfies all the conditions. As for the case $p=1$, again D is the null space of a nontrivial bounded functional on $L(R^n)$.

4.25 4.15 is useful here.

4.36 A function f which is BV on $I=[0,1]$ is said to be normalized, if $f(0)=0$ and f is right-continuous at each point x in the open interval $(0,1)$; the collection of normalized BV functions of I is denoted by NBV. The dual space of $C(I)$ can be identified with NBV, 4.26 in Chapter III is relevant here.

4.37 Given an open subset $\mathcal{O}$ of I begin by defining the set function $\psi(\mathcal{O}) = \sup\{Lf : f \in C(I) \text{ and } 0 \le f \le \chi_K$, where K is a compact subset of $\mathcal{O}\}$. Next set

$$\mu^*(E) = \inf\{\psi(\mathcal{O}) : \mathcal{O} \supseteq E, \mathcal{O} \text{ an open subset of } I\}, \quad \text{all } E \subseteq I.$$

The measure μ is now constructed by invoking 3.40 in Chapter V.

CHAPTER XV

6.3 Given $\varepsilon > 0$, consider $Y_n = \{y \in Y : \|y\| < 1/n\}$ and $A_n = \{x \in X : |L(x,y)| \le \varepsilon \text{ for all } y \in Y_n\}$. Since L is continuous in x, each A_n is closed, and since L is continuous in y, $X = \bigcup_n A_n$. By the Baire Category Theorem we get that $|L(x,y)| \le 2\varepsilon$ if $x \in \mathcal{O}$ and $y \in Y_N$, where $\mathcal{O}$ is a neighbourhood of the origin in X, and m is some integer.

6.5 A Hamel basis will get the job done.

6.11 Let $\eta = \inf_n \sqrt[n]{\|T^n\|}$; we begin by showing that actually $\eta = \lim_{n\to\infty} \sqrt[n]{\|T^n\|}$. Given $\varepsilon > 0$, choose m such that $\sqrt[m]{\|T^m\|} < \eta + \varepsilon$. Also let $M = \max\{1, \|T\|, \ldots, \|T^{m-1}\|\}$. Now consider any integer n and write it in the form $n = k_n m + l_n$, where k_n is a nonnegative integer, and $0 \le l_n \le m-1$. Then, since $\|T^n\| \le \|T\|^n$, it follows that $\sqrt[n]{\|T^n\|} \le \sqrt[n]{\|T^{l_n}\|\,\|T^m\|^{k_n}} \le M^{1/n}\|T^m\|^{k_n/n} < M^{1/n}(\eta+\varepsilon)^{(n-l_n)/n}$. Furthermore, since $\lim_{n\to\infty} M^{1/n}(\eta+\varepsilon)^{(n-l_n)/n} = \eta+\varepsilon$, there exists an integer N_ε such that for all $n \ge N_\varepsilon$, we have $M^{1/n}(\eta+\varepsilon)^{(n-l_n)/n} < \eta + 2\varepsilon$. Therefore, for those values of n we have $\eta \le \sqrt[n]{\|T^n\|} \le \eta + 2\varepsilon$, and consequently, as asserted, $\lim_{n\to\infty} \sqrt[n]{\|T^n\|}$ exists and it equals η. We now deduce that the series in question converges or diverges by applying the Cauchy test for the convergence of series to $\sum_{n=0}^{\infty} \|T^n\|$.

6.14 Let $x \in \overline{M}$, and suppose $\{x_n\}$ is a sequence of elements of M that converges to x. Consider now the sequence $\{T_0 x_n\} \subset Y$; since $\|T_0 x_n - T_0 x_m\| \le \|T_0\|\,\|x_n - x_m\|$, this is a Cauchy sequence, and since Y is complete, it converges to a limit that is independent of the choice of sequences of elements of M that converge to x. Now set $Tx = \lim_{n\to\infty} T_0 x_n$, where $\{x_n\}$ is any sequence of elements of M that converges to x. Obviously T is linear, and, taking limits in the inequality $\|T_0 x_n\| \le \|T_0\|\,\|x_n\|$, we find that $\|Tx\| \le \|T_0\|\,\|x\|$; that is, T is bounded with norm not exceeding $\|T_0\|$. The opposite inequality, as well as the uniqueness of T, are not hard to establish.

6.20 We construct a sequence $\{T_n\}$ of bounded linear operators on X which satisfies $\|T_n x\| \le k_x$ for $n = 1, 2, \ldots$, and yet for no constant c it is true that $\|T_n\| \le c$ for all n. We begin by writing polynomials as $x(t) = \sum_{n=0}^{\infty} a_n t^n$, where to each x there corresponds an integer N_x so that $a_n = 0$ for all $n \ge N_x$. As the sequence T_n we take the sequence of functionals L_n given by $L_n 0 = 0$ and for $x \ne 0$, $L_n x = a_0 + \cdots + a_{n-1}$. Clearly each L_n is linear, and, by the choice of norms, also bounded, with norm less than or equal to n. Now, for each x we have $|L_n x| \le (N_x + 1)\max_{0 \le j \le N_x} |a_j| = k_x$, but the polynomials $x(t) = 1 + t + \cdots + t^n$ are particularly troublesome because they satisfy $\|x\| = 1$ and $L_n x = n = n\|x\|$.

6.23 The construction of the embedding proceeds in stages. At stage one consider X^*, pick a nonzero $L_1 \in X^*$, and note that the null space X_1 of L_1 is nonempty, for otherwise L_1 would constitute an embedding of X, an infinite-dimensional space, into a finite-dimensional space. Note that also X_1 is infinite-dimensional, pick $x_1 \in X \setminus X_1, \|x_1\| = 1/2$, and once this is done proceed by induction as follows: Having chosen infinite-dimensional closed subspaces $X = X_0 \supset X_1 \supset \ldots \supset X_n$ and elements $x_i \in X_i \setminus X_{i-1}$ with $\|x_i\| \le 1/2^i$ for $i = 1, \ldots, n$, consider X_n^*, which, by the Hahn-Banach Theorem, is nonempty. Pick a nonzero $L_{n+1} \in X_n^*$, denote by X_{n+1} the null space of L_{n+1}, and pick $x_{n+1} \in X_n \setminus X_{n+1}$ with $\|x_{n+1}\| \le 1/2^{n+1}$. Having thus chosen the infinite sequence $\{x_n\} \subseteq X$ define a mapping $T: \ell^\infty \to X$ by $T((\lambda_n)) = \sum_{n=1}^{\infty} \lambda_n x_n$. Then T is a bounded linear mapping with norm less than or equal to 1, which, by construction, is also injective.

6.24 Assume that $\{x_n\}$ is an most countable dense subset of $\mathcal{B} = \{x \in X : \|x\| \le 1\}$, the unit ball of a separable Banach space B, and define a linear map T from ℓ^1 into B as follows: Given a sequence $(t_n) \in \ell^1$, let $T((t_n)) = \sum_{m=1}^{\infty} t_m x_m$. Since $\|T((t_n))\| \le \sum_{m=1}^{\infty} |t_m| \, \|x_m\| \le \|(t_n)\|_{\ell^1}$, T is continuous. Since T is linear, to show that T is onto it suffices to prove that T is onto $\mathcal{B}$. Given $x \in \mathcal{B}$, choose x_{n_1} in the dense set so that $\|x - x_{n_1}\| < 1/2$, and then x_{n_2}, also from the dense set, such that $\|2(x - x_{n_1}) - x_{n_2}\| < 1/2$, or $\|x - (x_{n_1} + \frac{1}{2}x_{n_2})\| < 1/2^2$, and $n_2 > n_1$. After $x_{n_1}, \ldots, x_{n_k}$, have been chosen, with $n_k > \ldots > n_1$, choose $x_{n_{k+1}}$ satisfying $\|x - \sum_{i=1}^{k+1} 2^{-i+1} x_{n_i}\| < 2^{-(k+1)}$, and $n_{k+1} > n_k$. Choose now a sequence in $(t_n) \in \ell^1$ as follows: Put $t_n = 0$ if $n \ne n_k$ for all k, and $t_n = 2^{-k+1}$ if $n = n_k$; clearly $T((t_n)) = x$. If N is the null space of T, the linear map T induces an isomorphism from ℓ^1/N onto B.

6.29 The proof follows essentially along the lines of that of the Open Mapping Theorem. In the notation of that theorem, we introduce the sequence $\{x_n\}$, and note that if $z_k = \sum_{n=1}^{k} x_n$, then $\lim_{k\to\infty} z_k = x$ exists, and, by (4.7), $\lim_{k\to\infty} Tz_k = y$. Since T is closed, x is in the domain of T, and $Tx = y$.

6.33 Suppose that g, f_n are $L^p(I)$ functions, $n = 1, 2, \ldots$, such that $\lim_{n\to\infty} \|f_n\|_p = 0$ and $\lim_{n\to\infty} \|Tf_n - g\|_p = 0$. Replacing these sequences with subsequences if necessary, by 4.12 in Chapter XII, we may assume that also $\lim_n f_n = 0$ a.e. and $|f_n| \le \phi$, where $\phi \in L^p(I)$, and $\lim_{n\to\infty} Tf_n = g$ a.e. In particular, off a null set of y's in I, for almost every $x \in I$ we have $\lim_{n\to\infty} |k(x,y)f_n(x)| = 0$, and $|k(x,y)f_n(y)| \le |k(x,y)\phi(y)| \in L(I)$. Thus, by LDCT, for almost every $x \in I$ we have $\lim_{n\to\infty} Tf_n = 0$, which in turn implies that $g = 0$ a.e. The result now follows by invoking the Closed Graph Theorem.

6.37 (iii) implies (i) requires no proof. As for (i) implies (ii), let T^* be a closed linear extension of T. If $y \in Y \setminus \{0\}$, then $y \notin G(T^*) \supset G(T)$. Moreover, since $G(T^*)$ is closed in $X \times Y$, $(0,y) \notin \overline{G(T)}$. Finally, we consider (ii) implies (iii): Define T^* as the linear operator whose graph is precisely $\overline{G(T)}$, i.e., $D(T^*) = \{x \in X : (x,z) \in \overline{G(T)} \text{ for some } z \in Y\}$. It is, then, not hard to check that T^* is a well-defined closed linear, minimal, extension of T.

CHAPTER XVI

5.21 The first element is $x_1 = (1/\|y_1\|)y_1$. Now, $y_2 = \langle y_2, x_1\rangle x_1 + z_2$, where z_2 is not the zero vector since the y_n's are linearly independent. Also $z_2 \perp x_1$, and so we can take $x_2 = (1/\|z_2\|)z_2$. Having chosen orthonormal vectors $x_1, \ldots, x_{n-1}$, note that the vector $z_n = y_n - \sum_{k=1}^{n-1}\langle y_n, x_k\rangle x_k$ is nonzero and orthogonal to $x_1, \ldots, x_{n-1}$. Pick now $x_n = (1/\|z_n\|)z_n$.

5.26 Since the sum $\sum_{m=1}^{\infty} \lambda_m(x_m - y_m)$ converges in X when the sum $\sum_{m=1}^{\infty} \lambda_m x_m$ converges, the mapping T given by $T(\sum_m \lambda_m x_m) = \sum_m \lambda_m(x_m - y_m)$ is well-defined, has norm $\|T\| \le A < 1$, and consequently, $T - I$ is invertible.

5.27 The sum in question is equal to $\langle \chi_{[a,x]}, \sum_n \langle \chi_{[a,x]}, \phi_n\rangle\phi_n\rangle$. The necessity follows then from Theorem 3.5. As for the sufficiency, note that the above identity, together with the case of equality in the Cauchy-Schwarz inequality imply that $\chi_{[a,x]} = \sum_n \langle \chi_{[a,x]}, \phi_n\rangle\phi_n$, where the equality is in the sense of $L^2(I)$. Pass now to the limit.

5.28 Let $\eta > 0$ be chosen so that $1 - M^2\eta > 0$. By Egorov's theorem there is a subset G of I such that $|I \setminus G| < \eta$ and $\varepsilon_n = \sup_{x\in G} |\lambda_n \phi_n(x)| \to$

0 as $n \to \infty$. Thus $\lambda_n^2 = \int_I (\lambda_n \phi_n(x))^2 dx \le \left(\int_G + \int_{I \setminus G}\right) (\lambda_n \phi_n(x))^2 dx \le \varepsilon^2 + M^2 \eta \lambda_n^2$, and consequently, $(1 - M^2\eta)\lambda_n^2 \le \varepsilon_n^2 \to 0$ as $n \to \infty$.

5.32 Since by the Riesz-Fischer theorem ℓ^2 embeds into every Hilbert space, a copy of the space I consisting of those sequences of the form $\{\lambda, \lambda^2, \lambda^3, \ldots\}$, with $0 < \lambda < 1$, sits inside every infinite-dimensional Hilbert space. Since card $I = c$, it only remains to verify that the sequences in I are linearly independent.

5.35 Put $T = I + U$, and note that $U^* f(x) = 2 \int_{[x,1]} e^{(t-x)} f(t)\, dt$. Using this observation it follows that if f is orthogonal to ϕ, then $U^*Uf = -Uf - U^*f$, and consequently, $T^*Tf = (I + U + U^* + U^*U)f = f$. Whence, $\langle f, \psi \rangle = \langle Tf, T\psi \rangle = \langle T^*Tf, \phi \rangle = \langle f, \phi \rangle = 0$, and $\|Tf\|_2^2 = \|f\|_2^2$, and (b) holds. As for (c), note that if g is orthogonal to ψ it follows that $T^*Tg = -Tg - T^*g$, and consequently in this case we have $T^*Tg = g$. Setting $f = T^*Tg$ we get that f is orthogonal to ϕ and $g = Tf$. Thus T is isometric as a map from $\{\phi\}^\perp$ onto $\{\psi\}^\perp$, and T is bijective. Indeed, if f is orthogonal to ϕ and $T(f + \lambda\phi) = Tf + \lambda\psi = 0$, then $Tf = 0$, $\lambda = 0$, and since $\|f\|_2 = \|Tf\|_2 = 0$, also $f + \lambda\phi = 0$. Given $g \in L^2(I)$, writting it as $g = g - (\langle g, \psi \rangle / \langle \psi, \psi \rangle)\psi + (\langle g, \psi \rangle / \langle \psi, \psi \rangle)\psi$, and setting $f = T^*g$ we have $Tf = g$ and further manipulations give that $f = T^*g - 2\langle g, \psi \rangle \phi$. This last relation may be written $f(x) = g(x) - 2\int_{[0,x]} e^{(-t+x)} g(t)\, dt$ which essentially corresponds to the inversion formula of $g(x) = f(x) + 2\int_{[0,x]} e^{(x-t)} f(t)\, dt$.

5.39 The Closed Graph Theorem does the job.

5.41 Since for $m \ge n$ we have that $0 \le T_m - T_n \le I$, and hence $\|T_m - T_n\| \le 1$, a computation using the generalized Cauchy-Schwarz inequality gives that

$$\|T_m x - T_n x\|^4 \le (\langle T_m x, x \rangle - \langle T_n x, x \rangle)\, \|x\|^2, \quad \text{all } x \in X\,.$$

But since the sequence $\{\langle T_n x, x \rangle\}$ is bounded and nondecreasing, and hence convergent, the above inequality implies that $\lim_{n\to\infty} T_n x = Tx$, say, exists for all $x \in X$. This mapping T is obviously linear and self-adjoint.

5.44 The estimate $\|f\|_2 \le \|f\|_\infty$ holds for every $f \in L^\infty(I)$. Thus, if f belongs to the $C(I)$ closure of X, it also belongs to the $L^2(I)$ closure of X, and since X is closed in $L^2(I)$, it follows that $f \in X$. That X is closed under these norms implies, in particular, that the restriction of these norms turn X into a Banach space, and that the identity map from

X normed by the $C(I)$ norm onto X normed by the $L^2(I)$ norm is continuous; the existence of c follows now from Corollary 4.3 in Chapter XV. As for the dimension, let $\{f_n\}_{n=1}^N$ be an ONS of X, and note that by the Pythagorean theorem it follows that $\|\sum_{n=1}^N \lambda_n f_n\|_\infty \le c\|\sum_{n=1}^N \lambda_n f_n\|_2 \le c$, for any choice of scalars $\lambda_1, \ldots, \lambda_N$ such that $\sum_{n=1}^N \lambda_n^2 = 1$. Thus, for almost every $x \in I$ we have $|\sum_{n=1}^N \lambda_n f_n(x)| \le c$, and, taking the sup over the $\{\lambda_n\}$'s, we get $(\sum_{n=1}^N f_n(x)^2)^{1/2} \le c$. Squaring this inequality and integrating over I it readily follows that $N = \|\sum_{n=1}^N f_n\|_2^2 \le c^2$.

5.46 We may assume that $x = 0$. Let $n_1 = 1$, and choose n_2 so that $|\langle x_{n_1}, x_{n_2}\rangle| \le 1$; this choice is possible since the x_n's converge weakly to 0. In general, having chosen $n_1 < \ldots < n_k$ and $x_{n_1}, \ldots, x_{n_k}$, choose n_{k+1} and $x_{n_{k+1}}$ such that $|\langle x_{n_1}; x_{n_{k+1}}\rangle|, \ldots, |\langle x_{n_k}, x_{n_{k+1}}\rangle| \le 1/k$. Furthermore, since the norms $\|x_n\|$ are uniformly bounded by c, say, it follows that $\|(x_{n_1} + \cdots + x_{n_k})/k\|^2 \le (c^2+2)/k \to 0$ as $k \to \infty$.

5.50 Assume that the sequence $\{x_n\}$ converges weakly to x in X; we must show that $\|T^*x_n - T^*x\| \to 0$ as $n \to \infty$. Now, on account of the continuity of T^* we know that the sequence $\{T^*x_n\}$ converges weakly to T^*x, so, by Proposition 4.2, it suffices to verify that $\lim_{n\to\infty} \|T^*x_n\| = \|T^*x\|$. Write $\|T^*x_n\|^2 = \langle T^*x_n, T^*x_n\rangle = \langle x_n, T(T^*x_n - T^*x)\rangle + \langle x_n, TT^*x\rangle = I + J$, say. Since the x_n's converge weakly, the sequence $\|x_n\|$ is bounded, and since $\{T^*x_n - T^*x\}$ converges weakly to 0 and T is compact, $\{T(T_n^x - T^*x)\}$ converges in norm to 0. Whence $\lim_{n\to\infty} I = 0$ and $\lim_{n\to\infty} J = \langle x, TT^*x\rangle = \|T^*x\|^2$.

Index